Design of Observer-based Compensators

Peter Hippe • Joachim Deutscher

Design of Observer-based Compensators

From the Time to the Frequency Domain

Springer

Peter Hippe, Dr.-Ing.
Lehrstuhl für Regelungstechnik
Universität Erlangen-Nürnberg
Cauerstraße 7
91058 Erlangen
Germany
peter.hippe@rt.eei.uni-erlangen.de

Joachim Deutscher, Dr.-Ing.
Lehrstuhl für Regelungstechnik
Universität Erlangen-Nürnberg
Cauerstraße 7
91058 Erlangen
Germany
joachim.deutscher@rt.eei.uni-erlangen.de

ISBN 978-1-84882-536-9 e-ISBN 978-1-84882-537-6
DOI 10.1007/978-1-84882-537-6
Springer Dordrecht Heidelberg London New York

British Library Cataloguing in Publication Data
A catalogue record for this book is available from the British Library

Library of Congress Control Number: 2009926141

Cover design: eStudioCalamar, Figueres/Berlin

Printed on acid-free paper

Springer is part of Springer Science+Business Media (www.springer.com)

Preface

The state-space approach in the time domain and the transfer-function approach in the frequency domain are the two major methods for the design of linear control systems. The frequency-domain approach allows an equivalent representation of the time-domain results and it can be formulated on the basis of polynomial matrix fraction descriptions. These polynomial matrix representations generalize the concept of transfer functions for single-input, single-output (SISO) systems to multivariable systems.

The motivation for formulating the results of the state-space approach in the frequency domain can be understood when considering the observer-based control in the time domain. First, a state feedback is designed to assign the eigenvalues of the resulting state feedback loop. In order to implement this control the states of the system have to be estimated by an observer. This leads to an observer-based compensator that assigns the eigenvalues of the state feedback loop and of the observer to the resulting closed-loop system. This property is the well-known separation principle of state feedback control. Since the corresponding compensator is driven by the input and the output of the system only the input–output behaviour of the controller influences the properties of the closed-loop system, and this input–output behaviour is completely characterized by the transfer functions of the compensator. Therefore, it seems reasonable to determine directly the transfer behaviour of the compensator when designing control systems. This has the advantage that the additional degrees of freedom that exist in choosing a suitable realization of the compensator can be used, for example, to achieve further robustness properties. In a time-domain design it is not at all obvious how the choice of the free parameters in the design of the state feedback and the observer will influence the internal realization of the resulting compensator.

In this book it is shown that the transfer behaviour of the observer-based compensator can be determined on the basis of the transfer behaviour of the system. This is achieved by introducing a parameterization of the state feedback loop and of the observer in the frequency domain. An important property of these parameterizations is that the number of independent design parameters coincides in the time-domain and the frequency-domain approaches and that connecting relations exist that allow an establishment of a one-to-one relationship between the time- and the frequency-domain parameterizations. Thus, every time-domain result can be transferred to an equivalent frequency-domain result and *vice versa*. As a consequence, each control problem can either be solved in the time domain or in the frequency domain, so that the best method for the solution can be chosen. For example, the decoupling of a reference transfer behaviour is easily stated and solved in the frequency domain since the design specification is to obtain a diagonal reference transfer matrix.

The general approach taken in this book is to formulate the time-domain solution first. Assuming that the reader is familiar with the basics of the state-space approach the presentation of the time-domain results is usually kept comparatively short. The development of the state-space methods given here is only more elaborate for those results which cannot be found in standard text books. Therefore, the book also provides a fast access to the controller design in the time domain.

Practical applications usually lead to system descriptions of high orders, so that the control design can only be carried out using computer assistance. In the last decades there have been important improvements in the algorithms for polynomial matrices, so that the polynomial approach can be used to design control systems for practical applications. All examples in the book were computed with the aid of the Polynomial Toolbox for MATLAB®. The m-files and the SIMULINK® files of the examples are available by writing an email to `polybook@rt.eei.uni-erlangen.de`.

Whereas the state-space descriptions of linear dynamic systems can be regarded as a standard tool, the frequency-domain representation of multiple-input, multiple-output systems on the basis of polynomial matrix fraction descriptions is not so widely known. Therefore, the *first chapter* of this book contains a short résumé of the basic facts on polynomial matrices used in the context of the design methods presented. A comprehensive introduction to polynomial methods, however, is not intended here.

The *second chapter* is devoted to the time-domain and the frequency-domain parameterizations of state feedback control. In the time domain the dynamics of the state feedback loop can be assigned by choosing the feedback gain K. It is shown that the resulting dynamics of the closed-loop system can be parameterized by the polynomial matrix $\tilde{D}(s)$ in the frequency domain. A relation is derived that establishes a one-to-one connection between the constant matrix K and the polynomial matrix $\tilde{D}(s)$, so that the time-domain parameterization can be obtained from the frequency-domain parameterization and *vice versa*.

In *Chapter 3* the design of reduced-order state observers is covered, which also includes the full-order observer. The time-domain formulation of the reduced-order observer used in this book was first proposed by Uttam and O'Halloran [62]. This particular form of the reduced-order observer has the advantage of allowing a parameterization by two gain matrices that completely characterize the observer dynamics influencing the closed-loop system. Consequently, a connecting relation can be determined between the observers gains and the polynomial matrix $\tilde{\bar{D}}(s)$ that characterizes the observer dynamics in the frequency domain.

In the time domain the observed-based compensator is specified by the gain matrices of the state feedback and the observer, and the state-space model of the system is used in the design. In the frequency domain one starts with the parameterizing polynomial matrices of the state feedback and the observer and needs to compute the transfer behaviour of the corresponding observer-based compensator. In *Chapter 4* it is demonstrated that this transfer behaviour can be computed on the basis of the transfer behaviour of the system. As the eigenvalues of the closed-loop system are arbitrarily assignable, compensators with high gain can result. Such compensators may give rise to output signals that pass the input limitations existing in every technical system. The resulting restrictions of the plant input signals can cause undesired effects in the transients of the closed-loop system and they can even lead to limit cycles. These undesired effects of input saturation are called windup. It is shown in this chapter that the frequency-domain representation of the observer-based compensator allows the formulation of systematic measures to prevent windup effects.

In single-input systems, a set of desired eigenvalues completely specifies the feedback gain K. If more than one input exists, only part of K is specified by the desired eigenvalues. Since the remaining parameters in K also influence the properties of the closed-loop system they need to be determined when designing the control. The so-called parametric approach allows a choice of the desired eigenvalues while assigning the additional degrees of freedom in K to meet other design specifications. The same approach can also be used to parameterize the observer gains. This parametric design method was originally formulated in the time domain. *Chapter 5* presents an equivalent frequency-domain approach that uses the poles and the so-called pole directions to obtain a complete parameterization of the observer-based controller. This parametric compensator design also leads to a new time-domain parameterization of the reduced-order observer of Uttam and O'Halloran.

In a system with p inputs the modification of one input usually affects all outputs, *i.e.*, there is a cross coupling between the inputs and outputs. When the set point of one controlled output is changed this coupling can lead to an undesired reaction in the remaining outputs. Using a decoupling control, the closed-loop system behaves to reference inputs as if p single-input, single-output systems were operated in parallel. Not all systems are decouplable by static state feedback. Therefore, also a partial decoupling has to be considered. In *Chapter 6* the frequency-domain conditions for the complete decoupling and the partial decoupling are derived followed by the frequency-domain design of the corresponding controllers.

An observer-based compensator as designed in Chapter 4 does not asymptotically compensate persistently acting disturbances. It is well known that constant disturbances are rejected by an integral controller action. This concept can be generalized to more general signal forms that are modelled as outputs of a linear time-invariant dynamical system. By incorporating a model of this signal process in the compensator, also called the internal model principle, such persistently acting disturbances can be asymptotically rejected. In *Chapter 7* the robust design introduced by Davison [10] is modified in the time domain by formulating the driven signal model as an observer. This, on the one hand, avoids undesired effects on the reference transfer behaviour if the signal forms of the references do not coincide with the signal forms of the disturbances. On the other hand, it allows a systematic prevention of windup. In the second part of this chapter this new approach is formulated in the frequency domain.

The time- and the frequency-domain designs of optimal state feedback controllers and observers are presented in *Chapter 8*. The state feedback control can be designed in an optimal way by minimizing a quadratic performance index. The solution of the corresponding optimization problem leads to an algebraic Riccati equation (ARE). By using the connecting relations between the time- and the frequency-domain parameterizations of state feedback a polynomial matrix equation is obtained for $\tilde{D}(s)$ that parameterizes the optimal state feedback control in the frequency domain. By solving this equation, the design of the optimal controller can be carried out directly in the frequency domain. Also, an optimal design of observers is possible and it yields optimal filters. Assuming that the system is driven by white noise and that its measurements are also corrupted by white noise with known covariance matrices, an asymptotic reconstruction of the state is not possible. However, an optimal state observer can be designed to yield a state estimate with minimal error variance in the stationary case. The frequency-domain design of this stationary Kalman filter also leads to a design equation for $\tilde{\bar{D}}(s)$ that directly follows from the corresponding Riccati equation when using the connecting relations between the time- and the frequency-domain parameterizations of observers. Given the polynomial matrices parameterizing the optimal state feedback and the Kalman filter, the resulting observer-based compensator solves the so-called linear quadratic Gaussian (LQG) control problem in the frequency domain.

The usual structure of observer-based state feedback control has the drawback that the disturbance behaviour cannot be designed independently from the reference transfer behaviour. In *Chapter 9* a two degrees of freedom structure for state feedback control is introduced. A feedforward control assures a reference behaviour of the system that coincides with the reference transfer behaviour of a controlled model of the system when no disturbances are present. Only in the presence of disturbances does an observer-based feedback controller become active to assure asymptotic tracking. As a consequence, the reference and the disturbance behaviour of the resulting closed-loop system can be designed independently, *i.e.*, the closed-loop system has two degrees of freedom. The time-domain design of such model-matching control systems in Chapter 9 is followed by an equivalent frequency-domain formulation.

Most control applications are based on computer control. Therefore, the discrete-time results for the design of observer-based controllers with disturbance rejection are presented in *Chapter 10*. The differences between the continuous-time and the discrete-time descriptions of observer-based compensators are minor, so that the presentation in Chapter 10 can be kept short. In *Chapter 11* the time- and the frequency-domain design of optimal observer-based controllers for discrete-time systems is covered.

The presentation of the design methods is always supplemented by accompanying examples, which can easily be replicated, since all necessary parameters are provided. Thus, any result not directly shown can be obtained by the corresponding simulations.

The authors are greatly indebted to their colleagues for their helpful feedback. In particular, they want to thank Christoph Wurmthaler for his perpetual interest and his never-tiring readiness to offer support. They are especially grateful to Bernhard Müller, because without his constant assistance and critical comments the book would not have materialized. Special thanks also to Günter Roppenecker for his encouragement and continuous support.

Erlangen
April 2009

Peter Hippe
Joachim Deutscher

Contents

1

Polynomial Matrix Fraction Descriptions

This chapter contains a short review of the polynomial description of linear time-invariant systems as far as it is needed in the design of observer-based compensators presented in this book. It is shown that the polynomial matrix fraction description is a straightforward generalization of transfer functions to multivariable systems. Similarly to the transfer function of a SISO system, the polynomial matrix fraction description displays two important properties of a multivariable system in a transparent manner, namely the poles and the zeros. For further details of the polynomial system description the interested reader is referred to [36] or to the more recent book [5]. Algorithms for handling polynomial matrices and polynomial matrix equations are not included in this presentation. Reliable algorithms are contained in the Polynomial Toolbox for MATLAB® (for further information see [53]). They allow the implementation directly on a computer of the polynomial methods for the analysis and design of linear control systems presented.

After introducing the right coprime matrix fraction description of a transfer matrix in Section 1.1 the dual representation of a right coprime matrix fraction description, namely the left coprime matrix fraction description, is presented in Section 1.2.

1.1 Right Coprime Matrix Fraction Description

Considered is a linear and time-invariant system described by the state equations

$$\dot{x}(t) = Ax(t) + Bu(t), \tag{1.1}$$

$$y(t) = Cx(t), \tag{1.2}$$

with state $x \in \mathbb{R}^n$, input $u \in \mathbb{R}^p$ and output $y \in \mathbb{R}^m$, $m \geq p$. In the following, it is assumed that the matrix B in (1.1) has full column rank and that the matrix C in (1.2) has full row rank. Also, let the pair (A, B) be controllable and the pair (C, A) be observable.

Remark 1.1. The assumptions for the matrices B and C do not constitute any loss of generality since on the one hand, full rank matrices can always be achieved by deleting dependent inputs and outputs. On the other hand, only the controllable and observable part of the system (1.1) and (1.2) is influenced by observer-based compensators, so that the unobservable and uncontrollable parts can be eliminated from the system description. The corresponding reduced system is attainable by a state transformation to *Kalman canonical form* (see, *e.g.*, [36]).

The transfer behaviour of the system (1.1) and (1.2) between the input u and the output y is given by

$$y(s) = C(sI - A)^{-1}Bu(s) = G(s)u(s). \tag{1.3}$$

An alternative representation of (1.3) can be obtained by expressing the $m \times p$ transfer matrix $G(s)$ as a fraction of polynomial matrices. This generalizes the representation of a scalar transfer function by a numerator and a denominator polynomial to the multivariable case. There are two different matrix fraction descriptions (MFD) of a transfer matrix, namely the right and the left MFDs.

First, the *right matrix fraction description*

$$G(s) = N(s)D^{-1}(s) \tag{1.4}$$

of the system is considered in which $N(s)$ is an $m \times p$ polynomial matrix and $D(s)$ is a $p \times p$ polynomial matrix. The order of the system (1.1) and the poles of the transfer behaviour (1.3) can be directly computed from the polynomial matrix $D(s)$, provided that certain conditions are satisfied. In the formulation of these conditions the following definitions are needed.

Definition 1.1 (Column degree). *The* ith column degree $\delta_{ci}[P(s)]$ *of a polynomial matrix* $P(s)$ *is given by the degree of the polynomial element of highest degree in the* ith *column of* $P(s)$.

Definition 1.2 (Highest column-degree-coefficient matrix). *The* $p \times p$ highest column-degree-coefficient matrix $\Gamma_c[P(s)]$ *of a* $p \times p$ *polynomial matrix* $P(s)$ *is the matrix of the coefficients of the polynomial elements with degree* $\delta_{ci}[P(s)]$ *in the* ith *column.*

It is shown in [67] that

$$\det D(s) = \det(\Gamma_c[D(s)])s^{\sum_{i=1}^{p}\delta_{ci}[D(s)]} + \text{lower degree terms in } s. \tag{1.5}$$

Thus, iff the highest column-degree-coefficient matrix $\Gamma_c[D(s)]$ is non-singular is the degree of $\det D(s)$ given by

$$\deg\det D(s) = \sum_{i=1}^{p}\delta_{ci}[D(s)]. \tag{1.6}$$

Since $\deg\det D(s) < \sum_{i=1}^{p}\delta_{ci}[D(s)]$ if $\det\Gamma_c[D(s)] = 0$ the following definition is obvious.

Definition 1.3 (Column-reduced polynomial matrix). *A $p \times p$ polynomial matrix $P(s)$ is called* column reduced *if* $\det \Gamma_c[P(s)] \neq 0$.

Example 1.1. Column-degree structure of a polynomial matrix
Consider the polynomial matrix

$$P(s) = \begin{bmatrix} s^2 - 3 & 1 & 2s \\ 4s + 2 & 2 & 0 \\ -s^2 & s + 3 & -3s + 2 \end{bmatrix}. \tag{1.7}$$

The column degrees of $P(s)$ are given by $\delta_{c1}[P(s)] = 2$, $\delta_{c2}[P(s)] = 1$ and $\delta_{c3}[P(s)] = 1$. Consequently, the highest column-degree-coefficient matrix of $P(s)$ has the form

$$\Gamma_c[P(s)] = \begin{bmatrix} 1 & 0 & 2 \\ 0 & 0 & 0 \\ -1 & 1 & -3 \end{bmatrix}. \tag{1.8}$$

The polynomial matrix $P(s)$ is not column reduced since $\det \Gamma_c[P(s)] = 0$. This can also be verified by computing

$$\det P(s) = 6s^3 + 44s^2 + 28s - 16, \tag{1.9}$$

which shows that $\deg \det P(s) < \delta_{c1}[P(s)] + \delta_{c2}[P(s)] + \delta_{c3}[P(s)] = 4$.

In order to relate the poles of the transfer behaviour (1.3) with the roots of $\det D(s)$ several properties of two polynomial matrices are introduced.

Definition 1.4 (Right divisor). *A non-singular polynomial matrix $R(s)$,* i.e., $\det R(s)$ *not identically zero, is a* right divisor *of a polynomial matrix $P(s)$ if there exists a polynomial matrix $\bar{P}(s)$, such that $P(s) = \bar{P}(s)R(s)$.*

Definition 1.5 (Greatest common right divisor). *A polynomial matrix $R(s)$ is a* greatest common right divisor *of two polynomial matrices $P_1(s)$ and $P_2(s)$ if*

1. *the polynomial matrix $R(s)$ is a common right divisor of $P_1(s)$ and $P_2(s)$,* i.e., $P_1(s) = \bar{P}_1(s)R(s)$ *and* $P_2(s) = \bar{P}_2(s)R(s)$ *and*
2. *any other common right divisor $Q(s)$ of $P_1(s)$ and $P_2(s)$ is also a right divisor of $R(s)$,* i.e., $R(s) = \bar{R}(s)Q(s)$.

Definition 1.6 (Right coprime polynomial matrices). *Two polynomial matrices $P_1(s)$ and $P_2(s)$ are* (relatively) right coprime *if their greatest common right divisor $U_R(s)$,* i.e., $P_1(s) = \bar{P}_1(s)U_R(s)$ *and* $P_2(s) = \bar{P}_2(s)U_R(s)$, *is a* unimodular matrix, i.e., $\det U_R(s)$ *is a non-zero constant independent of s.*

Example 1.2. Greatest common right divisor
The polynomial matrices

$$P_1(s) = \begin{bmatrix} s^2 & -1 \\ -s & s^2 \end{bmatrix} \quad \text{and} \quad P_2(s) = \begin{bmatrix} s & -s \\ 0 & 1 \end{bmatrix} \tag{1.10}$$

are not right coprime because

$$R(s) = \begin{bmatrix} s & 0 \\ 0 & 1 \end{bmatrix} \tag{1.11}$$

is a non-unimodular greatest common right divisor of $P_1(s)$ and $P_2(s)$, *i.e.*,

$$P_1(s) = \begin{bmatrix} s & -1 \\ -1 & s^2 \end{bmatrix} \begin{bmatrix} s & 0 \\ 0 & 1 \end{bmatrix} \quad \text{and} \quad P_2(s) = \begin{bmatrix} 1 & -s \\ 0 & 1 \end{bmatrix} \begin{bmatrix} s & 0 \\ 0 & 1 \end{bmatrix}, \tag{1.12}$$

with $\det R(s) = s$. This also means that the determinants of the polynomial matrices $P_1(s)$ and $P_2(s)$ have the common root $s = 0$, since $\det P_1(s) = \det \bar{P}_1(s) \det R(s)$ and $\det P_2(s) = \det \bar{P}_2(s) \det R(s)$.

The next theorem presents an alternative method for checking whether two polynomial matrices are right coprime.

Theorem 1.1 (Bezout identity for right coprime polynomial matrices). *The polynomial matrices $P_1(s)$ and $P_2(s)$ with the same number of columns are* right coprime *iff there exist polynomial matrices $X(s)$ and $Y(s)$, which are a solution of the* Bezout identity

$$X(s)P_1(s) + Y(s)P_2(s) = I. \tag{1.13}$$

Proof. A proof of this result can, *e.g.*, be found in [36]. □

Since the system (1.1) and (1.2) is supposed to be observable and controllable all eigenvalues of A in (1.1) are also poles of the transfer behaviour (1.3). It is shown in [36] that the roots of $\det D(s)$ are the poles of $G(s)$ provided that the polynomial matrices $N(s)$ and $D(s)$ in (1.4) are right coprime. If this is not the case, there exists a non-unimodular right divisor $R(s)$ of $N(s)$ and $D(s)$. This right divisor cancels in $G(s)$, so that the roots of $\det R(s)$, which are also roots of $\det D(s)$, are not poles of $G(s)$. For right coprime MFDs with $D(s)$ column reduced the order of the system can be obtained by inspection of $D(s)$ as

$$n = \sum_{i=1}^{p} \delta_{ci}[D(s)] \tag{1.14}$$

(see (1.6)). Furthermore, the relation

$$\det D(s) = c \det(sI - A), \tag{1.15}$$

with

$$c = \det \Gamma_c[D(s)] \neq 0 \tag{1.16}$$

(see (1.5)) is satisfied.

Remark 1.2. The assumption of a column-reduced $D(s)$ means no loss of generality since any *non-singular polynomial matrix* $D(s)$, *i.e.*, $\det D(s)$ does not vanish for all s, can be made column reduced by postmultiplication with an appropriate unimodular matrix $U_R(s)$ (see, *e.g.*, [67]).

Remark 1.3. In order to use the simple relation $\det D(s) = \det(sI - A)$ (see (1.15)) it is assumed in the following that a column-reduced polynomial matrix $D(s)$ satisfies $\det \Gamma_c[D(s)] = 1$. This can always be achieved in an MFD (1.4) by postmultiplying $N(s)$ and $D(s)$ with a suitable non-singular constant matrix.

Remark 1.4. Definition 1.6 also shows that a right coprime MFD (1.4), *i.e.*, a right MFD, where $N(s)$ and $D(s)$ are right coprime polynomial matrices, is unique up to a unimodular matrix $U_R(s)$ since

$$G(s) = N(s)D^{-1}(s) = N(s)U_R(s)U_R^{-1}(s)D^{-1}(s) = N^*(s)(D^*(s))^{-1}, \quad (1.17)$$

with $N^*(s) = N(s)U_R(s)$ and $D^*(s) = D(s)U_R(s)$.

Remark 1.5. Given a right coprime MFD $G(s) = N(s)D^{-1}(s)$ of a system, then the order of the minimal realization of $G(s)$ is $\deg\det D(s)$ [36]. Thus, by cancelling the greatest non-unimodular common right divisor $R(s)$ in the MFD $G(s) = N^*(s)(D^*(s))^{-1}$ with $N^*(s) = N(s)R(s)$ and $D^*(s) = D(s)R(s)$ the non-observable and the non-controllable parts of the system are removed in $G(s) = N(s)D^{-1}(s)$. Conversely, if the right coprime MFD $G(s) = N(s)D^{-1}(s)$ of a system of the order n is such that $\deg\det D(s) = n$ then the system is controllable and observable.

The transmission zeros (see, *e.g.*, [61]) of the transfer behaviour (1.3) are the μ values $s = s_{0i}$, $0 \leq \mu < n$, where the constant matrix $N(s_{0i})$ of a right coprime MFD is rank deficient. Thus, as with transfer functions the polynomial matrix $N(s)$ is called the *numerator matrix*. Accordingly, the polynomial matrix $D(s)$ is called the *denominator matrix* since in a right coprime MFD $G(s) = N(s)D^{-1}(s)$ of the transfer matrix $G(s)$ the poles are the roots of $\det D(s)$.

The direct feedthrough of a SISO system characterized by a proper transfer function can be determined as the quotient of the coefficients multiplying the highest power of s in the numerator and denominator polynomials. This result can be generalized to MFDs by using the following definition.

Definition 1.7 (Highest column-degree-coefficient matrix with respect to a given column degree). *Let $P(s)$ be an $m \times p$ polynomial matrix and $Q(s)$ be a $p \times p$ polynomial matrix. Then, the ith column of the $m \times p$ constant matrix $\Gamma_{\delta_c[Q(s)]}[P(s)]$ contains the coefficients of the monomials $s^{\delta_{ci}[Q(s)]}$ in the ith column of $P(s)$.*

Example 1.3. Highest column-degree-coefficient matrix with respect to a given column degree

Consider the polynomial matrices

$$Q(s) = \begin{bmatrix} s^2 & 1 \\ s & s \end{bmatrix} \quad \text{and} \quad P(s) = \begin{bmatrix} 3 & 2s^2 + s \\ 2 & 3 \end{bmatrix}, \tag{1.18}$$

then the highest column-degree-coefficient matrix of $P(s)$ with respect to the column degrees $\delta_{c1}[Q(s)] = 2$ and $\delta_{c2}[Q(s)] = 1$ of $Q(s)$ is given by

$$\Gamma_{\delta_c[Q(s)]}[P(s)] = \begin{bmatrix} 0 & 1 \\ 0 & 0 \end{bmatrix}. \tag{1.19}$$

Now, consider a transfer matrix

$$F(s) = N(s)D^{-1}(s) \tag{1.20}$$

that is proper (*i.e.*, the matrix $F(\infty)$ does not vanish but is finite) and assume that $D(s)$ is column reduced. Then, the relations

$$\delta_{ci}[N(s)] \leq \delta_{ci}[D(s)], \quad i = 1, 2, \ldots, p \tag{1.21}$$

are satisfied and the corresponding direct feedthrough of the transfer matrix $F(s)$ is given by

$$F(\infty) = \Gamma_{\delta_c[D(s)]}[N(s)]\Gamma_c^{-1}[D(s)]. \tag{1.22}$$

A proof of this result can be found in [9].

The following example demonstrates the concepts introduced so far.

Example 1.4. Right coprime matrix fraction description

Consider a system (1.1) and (1.2) whose transfer behaviour is represented by the right coprime MFD

$$G(s) = N(s)D^{-1}(s) = \begin{bmatrix} 1 & 0 \\ s & 1 \end{bmatrix} \begin{bmatrix} s^2 + 3s + 3 & -1 \\ -1 & s + 3 \end{bmatrix}^{-1}. \tag{1.23}$$

The column degrees of $D(s)$ are

$$\delta_{c1}[D(s)] = 2, \quad \delta_{c2}[D(s)] = 1 \tag{1.24}$$

and the highest column-degree-coefficient matrix is $\Gamma_c[D(s)] = I$. Thus, the denominator matrix $D(s)$ is column reduced and the order of the system can be obtained from

$$n = \delta_{c1}[D(s)] + \delta_{c2}[D(s)] = 3. \tag{1.25}$$

The poles of the transfer matrix $G(s)$ are the roots of

$$\det D(s) = (s + 2)^3, \tag{1.26}$$

so that the system has three poles at $s = -2$. The zeros of the transfer behaviour are the roots of $\det N(s) = 0$. Because $\det N(s) = 1$ for all s the system has no transmission zeros. In view of

$$\Gamma_{\delta_c[D(s)]}[N(s)] = \begin{bmatrix} 0 & 0 \\ 0 & 0 \end{bmatrix} \tag{1.27}$$

an application of (1.22) shows that the direct feedthrough of the transfer matrix (1.23) is vanishing.

The transfer behaviour of the system (1.1) between the input u and the state x

$$x(s) = (sI - A)^{-1}Bu(s) = N_x(s)D^{-1}(s)u(s), \tag{1.28}$$

where $N_x(s)$ is an $n \times p$ polynomial matrix and $N_x(s)$ and $D(s)$ are right coprime can be equivalently represented by a differential equation for the $p \times 1$ *partial state* $\pi(t)$ in the time domain. To this end, introduce the Laplace transform of the partial state

$$\pi(s) = [\pi_1(s) \;\; \pi_2(s) \;\; \ldots \;\; \pi_p(s)]^T = D^{-1}(s)u(s) \tag{1.29}$$

in (1.28). Then, a simple rearrangement of (1.28) and (1.29) yields the representation

$$D(s)\pi(s) = u(s), \tag{1.30}$$

$$x(s) = N_x(s)\pi(s). \tag{1.31}$$

By applying the inverse Laplace transform to (1.30) and (1.31), the *differential operator representation*

$$D(\tfrac{d}{dt})\pi(t) = u(t), \tag{1.32}$$

$$x(t) = N_x(\tfrac{d}{dt})\pi(t) \tag{1.33}$$

for the partial state $\pi(t)$ is obtained (see [67]).

Remark 1.6. The differential operator representation (1.32) and (1.33) is closely related to the flatness property that was first introduced for non-linear systems (see [18]). In fact, the partial state π is a flat output that differentially parameterizes the input u *via* (1.32) and the state x *via* (1.33). Further details on the flatness-based design of feedforward and tracking controllers using the differential operator representation (1.32) and (1.33) can be found in [14].

Also, the transfer behaviour (1.3) can be characterized by higher-order differential equations when setting

$$N(s) = CN_x(s) \tag{1.34}$$

(see (1.3), (1.4) and (1.28)), such that

$$D(s)\pi(s) = u(s), \tag{1.35}$$

$$y(s) = N(s)\pi(s). \tag{1.36}$$

When using the inverse Laplace transform this yields the differential operator representation

$$D(\tfrac{d}{dt})\pi(t) = u(t), \tag{1.37}$$

$$y(t) = N(\tfrac{d}{dt})\pi(t) \tag{1.38}$$

in the time domain.

The differential operator representation (1.32) and (1.33) of the system (1.1) can be used to determine a right coprime MFD (1.4) by inspection of the state equations. This is demonstrated in the following example.

Example 1.5. Computation of the right coprime MFD of a three-tank system using a differential operator representation

Three tanks with equal diameter are connected by tubes. Inputs are the flow rates of the two pumps supplying the first and the third tank. The three levels of the tanks are the state variables of the system. A Jacobian linearization of the differential equations for the tank levels leads to the state equations

$$\dot{x}_1(t) = -a\,x_1(t) + a\,x_2(t) + b\,u_1(t), \tag{1.39}$$

$$\dot{x}_2(t) = a\,x_1(t) - 2a\,x_2(t) + a\,x_3(t), \tag{1.40}$$

$$\dot{x}_3(t) = a\,x_2(t) - ka\,x_3(t) + b\,u_2(t), \tag{1.41}$$

where x and u are deviations from the operating point and

$$a = 0.5, \quad b = 1, \quad k = 2. \tag{1.42}$$

Since x_1 and x_3 are measured variables the outputs take the form

$$y_1(t) = x_1(t), \tag{1.43}$$

$$y_2(t) = x_3(t). \tag{1.44}$$

An inspection of the state equations shows that

$$\pi(t) = [\pi_1(t)\ \ \pi_2(t)]^T = [x_1(t)\ \ x_2(t)]^T \tag{1.45}$$

is a partial state, *i.e.*, it consists of the levels of the first and the second tank. This can be verified by expressing the remaining state x_3 and the inputs u by π. Solving (1.40) for x_3 and replacing x_2, $\dot{x}_2$ and x_1 by means of (1.45) yields

$$x_3(t) = \tfrac{1}{a}\left(\dot{\pi}_2(t) - a\,\pi_1(t) + 2a\,\pi_2(t)\right). \tag{1.46}$$

In view of (1.45) and (1.46) the differential operator representation

$$\begin{bmatrix} \frac{1}{b}\frac{d}{dt} + \frac{a}{b} & -\frac{a}{b} \\ -\frac{1}{b}\frac{d}{dt} - \frac{ka}{b} & \frac{1}{ab}\frac{d^2}{dt^2} + (k+2)\frac{1}{b}\frac{d}{dt} + (2k-1)\frac{a}{b} \end{bmatrix} \pi(t) = u(t), \tag{1.47}$$

$$y(t) = \begin{bmatrix} 1 & 0 \\ -1 & \frac{1}{a}\frac{d}{dt} + 2 \end{bmatrix} \pi(t) \tag{1.48}$$

(see (1.37) and (1.38)) is directly obtained from (1.39) and (1.41), where (1.48) is implied by (1.43), (1.44) and (1.46).

The partial state π can be eliminated in (1.47) and (1.48) by applying the Laplace transform with vanishing initial conditions $\pi(0) = 0$. This yields the transfer behaviour between u and y

$$\begin{aligned} y(s) &= N(s)D^{-1}(s)u(s) \\ &= \begin{bmatrix} 1 & 0 \\ -1 & \frac{1}{a}s + 2 \end{bmatrix} \begin{bmatrix} \frac{1}{b}s + \frac{a}{b} & -\frac{a}{b} \\ -\frac{1}{b}s - \frac{ka}{b} & \frac{1}{ab}s^2 + (k+2)\frac{1}{b}s + (2k-1)\frac{a}{b} \end{bmatrix}^{-1} u(s). \end{aligned} \tag{1.49}$$

The roots of $\det D(s)$ result by solving

$$\det D(s) = \frac{1}{ab^2}\, s^3 + \frac{k+3}{b^2}\, s^2 + \frac{3ka}{b^2}\, s + \frac{(k-1)a^2}{b^2} = 0, \tag{1.50}$$

and the roots of $\det N(s)$ are solutions of

$$\det N(s) = \tfrac{1}{a}s + 2 = 0. \tag{1.51}$$

Since (1.50) and (1.51) do not have a common root there does not exist a non-unimodular common right divisor of $N(s)$ and $D(s)$, *i.e.*, the MFD in (1.49) is right coprime (see Example 1.2). Consequently, one can use the results of Section 1.1 to determine the order of the system.

The highest column-degree-coefficient matrix and the column degrees of the denominator matrix $D(s)$ in (1.49) are

$$\Gamma_c[D(s)] = \begin{bmatrix} \frac{1}{b} & 0 \\ -\frac{1}{b} & \frac{1}{ab} \end{bmatrix}, \tag{1.52}$$

and

$$\delta_{c1}[D(s)] = 1, \quad \delta_{c2}[D(s)] = 2. \tag{1.53}$$

Since $\Gamma_c[D(s)]$ is non-singular in view of (1.42) and (1.52) the order n of the system is given by

$$n = \delta_{c1}[D(s)] + \delta_{c2}[D(s)] = 3. \tag{1.54}$$

The poles of (1.49) are the roots of (1.50) and the zeros of (1.49) are the roots of (1.51).

1.2 Left Coprime Matrix Fraction Description

A dual representation of the transfer matrix $G(s) = N(s)D^{-1}(s)$ can be obtained by converting the right coprime MFD to the *left coprime MFD*

$$G(s) = \bar{D}^{-1}(s)\bar{N}(s), \tag{1.55}$$

with an $m \times m$ *denominator matrix* $\bar{D}(s)$ and an $m \times p$ *numerator matrix* $\bar{N}(s)$. The right coprime MFD (1.4) is used for the design of state feedback controllers in the frequency domain (see Chapter 2), whereas the left coprime MFD (1.55) constitutes the basis of a design of observers in the frequency domain (see Chapter 3). In order to define the left coprimeness the following properties of two polynomial matrices are introduced.

Definition 1.8 (Left divisor). *A non-singular polynomial matrix* $L(s)$, i.e., $\det L(s)$ *not identically zero, is a* left divisor *of a polynomial matrix* $P(s)$ *if there exists a polynomial matrix* $\bar{P}(s)$*, such that* $P(s) = L(s)\bar{P}(s)$.

Definition 1.9 (Greatest common left divisor). *A polynomial matrix* $L(s)$ *is a* greatest common left divisor *of two polynomial matrices* $P_1(s)$ *and* $P_2(s)$ *if*

1. *the polynomial matrix* $L(s)$ *is a common left divisor of* $P_1(s)$ *and* $P_2(s)$, i.e., $P_1(s) = L(s)\bar{P}_1(s)$ *and* $P_2(s) = L(s)\bar{P}_2(s)$ *and*
2. *any other common left divisor* $Q(s)$ *of* $P_1(s)$ *and* $P_2(s)$ *is also a left divisor of* $L(s)$, i.e., $L(s) = Q(s)\bar{L}(s)$.

Definition 1.10 (Left coprime polynomial matrices). *Two polynomial matrices* $P_1(s)$ *and* $P_2(s)$ *are* (relatively) left coprime *if their greatest common left divisor* $U_L(s)$, i.e., $P_1(s) = U_L(s)\bar{P}_1(s)$ *and* $P_2(s) = U_L(s)\bar{P}_2(s)$*, is a* unimodular matrix, i.e., $\det U_L(s)$ *is a non-zero constant independent of* s.

Example 1.6. Greatest common left divisor

Consider the polynomial matrices

$$P_1(s) = \begin{bmatrix} s^2 & -1 \\ -s & s^2 \end{bmatrix} \quad \text{and} \quad P_2(s) = \begin{bmatrix} s & -s \\ 0 & 1 \end{bmatrix} \tag{1.56}$$

already introduced in Example 1.2. They are not right coprime. A greatest common left divisor is, for example, given by

$$L(s) = \begin{bmatrix} -1 & 0 \\ 0 & 1 \end{bmatrix}, \tag{1.57}$$

i.e.,

$$P_1(s) = \begin{bmatrix} -1 & 0 \\ 0 & 1 \end{bmatrix}\begin{bmatrix} -s^2 & 1 \\ -s & s^2 \end{bmatrix} \quad \text{and} \quad P_2(s) = \begin{bmatrix} -1 & 0 \\ 0 & 1 \end{bmatrix}\begin{bmatrix} -s & s \\ 0 & 1 \end{bmatrix}. \tag{1.58}$$

Since $L(s)$ is a unimodular matrix $P_1(s)$ and $P_2(s)$ are left coprime. This example shows that the left coprimeness of two polynomial matrices does not imply their right coprimeness.

By using the Bezout identity an alternative method can be obtained for checking whether two polynomial matrices are left coprime. The next theorem states this result, which is dual to Theorem 1.1.

Theorem 1.2 (Bezout identity for left coprime polynomial matrices). *The polynomial matrices $P_1(s)$ and $P_2(s)$ with the same number of rows are* left coprime *iff there exist polynomial matrices $\bar{X}(s)$ and $\bar{Y}(s)$, which are a solution of the* Bezout identity

$$P_1(s)\bar{X}(s) + P_2(s)\bar{Y}(s) = I. \tag{1.59}$$

Proof. A proof of this result can, *e.g.*, be found in [36]. □

Remark 1.7. The left coprime MFD (1.55) can also be computed on the basis of the state equations (1.1) and (1.2) when using the *right-to-left matrix fraction conversion*

$$C(sI - A)^{-1} = \bar{D}^{-1}(s)\bar{N}_x(s) \tag{1.60}$$

of the right MFD $C(sI - A)^{-1}$ with $\bar{D}(s)$ as in (1.55), so that

$$\bar{N}(s) = \bar{N}_x(s)B \tag{1.61}$$

obviously holds. Accordingly, the right MFD (1.4) is attainable by the *left-to-right matrix fraction conversion*

$$(sI - A)^{-1}B = N_x(s)D^{-1}(s) \tag{1.62}$$

of the left coprime MFD $(sI - A)^{-1}B$ with $D(s)$ as in (1.55), so that

$$N(s) = CN_x(s) \tag{1.63}$$

is obviously satisfied.

The quantities characterizing the degrees of the polynomial matrices in (1.55) and their degree structures are dual to the right coprime matrix fraction description. They can be obtained by applying the definitions of Section 1.1 to the transpose of the polynomial matrix in question. This leads to the following definitions.

Definition 1.11 (Row degree). *The* ith row degree $\delta_{ri}[P(s)]$ *of a polynomial matrix $P(s)$ is given by the degree of the polynomial element of highest degree in the ith row of $P(s)$.*

Definition 1.12 (Highest row-degree-coefficient matrix). *The $m \times m$* highest row-degree-coefficient matrix $\Gamma_r[P(s)]$ *of an $m \times m$ polynomial matrix $P(s)$ is the matrix of the coefficients of the polynomial elements with degree $\delta_{rj}[P(s)]$ in the jth row.*

Definition 1.13 (Row-reduced polynomial matrix). *An $m \times m$ polynomial matrix $P(s)$ is called* row reduced *if* $\det \Gamma_r[P(s)] \neq 0$.

Example 1.7. Row-degree structure of a polynomial matrix
Consider the polynomial matrix

$$P(s) = \begin{bmatrix} s^2 - 3 & 1 & 2s \\ 4s + 2 & 2 & 0 \\ -s^2 & s+3 & -3s + 2 \end{bmatrix} \tag{1.64}$$

already used in Example 1.1. The row degrees of this matrix are given by $\delta_{r1}[P(s)] = 2$, $\delta_{r2}[P(s)] = 1$ and $\delta_{r3}[P(s)] = 2$. Therefore, the highest row-degree-coefficient matrix of $P(s)$ has the form

$$\Gamma_r[P(s)] = \begin{bmatrix} 1 & 0 & 0 \\ 4 & 0 & 0 \\ -1 & 0 & 0 \end{bmatrix}. \tag{1.65}$$

The polynomial matrix $P(s)$ is not row reduced since $\det \Gamma_r[P(s)] = 0$. This can also be verified by computing

$$\det P(s) = 6s^3 + 44s^2 + 28s - 16, \tag{1.66}$$

which shows that $\deg \det P(s) < \delta_{r1}[P(s)] + \delta_{r2}[P(s)] + \delta_{r3}[P(s)] = 5$.

If and only if $\bar{D}(s)$ is row reduced can the order of the system (1.1) be obtained by inspection of the denominator matrix $\bar{D}(s)$ as

$$n = \sum_{j=1}^{m} \delta_{rj}[\bar{D}(s)] = \deg \det \bar{D}(s). \tag{1.67}$$

Furthermore, the relation

$$\det \bar{D}(s) = \bar{c} \det(sI - A), \tag{1.68}$$

with

$$\bar{c} = \det \Gamma_r[\bar{D}(s)] \neq 0 \tag{1.69}$$

holds. This shows that the poles of the transfer behaviour (1.3) (which coincide with the eigenvalues of A in (1.1) by assumption) are given by the roots of $\det \bar{D}(s)$.

Remark 1.8. The assumption of a row-reduced matrix $\bar{D}(s)$ means no loss of generality since any non-singular polynomial matrix $\bar{D}(s)$ (*i.e.*, $\det \bar{D}(s)$ does not vanish for all s) can be converted to a row-reduced polynomial matrix by pre-multiplication with an appropriate unimodular matrix $U_L(s)$ (see, *e.g.*, [67]).

Remark 1.9. In order to use the simple relation $\det \bar{D}(s) = \det(sI - A)$ (see (1.68)) it is assumed in the following that a row-reduced polynomial matrix $\bar{D}(s)$ satisfies $\det \Gamma_r[\bar{D}(s)] = 1$. This can always be achieved in an MFD (1.55) by pre-multiplying $\bar{N}(s)$ and $\bar{D}(s)$ with a suitable non-singular constant matrix.

Remark 1.10. Definition 1.10 also shows that a left coprime MFD (1.55) is unique up to a unimodular matrix $U_L(s)$ since

$$\begin{aligned} G(s) &= \bar{D}^{-1}(s)\bar{N}(s) \\ &= \bar{D}^{-1}(s)U_L^{-1}(s)U_L(s)\bar{N}(s) \\ &= (\bar{D}^*(s))^{-1}\bar{N}^*(s), \end{aligned} \tag{1.70}$$

with $\bar{N}^*(s) = U_L(s)\bar{N}(s)$ and $\bar{D}^*(s) = U_L(s)\bar{D}(s)$.

Remark 1.11. If the left MFD $G(s) = \bar{D}^{-1}(s)\bar{N}(s)$ of a system is coprime, then the minimal realization of $G(s)$ is of the order $\deg\det\bar{D}(s)$ [36]. Thus, by cancelling the greatest non-unimodular common left divisor $L(s)$ in $G(s) = (\bar{D}^*(s))^{-1}\bar{N}^*(s)$ with $\bar{N}^*(s) = L(s)\bar{N}(s)$ and $\bar{D}^*(s) = L(s)\bar{D}(s)$ the non-observable and the non-controllable parts of the system are removed in $G(s) = \bar{D}^{-1}(s)\bar{N}(s)$. If, on the other hand, the left coprime MFD $G(s) = \bar{D}^{-1}(s)\bar{N}(s)$ of a system of the order n is such that $\deg\det\bar{D}(s) = n$ then the system is controllable and observable.

Similar to the case of a right coprime MFD the (transmission) zeros of the transfer behaviour (1.3) are the μ values $s = s_{0i}$, $0 \leq \mu < n$, where the constant matrix $\bar{N}(s_{0i})$ of a left coprime MFD $G(s) = \bar{D}^{-1}(s)\bar{N}(s)$ is rank deficient.

Consider a proper transfer matrix

$$F(s) = \bar{D}^{-1}(s)\bar{N}(s) \tag{1.71}$$

(*i.e.*, the matrix $F(\infty)$ is not vanishing but finite), where $\bar{D}(s)$ is assumed to be row reduced. Then, the relations

$$\delta_{rj}[\bar{N}(s)] \leq \delta_{rj}[\bar{D}(s)], \quad j = 1, 2, \ldots, m \tag{1.72}$$

are satisfied. This is implied by (1.21) and by the duality between the left and the right coprime MFD. The direct feedthrough of $F(s)$ in (1.71) can be determined on the basis of the following definition.

Definition 1.14 (Highest row-degree-coefficient matrix with respect to a given row degree). *Let $Q(s)$ be an $m \times m$ polynomial matrix and $P(s)$ be an $m \times p$ polynomial matrix. Then, the* jth *row of the $m \times p$ constant matrix $\Gamma_{\delta_r[Q(s)]}[P(s)]$ contains the coefficients of the monomials $s^{\delta_{rj}[Q(s)]}$ in the* jth *row of $P(s)$.*

Example 1.8. Highest row-degree-coefficient matrix with respect to a given row degree

Consider the polynomial matrices

$$Q(s) = \begin{bmatrix} s^2 & 1 \\ 3 & s \end{bmatrix} \quad \text{and} \quad P(s) = \begin{bmatrix} 3 & 2s^2+s \\ 2 & 3 \end{bmatrix}, \tag{1.73}$$

which were already used in the Example 1.3. When applying Definition 1.14 the highest row-degree-coefficient matrix of $P(s)$ with respect to the row degrees $\delta_{r1}[Q(s)] = 2$ and $\delta_{r2}[Q(s)] = 1$ of $Q(s)$ is given by

$$\Gamma_{\delta_r[Q(s)]}[P(s)] = \begin{bmatrix} 0 & 2 \\ 0 & 0 \end{bmatrix}. \tag{1.74}$$

By using the matrix introduced in Definition 1.14 the direct feedthrough of the transfer matrix $F(s)$ in (1.71) is given by

$$F(\infty) = \Gamma_r^{-1}[\bar{D}(s)]\Gamma_{\delta_r[\bar{D}(s)]}[\bar{N}(s)]. \tag{1.75}$$

This result can be directly obtained when applying (1.22) to the transpose of (1.71).

If the transfer behaviour (1.3) is characterized by a left coprime MFD (1.55), *i.e.*,

$$y(s) = \bar{D}^{-1}(s)\bar{N}(s)u(s) \tag{1.76}$$

a representation of the system by higher-order differential equations can be computed by introducing the Laplace transform

$$\bar{\pi}(s) = \bar{N}(s)u(s) \tag{1.77}$$

of the $m \times 1$ *partial state* $\bar{\pi}(t)$. After a simple rearrangement of (1.76) and (1.77) one obtains the representation

$$\bar{\pi}(s) = \bar{N}(s)u(s), \tag{1.78}$$

$$\bar{D}(s)y(s) = \bar{\pi}(s). \tag{1.79}$$

An application of the inverse Laplace transform to (1.78) and (1.79) yields the *differential operator representation*

$$\bar{\pi}(t) = \bar{N}(\tfrac{d}{dt})u(t), \tag{1.80}$$

$$\bar{D}(\tfrac{d}{dt})y(t) = \bar{\pi}(t) \tag{1.81}$$

of the system in the time domain. This is a dual form of the representation (1.37) and (1.38).

Example 1.9. Left coprime matrix fraction description

Consider the right coprime MFD of the transfer matrix already discussed in Example 1.4. Applying the right-to-left matrix fraction conversion to this MFD the left coprime MFD

$$G(s) = \bar{D}^{-1}(s)\bar{N}(s) = \begin{bmatrix} s^2+4s+3 & -1 \\ s+2 & s+2 \end{bmatrix}^{-1} \begin{bmatrix} 1 & 0 \\ 1 & 1 \end{bmatrix} \tag{1.82}$$

is obtained. The row degrees of $\bar{D}(s)$ are

$$\delta_{r1}[\bar{D}(s)] = 2, \quad \delta_{r2}[\bar{D}(s)] = 1, \tag{1.83}$$

so that the highest row-degree-coefficient matrix of the denominator matrix $\bar{D}(s)$ has the from

$$\Gamma_r[\bar{D}(s)] = \begin{bmatrix} 1 & 0 \\ 1 & 1 \end{bmatrix}. \tag{1.84}$$

Because this matrix has full rank, the denominator matrix $\bar{D}(s)$ is row reduced, so that the order of the system can be obtained from the sum of its row degrees, *i.e.*,

$$n = \delta_{r1}[\bar{D}(s)] + \delta_{r2}[\bar{D}(s)] = 3. \tag{1.85}$$

As the factorization (1.82) is coprime the poles of the transfer matrix $G(s)$ are the roots of

$$\det \bar{D}(s) = (s+2)^3. \tag{1.86}$$

Therefore, the system has three poles at $s = -2$.

Since $\bar{N}(s)$ is a square matrix, the zeros of the transfer behaviour are the roots of $\det \bar{N}(s) = 0$. As

$$\det \bar{N}(s) = 1 \tag{1.87}$$

for all s the system has no transmission zeros.

A comparison of these results with those of Example 1.4 demonstrates that the left coprime MFD is a dual representation of the right coprime MFD (1.4). Since

$$\Gamma_{\delta_r[\bar{D}(s)]}[\bar{N}(s)] = \begin{bmatrix} 0 & 0 \\ 0 & 0 \end{bmatrix} \tag{1.88}$$

an inspection of (1.75) shows that the transfer matrix (1.82) has no direct feedthrough.

2

State Feedback Control

If a system is controllable all eigenvalues can be placed at arbitrary locations by static state feedback. Therefore, state feedback control can be used to stabilize the closed-loop system and to achieve further design specifications. Hence, the parameterization of state feedback controllers is an important aspect of control theory.

In a system of the order n with one input u (*i.e.*, a single-input system), the desired eigenvalues of the closed-loop system specify all n elements of the $1 \times n$ state feedback gain k^T. If the system has $p > 1$ inputs (*i.e.*, in a multiple-input system), the desired eigenvalues again specify n elements of the $p \times n$ state feedback gain K. Therefore, after the assignment of the eigenvalues there remain $(p-1)n$ degrees of freedom parameterizing various properties of the closed-loop system as, *e.g.*, the zeros in the elements of its transfer matrix.

In the frequency domain, the state feedback is parameterized by a polynomial matrix $\tilde{D}(s)$. In the single-input case, all free coefficients in $\tilde{D}(s)$ are specified by the desired eigenvalues of the closed-loop system, whereas in the multiple-input case additional $(p-1)n$ degrees of freedom also exist.

If the system described by its transfer behaviour has n linearly independent measurable outputs, a pole-placing output feedback controller can be computed from the parameterizing polynomial matrix $\tilde{D}(s)$ without recourse to any state-space description.

The main results related to a time-domain representation of state feedback control are summarized briefly in Section 2.1. A more comprehensive treatment of the subject can be found in many text books on the control design of linear systems as, *e.g.*, [36]. Section 2.2 describes the frequency-domain design of state feedback control. It is based on the right coprime MFD of the system, and it is parameterized by the coefficients of the denominator matrix $\tilde{D}(s)$. It is shown that this polynomial matrix contains the same number of free parameters as the state feedback gain K. A connecting relation is also derived that, given a time-domain parameterization of state feedback, enables the designer to compute an equivalent frequency-domain parameterization and *vice versa*.

2.1 State Feedback in the Time Domain

Considered is the control of linear, time-invariant systems with the state equations

$$\dot{x}(t) = Ax(t) + Bu(t), \tag{2.1}$$

$$y(t) = Cx(t), \tag{2.2}$$

where $x \in \mathbb{R}^n$ is the state, $u \in \mathbb{R}^p$ is the input, and $y \in \mathbb{R}^m$ with $m \geq p$ is the measurement. It is assumed that the inputs and outputs are linearly independent, *i.e.*, rank $B = p$ and rank $C = m$.

The controlled output $y_c \in \mathbb{R}^p$ is assumed to be measurable. Therefore, a $p \times m$ selection matrix

$$\Xi = \begin{bmatrix} e_{i_1}^T \\ e_{i_2}^T \\ \vdots \\ e_{i_p}^T \end{bmatrix}, \; i_\nu \in \{1, 2, \ldots, m\}, \; \nu = 1, 2, \ldots, p \tag{2.3}$$

exists with e_{i_ν} denoting the i_νth unit vector, so that

$$y_c(t) = \Xi y(t) = \Xi C x(t) = C_c x(t), \tag{2.4}$$

i.e., one has $y_c^1 = y^{i_1}, \ldots, y_c^p = y^{i_p}$, where y^j denotes the jth element of the vector y.

Remark 2.1. In view of designing reduced-order observers the $m \times 1$ vector y of the measurements will be subdivided into an $(m - \kappa) \times 1$ vector y_1 and a $\kappa \times 1$ vector y_2 in many parts of this book. Therefore, the unusual notion y^j is adopted for the jth component of y.

On the one hand, only the controllable and observable part of a system can be influenced by output feedback (as, for example, observer-based) control. On the other hand, the coprime MFDs of the system exactly describe its controllable and observable part. Therefore, the time-domain representation (C, A, B) of the model of the system used for controller design is always assumed to be such that the pair (A, B) is controllable and the pair (C, A) is observable (see also Remark 1.1).

The stabilizing state feedback has the form

$$u(t) = -Kx(t) + Mr(t), \tag{2.5}$$

where $r \in \mathbb{R}^p$ is the reference input. In (2.5) the $p \times n$ feedback gain K must be chosen such that the eigenvalues $\tilde{s}_\nu$, $\nu = 1, 2, \ldots, n$, of the controlled system, *i.e.*, the zeros of the characteristic polynomial $\det(sI - A + BK)$,

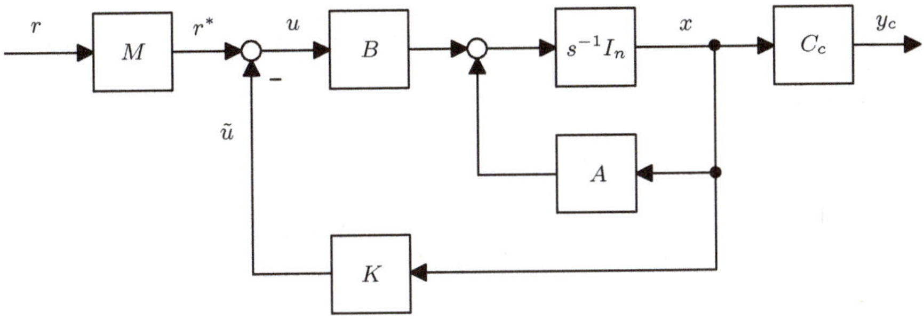

Figure 2.1. State feedback control in the time domain

are located in the complex open left-half plane. The remaining $(p-1)n$ free parameters in K also influence the properties of the closed-loop system. They can, for example, be exploited to assure a reference transfer behaviour that is decoupled (see also Chapter 6).

By applying the control (2.5) to (2.1) and (2.2) the closed-loop system is described by the state equations

$$\dot{x}(t) = (A - BK)x(t) + BMr(t), \tag{2.6}$$

$$y_c(t) = C_c x(t). \tag{2.7}$$

A block diagram representation of the closed-loop system (2.6) and (2.7) is shown in Figure 2.1.

A reference transfer behaviour

$$y_c(s) = C_c(sI - A + BK)^{-1}BMr(s) = G_r(s)r(s) \tag{2.8}$$

with vanishing steady-state error $y_c(\infty) - r(\infty)$ for stationary constant reference signals (*i.e.*, $r(\infty) = const$) requires $G_r(0) = I$. For all stabilizing gains K, *i.e.*, the inverse of $A - BK$ exists, this is assured by the constant $p \times p$ matrix

$$M = \left[C_c(-A + BK)^{-1}B\right]^{-1} \tag{2.9}$$

(see (2.5)). The matrix in square brackets in (2.9) has full rank iff the system (C_c, A, B) has no invariant zero at $s = 0$. This can be shown by elementary operations applied to Rosenbrock's system matrix (see [58]) defined for the system (2.6) and (2.7). In the following, it will always be assumed that no invariant zero of the system (C_c, A, B) is located at $s = 0$.

2.2 Parameterization of the State Feedback in the Frequency Domain

The system (2.1) and (2.2) is now assumed to be described by its transfer behaviour $y(s) = G(s)u(s)$, where the $m \times p$ transfer matrix $G(s)$ is related to the time-domain quantities by

$$G(s) = C(sI - A)^{-1}B. \tag{2.10}$$

The transfer matrix $G(s)$ of the system is represented with the aid of an $m \times p$ numerator matrix $N(s)$ and a $p \times p$ denominator matrix $D(s)$ by a right coprime MFD

$$G(s) = N(s)D^{-1}(s). \tag{2.11}$$

The controlled output $y_c = \Xi y$ (see also (2.4)) is characterized by its transfer behaviour $y_c(s) = G_c(s)u(s)$ and the transfer matrix $G_c(s)$ is represented by the right MFD

$$G_c(s) = N_c(s)D^{-1}(s), \tag{2.12}$$

where $N_c(s)$ is a $p \times p$ polynomial matrix. This polynomial matrix has the form $N_c(s) = \Xi N(s)$ (see also (2.4)).

Remark 2.2. In general, the MFD (2.12) need not be coprime, *i.e.*, the system can be such that the order of $y_c(s) = G_c(s)u(s)$ is smaller than the order of $y(s) = G(s)u(s)$. A right coprime MFD of $G_c(s)$ then has the form $G_c(s) = N_c^*(s)D^{*-1}(s)$ with $\deg(\det D^*(s)) < \deg(\det D(s))$. Throughout this book, the MFD of $G_c(s)$ is assumed to have the form (2.12) with $D(s)$ as defined in (2.11) and $N_c(s) = \Xi N(s)$.

The MFDs (2.11) and (2.12) are assumed to be such that $D(s)$ is column reduced. The poles of $G(s)$ are the zeros of $\det D(s)$ and the invariant zeros of the system (2.1)–(2.2) are the zeros of $\det N_c(s)$. If the MFD (2.12) is coprime the zeros of $\det N_c(s)$ are the transmission zeros of $G_c(s)$.

In the discussion of state feedback control it is always assumed that the state x of the system is directly measurable. Therefore, the right coprime MFD in the transfer behaviour

$$x(s) = N_x(s)D^{-1}(s)u(s) = (sI - A)^{-1}Bu(s) \tag{2.13}$$

is also considered with $D(s)$ as defined in (2.11) and $N_x(s)$ being an $n \times p$ polynomial matrix. Obviously, the equalities

$$N(s) = CN_x(s) \quad \text{and} \quad N_c(s) = C_cN_x(s) \tag{2.14}$$

then also hold.

Using the MFDs (2.12) and (2.13), the transfer behaviour of the system in Figure 2.1 can be equally represented by the block diagram in Figure 2.2 with a feedback

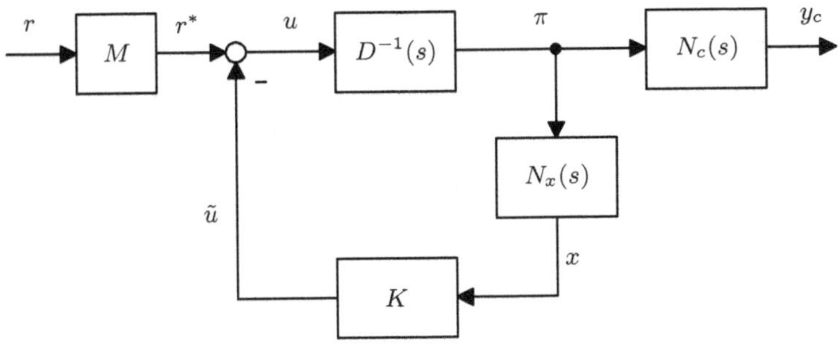

Figure 2.2. State feedback control in the frequency domain

$$\tilde{u}(s) = KN_x(s)\pi(s) \tag{2.15}$$

of the partial state π (see (1.29)).

In the closed-loop system of Figure 2.1 the transfer behaviour between the modified reference input r^* and the controlled output y_c is

$$y_c(s) = C_c(sI - A + BK)^{-1}Br^*(s). \tag{2.16}$$

In the closed-loop system of Figure 2.2 this transfer behaviour has the form

$$y_c(s) = N_c(s)\left[D(s) + KN_x(s)\right]^{-1} r^*(s). \tag{2.17}$$

Thus, the $p \times p$ denominator matrix

$$\tilde{D}(s) = D(s) + KN_x(s) \tag{2.18}$$

in the MFD (2.17) characterizes the dynamics of the closed-loop system in Figure 2.2.

By using the substitution (2.18) the Equations (2.16) and (2.17) can be written as

$$y_c(s) = \frac{C_c\text{adj}\{sI - A + BK\}B}{\det(sI - A + BK)}r^*(s) = \frac{N_c(s)\text{adj}\{\tilde{D}(s)\}}{\det \tilde{D}(s)}r^*(s), \tag{2.19}$$

where adj$\{\cdot\}$ denotes the adjoint of a matrix. This shows that

$$\det \tilde{D}(s) = \det(sI - A + BK). \tag{2.20}$$

By inserting (2.18) and $r^* = Mr$ in (2.17) one obtains the reference transfer behaviour

$$y_c(s) = G_r(s)r(s) = N_c(s)\tilde{D}^{-1}(s)Mr(s) \tag{2.21}$$

of the state feedback control in the frequency domain. This also shows that the invariant zeros of the system (*i.e.*, the zeros of $\det N_c(s)$) are not influenced by state feedback.

A vanishing steady-state error $y_c(\infty) - r(\infty)$ for stationary constant reference signals (*i.e.*, $r(\infty) = const$) is assured by $G_r(0) = I$ which yields

$$M = \tilde{D}(0)N_c^{-1}(0), \tag{2.22}$$

in view of (2.21). The matrix M in (2.22) only exists if $\det N_c(s)$ has no root at $s = 0$ and this is the frequency-domain equivalent to the corresponding results in Section 2.1.

Given a frequency-domain parameterization of the state feedback control by $\tilde{D}(s)$, the Relation (2.18) can be used to obtain an equivalent feedback gain K or *vice versa.* When multiplying (2.18) from the right by $D^{-1}(s)$ and using (2.13) one obtains the well-known connecting relation

$$\tilde{D}(s)D^{-1}(s) = I + K(sI - A)^{-1}B \tag{2.23}$$

between the time- and the frequency-domain representations of state feedback control (see, *e.g.*, [2, 36]).

In the following discussion of the properties of $\tilde{D}(s)$ the notion of the polynomial part of a rational matrix is required. Recall that given a rational matrix $G(s)$ and its corresponding limit value $G_\infty = \lim_{s\to\infty} G(s)$, this matrix is called *strictly proper* if $G_\infty = 0$, *proper* if G_∞ is finite and not vanishing, and *improper* if at least one element of G_∞ is not finite.

Definition 2.1 (Polynomial part and strictly proper part of a rational matrix). *Any given transfer matrix* $G(s)$ *can be represented as*

$$G(s) = \Pi\{G(s)\} + SP\{G(s)\}, \tag{2.24}$$

where the polynomial part $\Pi\{\cdot\}$ *is a polynomial matrix and the* strictly proper part $SP\{\cdot\}$ *is a strictly proper rational matrix.*

Therefore, in a proper rational matrix $G(s)$ the polynomial part $\Pi\{G(s)\}$ coincides with the direct feedthrough G_∞ from the input u to the output y (see also (1.22) and (1.75)).

Example 2.1. Decomposition of a rational matrix into its polynomial and strictly proper parts

For the rational matrix

$$G(s) = \frac{1}{s+1}\begin{bmatrix} s+3 & 2s+4 \\ 3 & s^2+3s+2 \end{bmatrix} \tag{2.25}$$

the polynomial part is given by

$$\Pi\{G(s)\} = \begin{bmatrix} 1 & 2 \\ 0 & s+2 \end{bmatrix}, \tag{2.26}$$

and the strictly proper part by

$$SP\{G(s)\} = \frac{1}{s+1}\begin{bmatrix} 2 & 2 \\ 3 & 0 \end{bmatrix}. \tag{2.27}$$

In the time domain the state feedback is parameterized by the $p \times n$ feedback gain K, which contains pn free parameters. The following theorem shows that the $p \times p$ polynomial matrix $\tilde{D}(s)$ plays the corresponding role in the frequency domain.

Theorem 2.1 (Parameterizing polynomial matrix of state feedback). *The $p \times p$ polynomial matrix $\tilde{D}(s)$ characterizing the dynamics of the state feedback loop in the frequency domain has the properties*

$$\delta_{ci}[\tilde{D}(s)] = \delta_{ci}[D(s)], \quad i = 1, 2, \ldots, p, \tag{2.28}$$

and

$$\Gamma_c[\tilde{D}(s)] = \Gamma_c[D(s)], \tag{2.29}$$

and it has exactly the same number of free parameters as the state feedback gain K, namely pn.

Proof. Since $K(sI - A)^{-1}B$ is strictly proper the right-hand side of (2.23) indicates that the polynomial part $\Pi\{\tilde{D}(s)D^{-1}(s)\}$ is the identity matrix. This implies that $\tilde{D}(s)D^{-1}(s)$ is proper and as $D(s)$ is column reduced (*i.e.*, $\det \Gamma_c[D(s)] \neq 0$) this further implies that $\delta_{ci}[\tilde{D}(s)] \leq \delta_{ci}[D(s)]$, $i = 1, 2, \ldots, p$ (see Section 1.1). Applying (1.22) one obtains

$$\Pi\{\tilde{D}(s)D^{-1}(s)\} = \Gamma_{\delta_c[D(s)]}[\tilde{D}(s)]\Gamma_c^{-1}[D(s)] = I. \tag{2.30}$$

Postmultiplying this by $\Gamma_c[D(s)]$ leads to

$$\Gamma_{\delta_c[D(s)]}[\tilde{D}(s)] = \Gamma_c[D(s)]. \tag{2.31}$$

Since $D(s)$ is column reduced also $\det \Gamma_{\delta_c[D(s)]}[\tilde{D}(s)] \neq 0$. This, on the other hand, shows that (2.28) is satisfied, because if $\delta_{ci}[\tilde{D}(s)] < \delta_{ci}[D(s)]$ for any i, the corresponding column of $\Gamma_{\delta_c[D(s)]}[\tilde{D}(s)]$ would vanish, contradicting $\det \Gamma_{\delta_c[D(s)]}[\tilde{D}(s)] \neq 0$. With (2.28) satisfied, (2.29) directly follows from (2.31).

Since $D(s)$ is column reduced the order of the system is

$$n = \sum_{i=1}^{p} \delta_{ci}[D(s)] = \sum_{i=1}^{p} \delta_{ci}[\tilde{D}(s)] \tag{2.32}$$

in view of (2.28). Because of (2.29) the number n_{fi} of free parameters in each column of $\tilde{D}(s)$ is $n_{fi} = p\delta_{ci}[\tilde{D}(s)]$ and consequently, the total number n_f of free parameters in $\tilde{D}(s)$ is

$$n_f = \sum_{i=1}^{p} p\,\delta_{ci}[\tilde{D}(s)] = pn \tag{2.33}$$

in view of (2.32). □

Remark 2.3. Representing $\tilde{D}(s)$, which has the properties (2.28) and (2.29), in the form

$$\tilde{D}(s) = \Gamma_c[D(s)]\mathrm{diag}\left(s^{\delta_{ci}[D(s)]}\right) + \tilde{D}_c S(s), \tag{2.34}$$

with

$$S(s) = \mathrm{diag}\left([s^{\delta_{c1}[D(s)]-1} \dots s\ 1]^T, \dots, [s^{\delta_{cp}[D(s)]-1} \dots s\ 1]^T\right) \tag{2.35}$$

it becomes obvious that the degrees of freedom in $\tilde{D}(s)$ are contained in the constant matrix $\tilde{D}_c$ of freely assignable coefficients in (2.34). It has the same dimensions as K, *i.e.*, it is a $p \times n$ matrix since $\sum_{i=1}^{p} \delta_{ci}[D(s)] = n$ (see (2.32)).

Therefore, the linear state feedback control can either be parameterized in the time domain by the $p \times n$ constant matrix K or in the frequency domain by the $p \times p$ polynomial matrix $\tilde{D}(s)$. In both cases, the same number pn of free parameters exists.

Given a set of desired eigenvalues of the closed-loop system, the parameterization via $\tilde{D}(s)$ or K is uniquely defined in the case of single-input systems. For multiple-input systems, there exist $(p-1)n$ additional degrees of freedom in $\tilde{D}(s)$ and K, so that the parameterization of a state feedback control yielding desired eigenvalues of the closed-loop system is more complicated in this case.

Remark 2.4. In the frequency domain there exists a simple way to solve the eigenvalue-assignment problem by using the parameterizing matrix

$$\tilde{D}(s) = \Gamma_c[D(s)] \begin{bmatrix} \tilde{d}_1(s) & & \\ & \ddots & \\ & & \tilde{d}_p(s) \end{bmatrix}, \tag{2.36}$$

with

$$\deg \tilde{d}_i(s) = \delta_{ci}[D(s)], \quad i = 1, 2, \dots, p, \tag{2.37}$$

so that one obtains

$$\det \tilde{D}(s) = \tilde{d}_1(s) \cdot \ldots \cdot \tilde{d}_p(s). \tag{2.38}$$

More degrees of freedom can possibly be used when choosing the free coefficients in $\tilde{D}(s)$ in an appropriate way, so that (2.38) is satisfied for given polynomials $\tilde{d}_i(s)$ (see, *e.g.*, Example 2.2). If all degrees of freedom need to be used, the parametric approach presented in Chapter 5 can be used.

Equation (2.18) is a special form of the Diophantine equation

$$P(s)N_x(s) + Q(s)D(s) = \tilde{D}(s) \tag{2.39}$$

with polynomial matrices $P(s)$ and $Q(s)$. A solution of this Diophantine equation is given by the pair of constant matrices

$$P(s) = P = K \text{ and } Q(s) = Q = I, \tag{2.40}$$

which becomes obvious by comparison of (2.39) and (2.18).

Remark 2.5. If the pair $(N_x(s), D(s))$ is right coprime (*i.e.*, if the pair (A, B) is controllable) the Diophantine equation (2.39) can be solved for arbitrary right-hand sides $\tilde{D}(s)$ having the properties (2.28) and (2.29), *i.e.*, the dynamics of the closed-loop system can be assigned arbitrarily. If a controllability defect occurs, the matrices $N_x(s)$ and $D(s)$ have a greatest common right divisor $R(s)$, which is no longer a unimodular matrix. Then, the Diophantine equation (2.39) takes the form

$$P(s)N_x^*(s)R(s) + Q(s)D^*(s)R(s) = \tilde{D}(s), \tag{2.41}$$

and this makes it obvious that a solution only exists if the polynomial matrix $\tilde{D}(s)$ also has the form $\tilde{D}(s) = \tilde{D}^*(s)R(s)$. This immediately shows which part of the system dynamics cannot be changed by state feedback control.

If n linearly independent measured outputs y exist, an arbitrary assignment of all eigenvalues of the closed-loop system is also possible by an output feedback $u = -Py$ with the constant feedback gain P. This output feedback can be directly computed in the frequency domain without recourse to a time-domain representation of the system by replacing the polynomial matrix $N_x(s)$ by $N(s)$ in the Diophantine equation (2.39).

Example 2.2. Parameterization of a state feedback control in the frequency domain

Considered is a system with two inputs u and three measured outputs y. The right coprime MFD of its transfer matrix $G(s)$ has the form

$$G(s) = N(s)D^{-1}(s) = \begin{bmatrix} 1 & 0 \\ s & 1 \\ 0 & 1 \end{bmatrix} \begin{bmatrix} s^2+3s+3 & -1 \\ -1 & s+3 \end{bmatrix}^{-1}. \tag{2.42}$$

The characteristic polynomial of the system is $\det D(s) = (s+2)^3$, *i.e.*, it has the order $n = 3$ and its three poles are located at $s = -2$. As no constant left annihilator $N^{\perp}$ of $N(s)$ exists the $m = 3$ outputs are linearly independent, which means that no output can be represented as a linear combination of the other two.

It is assumed that the controlled outputs y_c coincide with the first two measurements. Therefore, the selection matrix in $y_c = \Xi y$ (see (2.4)) is

$$\Xi = \begin{bmatrix} 1 & 0 & 0 \\ 0 & 1 & 0 \end{bmatrix}, \tag{2.43}$$

giving the corresponding numerator matrix

$$N_c(s) = \Xi N(s) = \begin{bmatrix} 1 & 0 \\ s & 1 \end{bmatrix}. \tag{2.44}$$

Since $\det N_c(s) = 1$ the transfer matrix $G_c(s) = N_c(s)D^{-1}(s)$ has no zero.

The control is intended to place all three eigenvalues of the system at $\tilde{s}_i = -5$, $i = 1, 2, 3$, and since n linearly independent measurements exist, it can be realized by an output feedback

$$u(s) = -Py(s) + Mr(s). \tag{2.45}$$

The column degrees of $D(s)$ are $\delta_{c1}[D(s)] = 2$, $\delta_{c2}[D(s)] = 1$ and the highest column-degree-coefficient matrix is $\Gamma_c[D(s)] = I$. Therefore, the parameterizing matrix $\tilde{D}(s)$ has the general form

$$\tilde{D}(s) = \begin{bmatrix} s^2 + \alpha s + \beta & \gamma \\ \delta s + \varepsilon & s + \varphi \end{bmatrix}, \tag{2.46}$$

i.e., there are $pn = 6$ free parameters α, β, γ, δ, ε, and φ.

A choice of $\alpha = 10$, $\beta = 25$, $\gamma = 0$, $\delta = 2$, $\varepsilon = 10$, and $\varphi = 5$ assures the desired characteristic polynomial $\det \tilde{D}(s) = (s+5)^3$ and leads to

$$\tilde{D}(s) = \begin{bmatrix} s^2+10s+25 & 0 \\ 2s + 10 & s + 5 \end{bmatrix}. \tag{2.47}$$

If vanishing steady-state errors for constant reference inputs are required this is assured by the constant matrix

$$M = \tilde{D}(0)N_c^{-1}(0) = \begin{bmatrix} 25 & 0 \\ 10 & 5 \end{bmatrix}. \tag{2.48}$$

In order to obtain the feedback gain P in (2.45) the Diophantine equation (2.39) with $N_x(s) = N(s)$, *i.e.*,

$$P(s)N(s) + Q(s)D(s) = \tilde{D}(s) \tag{2.49}$$

has to be solved. By inserting the above matrices $N(s)$, $D(s)$ and $\tilde{D}(s)$ one obtains $Q(s) = Q = I$ and

$$P(s) = P = \begin{bmatrix} 22 & 7 & -6 \\ 11 & 2 & 0 \end{bmatrix}. \tag{2.50}$$

To assign the desired eigenvalues of the closed-loop system it is obviously much easier to look for a polynomial matrix $\tilde{D}(s)$ in the frequency domain than to determine K from $\det(sI - A + BK) = \prod_{i=1}^{n}(s - \tilde{s}_i)$ in a time-domain approach.

3

State Observers

State feedback control for an arbitrary assignment of the eigenvalues is only realizable if all the states of the system are measurable, *i.e.*, if the number m of measurements coincides with the order n of the system. For $m < n$ an observer must be used and the feedback of the estimated states defines the observer-based compensator.

Given m linearly independent measurements y, then the state can be estimated with the aid of an observer of the order $n_O = n - \kappa$ with $0 \leq \kappa \leq m$. This includes as special cases the full-order or identity observer ($n_O = n$) and the completely reduced-order observer ($n_O = n - m$). If additional properties of the observer-based compensator have to be assured, such as asymptotic disturbance rejection, observers of the order $n_O > n$ are sometimes also used. For state reconstruction, however, the identity observer with $n_O = n$ usually constitutes the upper limit.

The best-known approach to state observers is the so-called Luenberger observer (see [46]) that can be designed by solving a Sylvester equation. Unfortunately, in a Luenberger observer of the order $n_O < n$ it is not obvious which of the parameters influence the properties of the closed-loop system when applying observer-based control. Since a connecting relation between the time- and the frequency-domain parameterizations of an observer can only be established if the influential parameters can also be identified in the time domain, Luenberger's formulation cannot be used as a basis for the frequency-domain design of reduced-order observers.

A different approach was presented by Uttam and O'Halloran in [62], which has also been discussed in [21]. This approach uses a characterization of reduced-order observers by two matrices, which allow to identify the $m(n-\kappa)$ influential parameters in the above defined sense. This representation will be used throughout the book. If an optimization approach (see Chapter 8) or a parametric design (see Chapter 5) are applied, the characterizing matrices directly result. They can also be obtained from the frequency-domain parameterization of the observer.

In the frequency domain all degrees of freedom of the observer are characterized by an $m \times m$ polynomial matrix $\tilde{\bar{D}}(s)$, so that for any given parameterizing $\tilde{\bar{D}}(s)$ a corresponding time-domain realization of the observer can be obtained.

In Section 3.1 the time-domain version of the (reduced-order) observer of Uttam and O'Halloran is presented. It also allows a non-minimal realization of reduced-order observers in the time domain to be derived. This will play a crucial role in deriving the relations between the time-domain and the frequency-domain representations of such observers. Since the manipulations in the derivation of the reduced-order observer are quite involved, the frequency-domain parameterization of the full-order observer or identity observer is first presented in Section 3.2. This section can be regarded as a motivation for the steps necessary to define the parameterizing polynomial matrix $\tilde{\bar{D}}(s)$ of a state observer of the general order $n_O = n - \kappa$, $0 \leq \kappa \leq m$, as described in Section 3.3.

3.1 The Reduced-order Observer in the Time Domain

Considered are systems with the state equations

$$\dot{x}(t) = Ax(t) + Bu(t), \tag{3.1}$$

$$y(t) = Cx(t), \tag{3.2}$$

where $x \in \mathbb{R}^n$ is the state, $u \in \mathbb{R}^p$ is the input, and $y \in \mathbb{R}^m$ with $m \geq p$ is the measurement. It is assumed that the pair (A, B) is controllable and the pair (C, A) is observable and that the inputs and outputs are linearly independent, *i.e.*, rank $B = p$ and rank $C = m$. In view of designing reduced-order observers of the order $n_O = n - \kappa$, $0 \leq \kappa \leq m$, the $m \times 1$ output vector y of these systems is arranged according to

$$y(t) = \begin{bmatrix} y_1(t) \\ y_2(t) \end{bmatrix} = \begin{bmatrix} C_1 \\ C_2 \end{bmatrix} x(t). \tag{3.3}$$

Here $y_2 \in \mathbb{R}^\kappa$ with $0 \leq \kappa \leq m$, contains the measurements directly used in the construction of the estimate $\hat{x}$ and $y_1 \in \mathbb{R}^{m-\kappa}$ contains the remaining $m - \kappa$ measurements. Therefore, y_2 and C_2 vanish for $\kappa = 0$ (characterizing the *full-order* or *identity observer*) and y_1 and C_1 vanish for $\kappa = m$ (characterizing the *completely reduced-order observer*).

This allows all observers of the orders $n - m \leq n_O \leq n$ to be included in one common scheme.

When using y_2 to reconstruct the state of the system, only $n - \kappa$ linear combinations

$$\zeta(t) = Tx(t) \tag{3.4}$$

of the state are needed to reconstruct x. Provided that the $(n-\kappa)\times n$ matrix T has full row rank and the rows of T and C_2 are linearly independent, the inverse of the composite matrix

$$\begin{bmatrix} C_2 \\ T \end{bmatrix}^{-1} = \begin{bmatrix} \Psi_2 & \Theta \end{bmatrix} \tag{3.5}$$

exists and it can be split into the $n \times \kappa$ matrix Ψ_2 and the $n \times (n-\kappa)$ matrix Θ. As a direct consequence of (3.5) the following two relations

$$\begin{bmatrix} C_2\Psi_2 & C_2\Theta \\ T\Psi_2 & T\Theta \end{bmatrix} = \begin{bmatrix} I_\kappa & 0 \\ 0 & I_{n-\kappa} \end{bmatrix}, \tag{3.6}$$

and

$$\Psi_2 C_2 + \Theta T = I_n \tag{3.7}$$

are satisfied. Using the output y_2 and the state ζ as defined in (3.4) the state x of the system can be expressed as

$$x(t) = \begin{bmatrix} C_2 \\ T \end{bmatrix}^{-1} \begin{bmatrix} y_2(t) \\ \zeta(t) \end{bmatrix} = \Psi_2\, y_2(t) + \Theta\, \zeta(t), \tag{3.8}$$

because of (3.5). Therefore, the state x can be reconstructed from the κ measurements y_2 and the $n-\kappa$ states ζ.

There remains the task of estimating the state ζ. By successively using (3.4), (3.1) and (3.8) one obtains the differential equation

$$\begin{aligned} \dot{\zeta}(t) &= T\dot{x}(t) = TA\,x(t) + TB\,u(t) \\ &= TA\Theta\,\zeta(t) + TA\Psi_2\, y_2(t) + TB\,u(t) \end{aligned} \tag{3.9}$$

for the state ζ. This is a system of the order $n-\kappa$ and the identity observer for this reduced-order system is

$$\dot{\hat{\zeta}}(t) = TA\Theta\,\hat{\zeta}(t) + TA\Psi_2\, y_2(t) + L_1^*\,(y_1(t) - \hat{y}_1(t)) + TB\,u(t). \tag{3.10}$$

To justify the structure of the observer in (3.10) insert the estimate $\hat{\zeta}$ in (3.8) to obtain the state estimate $\hat{x}$ and consequently

$$\hat{y}_1(t) = C_1\hat{x}(t) = C_1\Theta\,\hat{\zeta}(t) + C_1\Psi_2\, y_2(t). \tag{3.11}$$

Obviously $\hat{y}_1$ depends on $\hat{\zeta}$, so that it makes sense to use the correction term $L_1^*\,(y_1(t) - \hat{y}_1(t))$ in the observer. Because of (3.6) and (3.8) the estimate $\hat{y}_2$ is given by

$$\hat{y}_2(t) = C_2\hat{x}(t) = C_2\Psi_2\, y_2(t) + C_2\Theta\,\hat{\zeta}(t) = y_2(t), \tag{3.12}$$

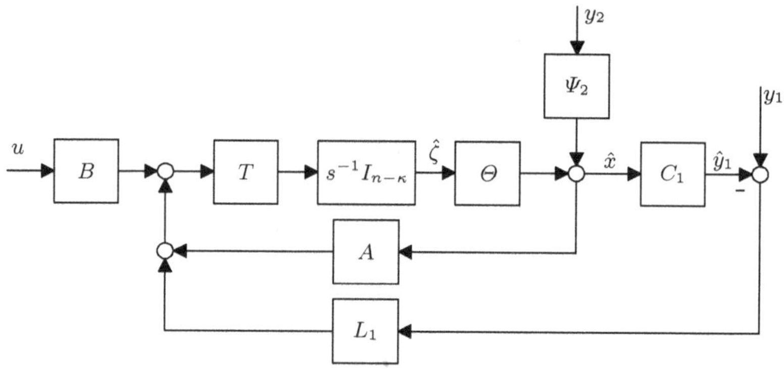

Figure 3.1. Block diagram representation of the reduced-order observer

i.e., it is independent of $\hat{\zeta}$, so that it would be pointless to incorporate $y_2 - \hat{y}_2$ in the correction term.

Because $T\Theta = I_{n-\kappa}$ (see (3.6)), the state equation (3.10) of the observer can be simplified by introducing

$$L_1 = \Theta L_1^*, \tag{3.13}$$

such that

$$TL_1 = T\Theta L_1^* = L_1^*. \tag{3.14}$$

Figure 3.1 shows a block diagram of the reduced-order observer.

Using (3.11) and (3.13) in (3.10) one obtains the final form

$$\dot{\hat{\zeta}}(t) = T(A-L_1C_1)\Theta\hat{\zeta}(t) + [TL_1 \quad T(A-L_1C_1)\Psi_2]\begin{bmatrix} y_1(t) \\ y_2(t) \end{bmatrix} + TBu(t), \tag{3.15}$$

$$\hat{x}(t) = \Theta\hat{\zeta}(t) + \Psi_2 y_2(t) \tag{3.16}$$

of the (reduced-order) observer and it is obvious that the observed state $\hat{x}$ and the observer state $\hat{\zeta}$ are related by

$$\hat{\zeta}(t) = T\hat{x}(t). \tag{3.17}$$

It was shown in [15] that the degrees of freedom contained in the matrices L_1 and Ψ_2 parameterize the transfer behaviour of the observer, whereas T and Θ only influence its internal structure. It can also be shown (see, *e.g.*, [15]) that there exist matrices L_1 and Ψ_2 such that the observer converges provided that the pair (C, A) is observable.

As the system is assumed to be observable, the error dynamics

$$\dot{\zeta}(t) - \dot{\hat{\zeta}}(t) = T(A - L_1C_1)\Theta(\zeta(t) - \hat{\zeta}(t)) \tag{3.18}$$

can always be made asymptotically stable.

All nm entries of L_1 and Ψ_2, however, cannot be chosen arbitrarily. Due to $C_2\Theta = 0$, which is a consequence of (3.6), the relation

$$C_2 L_1 = 0_{\kappa,m-\kappa} \tag{3.19}$$

is satisfied by the $n \times (m-\kappa)$ matrix L_1 (see (3.13)).

Furthermore, the restriction

$$C_2 \Psi_2 = I_\kappa \tag{3.20}$$

(see also (3.6)) has to be satisfied. By (3.19) $\kappa(m-\kappa)$ and by (3.20) κ^2 parameters are fixed, so that only $m(n-\kappa)$ degrees of freedom parameterize the transfer behaviour of an observer of the order $n_O = n - \kappa$, $0 \le \kappa \le m$.

Given the constant matrices L_1 and Ψ_2, either resulting from a parameterization described in Section 5.4, an optimal estimation scheme (see Section 8.2) or a given frequency-domain parameterization (see Section 3.3), the remaining matrices characterizing the observer (3.15) and (3.16) can be obtained *via*

$$T\Psi_2 = 0, \tag{3.21}$$

(see (3.6)) where T is a solution with full row rank $n-\kappa$. The rows of this T are linearly independent of the rows of C_2 in (3.20). If this was not the case, some row t_i^T of T could be represented as a linear combination of rows of C_2, *i.e.*, $t_i^T = m_i^T C_2$ for some vector $m_i^T \neq 0^T$. Then $t_i^T \Psi_2 = m_i^T C_2 \Psi_2 = m_i^T \neq 0^T$ in view of (3.20), which contradicts (3.21). Therefore, the matrix Θ is given by

$$\Theta = \begin{bmatrix} C_2 \\ T \end{bmatrix}^{-1} \begin{bmatrix} 0 \\ I_{n-\kappa} \end{bmatrix} \tag{3.22}$$

(see (3.5)).

For $n_O = n$ or $\kappa = 0$, $C_1 = C$ and C_2 vanishes. Therefore, also Ψ_2 vanishes. Setting $L_1 = L$, $T = \Theta = I$ and consequently $\hat{\zeta} = \hat{x}$ the Equation (3.15) of the observer takes the form

$$\dot{\hat{x}}(t) = (A - LC)\hat{x}(t) + Ly(t) + Bu(t), \tag{3.23}$$

i.e., one obtains the well-known identity or full-order observer.

For $n_O = n - m$ or $\kappa = m$, $C_2 = C$ and C_1 vanishes. Consequently, L_1 also vanishes. Setting $\Psi_2 = \Psi$ Equation (3.15) of the observer takes the form

$$\dot{\hat{\zeta}}(t) = TA\Theta\hat{\zeta}(t) + TA\Psi y(t) + TBu(t), \tag{3.24}$$

with $\hat{x} = \Theta\hat{\zeta} + \Psi y$. This is the completely reduced-order observer, *i.e.*, the observer with the smallest order considered.

3.2 Parameterization of the Full-order Observer in the Frequency Domain

In Section 3.3 the polynomial matrix will be derived that parameterizes observers of the general order $n_O = n - \kappa$, $0 \leq \kappa \leq m$, in the frequency domain. A motivation for the steps taken in this derivation can best be presented when briefly discussing the full-order observer with $n_O = n$.

A frequency-domain representation of an observer cannot yield a state estimate, because the state is not defined by the input–output behaviour of the system in terms of its transfer matrix $G(s)$. Instead, the goal of the following sections is to look for a frequency-domain parameterization of the dynamics of the observation error that contains all degrees of freedom of the observer and that is based on the left coprime MFD of $G(s)$. Given such a frequency-domain parameterization one can easily obtain a corresponding time-domain realization of the observer.

Figure 3.2 shows the block diagram of the full-order observer (*i.e.*, the case $\kappa = 0$) in the time domain. This observer is completely parameterized by the constant $n \times m$ matrix L.

To obtain a frequency-domain parameterization of all degrees of freedom of the observer the transfer matrix $G(s)$ in $y(s) = G(s)u(s)$ of the system is represented by the left coprime MFD

$$G(s) = \bar{D}^{-1}(s)\bar{N}(s), \tag{3.25}$$

where the $m \times m$ polynomial matrix $\bar{D}(s)$ is supposed to be row reduced and $\bar{N}(s)$ is an $m \times p$ polynomial matrix. Since the output error $\varepsilon = y - \hat{y}$ is fed back to the input of the integrator, also the left coprime MFD

$$C(sI - A)^{-1} = \bar{D}^{-1}(s)\bar{N}_x(s) \tag{3.26}$$

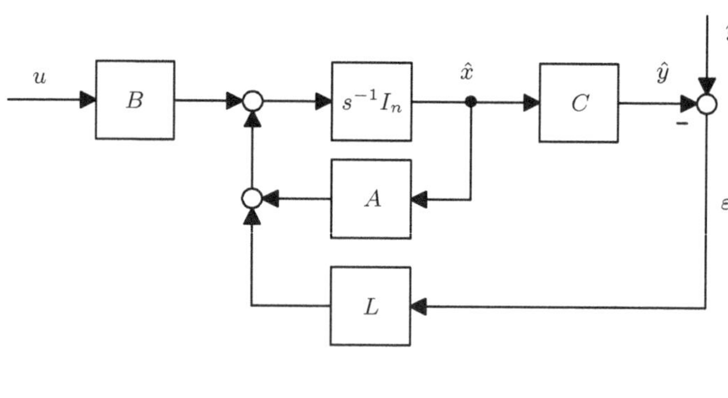

Figure 3.2. Block diagram of the full-order observer in the time domain

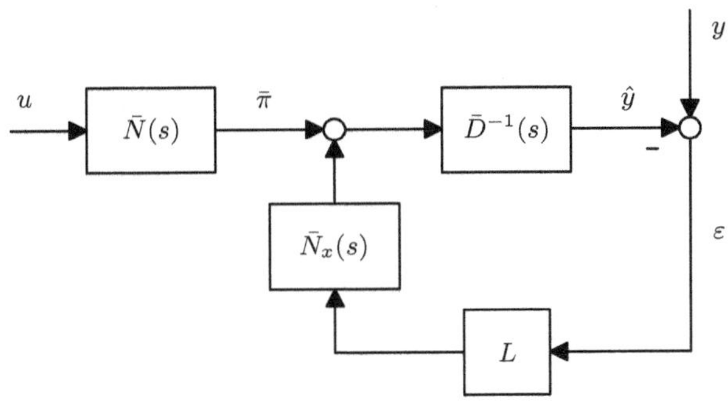

Figure 3.3. Block diagram of the full-order observer in the frequency domain

is considered with $\bar{D}(s)$ as defined in (3.25) and $\bar{N}_x(s)$ being an $m \times n$ polynomial matrix.

Obviously, in view of $G(s) = C(sI - A)^{-1}B$ and its left coprime MFD (3.25) the equality

$$\bar{N}(s) = \bar{N}_x(s)B \tag{3.27}$$

then also holds. Using the MFDs (3.25) and (3.26) the transfer behaviour of the system in Figure 3.2 is equivalently represented by the block diagram in Figure 3.3, where the weighted output error is added to the partial state $\bar{\pi}$ (see (1.77)).

The transfer behaviour of the observer in Figure 3.2 between its inputs u and y and the observation error $\varepsilon = y - \hat{y}$ has the form

$$\varepsilon(s) = (I - C(sI - A + LC)^{-1}L)y(s) - C(sI - A + LC)^{-1}Bu(s). \tag{3.28}$$

In the observer representation of Figure 3.3 the same transfer behaviour can be expressed as

$$\varepsilon(s) = [\bar{N}_x(s)L + \bar{D}(s)]^{-1}(\bar{D}(s)y(s) - \bar{N}(s)u(s)). \tag{3.29}$$

Since the transfer behaviours of the observer representations in Figures 3.2 and 3.3 coincide, a comparison of (3.28) and (3.29) shows that the $m \times m$ denominator matrix

$$\tilde{\bar{D}}(s) = \bar{N}_x(s)L + \bar{D}(s) \tag{3.30}$$

in (3.29) characterizes the dynamics of the observer. This becomes obvious when observing that (3.28) and (3.29) with the abbreviation (3.30) can be written as

$$\varepsilon(s) = -\frac{C\mathrm{adj}\{sI - A + LC\}}{\det(sI - A + LC)}(Ly(s) + Bu(s)) + y(s) \tag{3.31}$$

$$= \frac{\mathrm{adj}\{\bar{\tilde{D}}(s)\}}{\det \bar{\tilde{D}}(s)}(\bar{D}(s)y(s) - \bar{N}(s)u(s)),$$

where adj$\{\cdot\}$ denotes the adjoint of a matrix. Comparing the denominators of both terms shows that

$$\det \bar{\tilde{D}}(s) = \det(sI - A + LC). \tag{3.32}$$

Given a frequency-domain parameterization of the observer *via* the polynomial matrix $\bar{\tilde{D}}(s)$, then (3.30) can be used to obtain an equivalent time-domain parameterization *via* the constant matrix L and consequently also a realization of the observer or *vice versa.*

Pre-multiplying (3.30) by $\bar{D}^{-1}(s)$ and using (3.26) one obtains the connecting relation

$$\bar{D}^{-1}(s)\bar{\tilde{D}}(s) = I + C(sI - A)^{-1}L \tag{3.33}$$

between the time-domain parameterization of the observer *via* the gain L and its frequency-domain parameterization *via* the $m \times m$ polynomial matrix $\bar{\tilde{D}}(s)$ (see, *e.g.*, [2]).

By a dual version of the arguments in Section 2.2 (see Theorem 2.1) it can be shown that (3.33) specifies the structure of $\bar{\tilde{D}}(s)$ according to

$$\delta_{rj}[\bar{\tilde{D}}(s)] = \delta_{rj}[\bar{D}(s)], \quad j = 1, 2, \ldots, m, \tag{3.34}$$

and

$$\Gamma_r[\bar{\tilde{D}}(s)] = \Gamma_r[\bar{D}(s)]. \tag{3.35}$$

As a consequence of this, the number of free parameters in $\bar{\tilde{D}}(s)$ is mn and this coincides with the number of elements in L, *i.e.*, with the number of degrees of freedom in the observer. A formal proof of this will be presented in Section 3.3, where the general case of observers of the order $n_O = n - \kappa$, $0 \leq \kappa \leq m$, is discussed, which contains the full-order observer as a special case for $\kappa = 0$.

Example 3.1. Frequency-domain parameterization of a full-order observer
Considered is a system of the order $n = 3$ with $p = 1$ input u and $m = 2$ outputs y. Its left coprime MFD is characterized by

$$\bar{D}^{-1}(s)\bar{N}(s) = \begin{bmatrix} s+3 & 1 \\ -2s-6 & s^2+6s+7 \end{bmatrix}^{-1} \begin{bmatrix} 1 \\ s+2 \end{bmatrix}. \tag{3.36}$$

The row degrees of $\bar{D}(s)$ are

$$\delta_{r1}[\bar{D}(s)] = 1 \text{ and } \delta_{r2}[\bar{D}(s)] = 2. \tag{3.37}$$

The highest row-degree-coefficient matrix of $\bar{D}(s)$ is

$$\Gamma_r[\bar{D}(s)] = \begin{bmatrix} 1 & 0 \\ 0 & 1 \end{bmatrix}. \tag{3.38}$$

Therefore, the polynomial matrix that parameterizes a full-order observer for this system is given by

$$\tilde{\bar{D}}(s) = \begin{bmatrix} s+\alpha & \beta \\ \gamma s+\delta & s^2+\varepsilon s+\varphi \end{bmatrix}, \tag{3.39}$$

and it has $mn = 6$ free parameters. An observer with eigenvalues at $s = -5$ is, for example, obained by choosing $\alpha = 5$, $\beta = 0$, $\gamma = -2$, $\delta = -10$, $\varepsilon = 10$ and $\varphi = 25$, which leads to

$$\tilde{\bar{D}}(s) = \begin{bmatrix} s+5 & 0 \\ -2s-10 & s^2+10s+25 \end{bmatrix}. \tag{3.40}$$

An equivalent time-domain parameterization of this observer can be obtained by choosing some state equations for the system considered. The system

$$\begin{aligned} \dot{x}(t) &= Ax(t) + Bu(t), \\ y(t) &= Cx(t), \end{aligned} \tag{3.41}$$

with

$$A = \begin{bmatrix} 0 & 1 & 0 \\ 0 & 0 & 1 \\ -27 & -27 & -9 \end{bmatrix}, \quad B = \begin{bmatrix} 0 \\ 0 \\ 1 \end{bmatrix}, \quad C = \begin{bmatrix} 5 & 5 & 1 \\ 12 & 7 & 1 \end{bmatrix} \tag{3.42}$$

has the transfer behaviour described by (3.36). The MFD (3.26) is characterized by the above $\bar{D}(s)$ and

$$\bar{N}_x(s) = \begin{bmatrix} 5 & 5 & 1 \\ 12s+35 & 7s+17 & s+2 \end{bmatrix}. \tag{3.43}$$

Solving (3.30) for the feedback gain L yields the parameterization of the full-order observer in the time domain, namely

$$L = \begin{bmatrix} -2 & 5 \\ 6 & -15 \\ -18 & 49 \end{bmatrix}. \tag{3.44}$$

In a time-domain approach the state equations of the observer-based compensator are completely specified when the state feedback gain K and the feedback gain L of the output estimation error are known. The polynomial matrices $\tilde{D}(s)$ and $\tilde{\bar{D}}(s)$ parameterize the state feedback and the observer in the frequency domain, so that the transfer behaviour of the corresponding observer-based compensator can be computed. This will be discussed in Chapter 4.

3.3 Parameterization of the Reduced-order Observer in the Frequency Domain

In the frequency domain the parameterization of the state observer is based on the left coprime MFD of the system. This MFD characterizes a system of the order n. Since the identity observer is also of the order n, its frequency-domain parameterization is a comparatively easy task (see Section 3.2).

A reduced-order observer, however, is of the order $n - \kappa$ with $0 < \kappa \leq m$. If its frequency-domain representation is based on the left coprime MFD of the system, a straightforward time-domain realization is therefore non-minimal. Thus, a connecting relation between the time- and the frequency-domain parameterizations of reduced-order observers can only be formulated if there also exists a non-minimal realization of such observers based on a full-order model in the time domain. In the following, it will be shown that the block diagram in Figure 3.4 represents such a non-minimal realization of the reduced-order observer on the basis of a full-order model of the system (3.1) and (3.2). This is established by verifying that the transfer behaviour between the inputs y_1, y_2 and u and its output $\hat{x}$ can be described by $n - \kappa$ differential equations of the form (3.15) and an algebraic output equation of the form (3.16). Furthermore, $y_2 = \hat{y}_2$ for all t, which is required in the light of (3.12).

To show this, the block diagram of Figure 3.4 is redrawn as the equivalent block diagram of Figure 3.5. The two diagrams are related as follows.

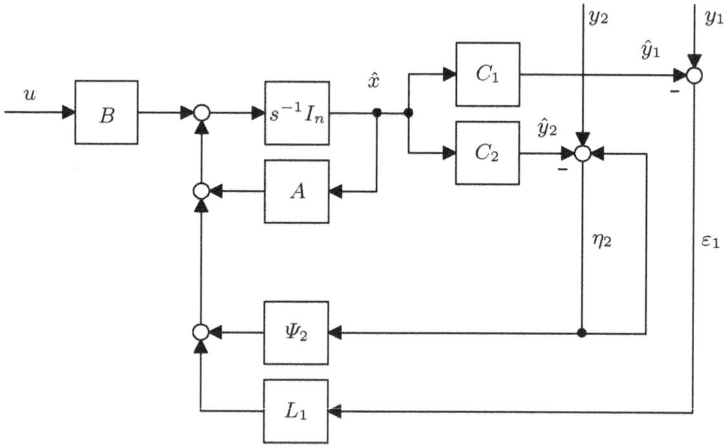

Figure 3.4. Non-minimal realization of the reduced-order observer based on the time-domain parameterization

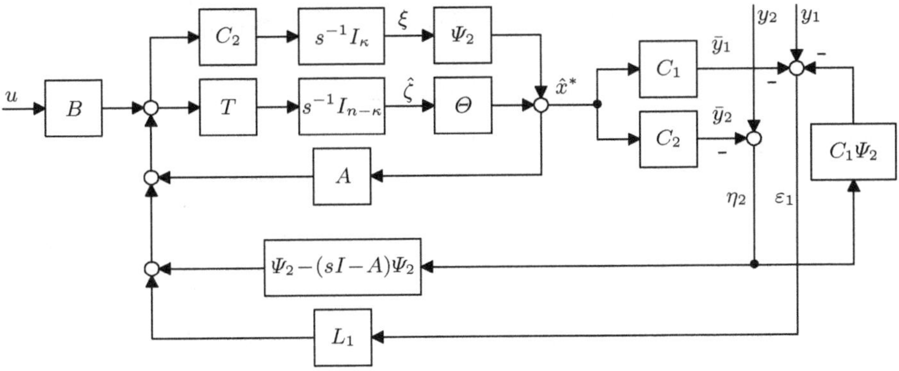

Figure 3.5. Modified version of the block diagram in Figure 3.4

- The integrator $s^{-1}I_n$ in Figure 3.4 is replaced by $s^{-1}(\Psi_2 C_2 + \Theta T)$ in Figure 3.5, based on (3.7).
- The positive unity feedback of η_2 onto itself in Figure 3.4 is replaced by a feedback of η_2 *via* $-(sI - A)\Psi_2\eta_2$ to the input of the integrator in Figure 3.5. This yields $C_2(sI-A)^{-1}(sI-A)\Psi_2\eta_2 = C_2\Psi_2\eta_2 = \eta_2$ at the summation $y_2 - \bar{y}_2$ in Figure 3.5, *i.e.*, it is equivalent to the positive feedback of η_2 onto itself. This feedback, however, also gives rise to an additional quantity $C_1(sI - A)^{-1}(sI - A)\Psi_2\eta_2$ at the summation $y_1 - \bar{y}_1$ that is compensated by the additional feedback $-C_1\Psi_2\eta_2$ onto this summation in Figure 3.5.

Because of the additional feedback $-(sI - A)\Psi_2\eta_2$ in Figure 3.5 the signal $\hat{x}^*$ is given by

$$\begin{aligned}\hat{x}^*(s) &= \hat{x}(s) - (sI - A)^{-1}(sI - A)\Psi_2\eta_2(s)\\ &= \hat{x}(s) - \Psi_2\eta_2(s) = \hat{x}(s) - \Psi_2(y_2(s) - \bar{y}_2(s)).\end{aligned} \tag{3.45}$$

With $\hat{x}^* = \Psi_2\xi + \Theta\hat{\zeta}$ and $\bar{y}_2 = C_2\hat{x}^*$ (see Figure 3.5) one obtains

$$\bar{y}_2(s) = C_2(\Psi_2\xi(s) + \Theta\hat{\zeta}(s)) = \xi(s), \tag{3.46}$$

because of $C_2\Psi_2 = I_\kappa$ and $C_2\Theta = 0$ (see (3.6)). Introducing (3.46) on the right-hand side of (3.45) leads to

$$\hat{x}^*(s) = \hat{x}(s) - \Psi_2(y_2(s) - \xi(s)). \tag{3.47}$$

Because

$$\hat{x}^*(s) = \Psi_2\xi(s) + \Theta\hat{\zeta}(s) \tag{3.48}$$

(see Figure 3.5) this yields

$$\hat{x}(s) - \Psi_2 y_2(s) + \Psi_2 \xi(s) = \Psi_2 \xi(s) + \Theta\hat{\zeta}(s), \tag{3.49}$$

so that

$$\hat{x}(s) = \Psi_2 y_2(s) + \Theta\hat{\zeta}(s) \tag{3.50}$$

when solving for $\hat{x}$. This is equivalent to

$$\hat{x}(t) = \Psi_2 y_2(t) + \Theta\hat{\zeta}(t) \tag{3.51}$$

in the time domain, which is exactly the static relation (3.16). Because $\bar{y}_1 = C_1(\Theta\hat{\zeta} + \Psi_2\xi)$ (see Figure 3.5) and $\bar{y}_2 = \xi$ (see (3.46)) the input to the integrator $s^{-1}I_{n-\kappa}$ in Figure 3.5 is given by

$$\begin{aligned} s\hat{\zeta}(s) = T\,\{&Bu(s) + A\hat{x}^*(s) \\ &+L_1\left[y_1(s) - \bar{y}_1(s) - C_1\Psi_2\eta_2(s)\right] \\ &+\Psi_2\eta_2(s) - (sI - A)\Psi_2\eta_2(s)\} \\ = T\Big\{&Bu(s) + A(\Psi_2\xi(s) + \Theta\hat{\zeta}(s)) \\ &+L_1\Big[y_1(s) - C_1\Theta\hat{\zeta}(s) - C_1\Psi_2\xi(s) - C_1\Psi_2 y_2(s) + C_1\Psi_2\xi(s)\Big] \\ &+(\Psi_2 - s\Psi_2 + A\Psi_2)[y_2(s) - \xi(s)]\big\}, \end{aligned} \tag{3.52}$$

in which (3.48) and $\eta_2 = y_2 - \bar{y}_2$ were used. Because of $T\Psi_2 = 0$ (see (3.6)) this takes the form

$$\dot{\hat{\zeta}}(t) = T(A - L_1C_1)\Theta\hat{\zeta}(t) + \begin{bmatrix} TL_1 & T(A - L_1C_1)\Psi_2 \end{bmatrix} \begin{bmatrix} y_1(t) \\ y_2(t) \end{bmatrix} + TBu(t) \tag{3.53}$$

in the time domain, which is exactly (3.15).

Since $\hat{y}_2 = C_2\hat{x}$ (see Figure 3.4) one obtains

$$\hat{y}_2(t) = C_2(\Psi_2 y_2(t) + \Theta\hat{\zeta}(t)) = y_2(t), \tag{3.54}$$

because of $C_2\Psi_2 = I_\kappa$ and $C_2\Theta = 0$. Thus, (3.51)–(3.54) verify that Figure 3.4 is indeed a realization of the reduced-order observer.

Remark 3.1. The positive unity feedback of η_2 onto itself may be interpreted as an order reduction of the entire system consisting of the plant and the observer of Figure 3.4. This can be shown when inspecting the input to the integrator $s^{-1}I_\kappa$ in Figure 3.5. Observing again that $\bar{y}_1 = C_1(\Theta\hat{\zeta} + \Psi_2\xi)$ and $\bar{y}_2 = \xi$ one obtains

$$\begin{aligned} s\hat{\xi}(s) = C_2\,\{&Bu(s) + A\hat{x}^*(s) \\ &+L_1\left[y_1(s) - \bar{y}_1(s) - C_1\Psi_2\eta_2(s)\right] \\ &+\Psi_2\eta_2(s) - (sI - A)\Psi_2\eta_2(s)\} \\ = C_2\Big\{&Bu(s) + A(\Psi_2\xi(s) + \Theta\hat{\zeta}(s)) \\ &+L_1\Big[y_1(s) - C_1\Theta\hat{\zeta}(s) - C_1\Psi_2\xi(s) - C_1\Psi_2 y_2(s) + C_1\Psi_2\xi(s)\Big] \\ &+(\Psi_2 - s\Psi_2 + A\Psi_2)[y_2(s) - \xi(s)]\big\}, \end{aligned} \tag{3.55}$$

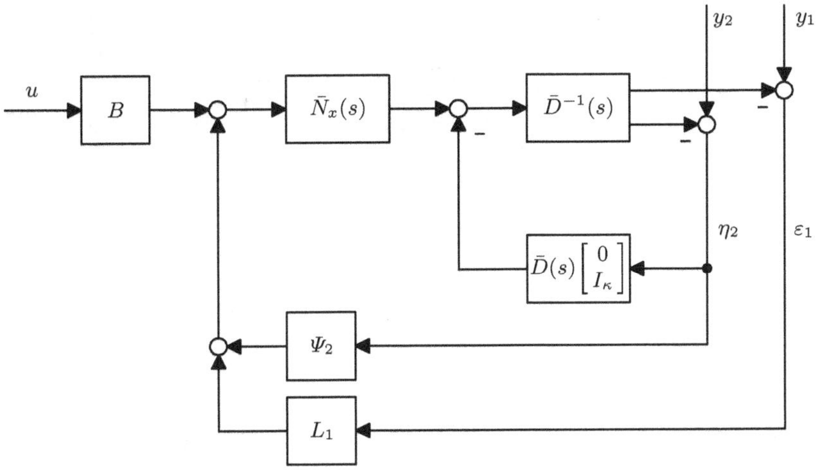

Figure 3.6. Frequency-domain representation of the block diagram in Figure 3.4

in which (3.48) and $\eta_2 = y_2 - \bar{y}_2$ were used. And because of $C_2L_1 = 0$ and $C_2\Psi_2 = I_\kappa$ (see (3.19) and (3.6)) this takes the form

$$\xi(t) = y_2(t) - \dot{y}_2(t) + C_2A\Psi_2y_2(t) + C_2A\Theta\hat{\zeta}(t) + C_2Bu(t) \tag{3.56}$$

in the time domain. By using $y_2 = C_2x$, $\dot{y}_2 = C_2\dot{x} = C_2Ax + C_2Bu$ and $I - \Psi_2C_2 = \Theta T$ (3.56) can further be simplified to

$$\xi(t) = C_2[I_n - A\Theta T]x(t) + C_2A\Theta\hat{\zeta}(t). \tag{3.57}$$

In view of (3.57) the output ξ of the κ integrators can be expressed as a linear combination of the states x of the system and the states $\hat{\zeta}$ of the observer. As the outputs ξ of the integrators are not independent variables they are consequently not states of the system consisting of the plant (3.1) and (3.2) and the observer shown in Figure 3.4. The order of this system is therefore given by $\dim(x) + \dim(\hat{\zeta}) = n + (n - \kappa)$.

In order to parameterize the reduced-order observer in the frequency domain, the transfer matrix $G(s)$ is again represented by the $m \times p$ numerator matrix $\bar{N}(s)$ and the $m \times m$ denominator matrix $\bar{D}(s)$ in the left coprime MFD (3.25). Also considered is the left coprime MFD (3.26). When representing the system model by the MFD (3.26) and relocating the positive feedback of the κ-dimensional signal vector η_2 onto itself to the input of the block with transfer behaviour $\bar{D}^{-1}(s)$, the block diagram of Figure 3.4 takes the form shown in Figure 3.6. When describing the input–output behaviour of the system by the MFD (3.25) the block diagram of Figure 3.6 can also be

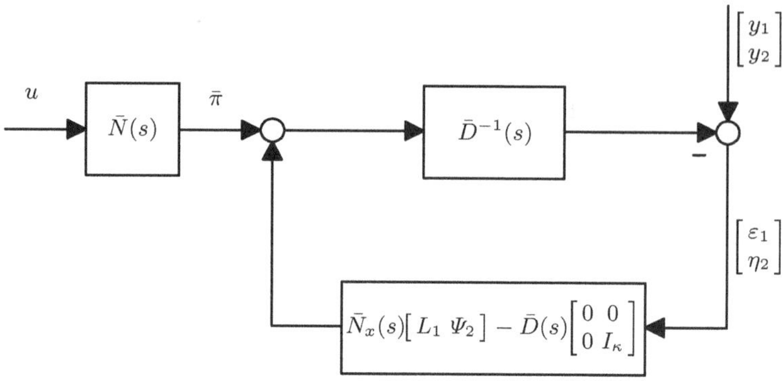

Figure 3.7. Frequency-domain representation of the reduced-order observer

represented as shown in Figure 3.7. The transfer behaviour of this system is the same as that of the non-minimal representation of the reduced-order observer shown in Figure 3.4.

First, consider the transfer behaviour between the input signals y and u and the output signals ε_1 and η_2 in the time-domain representation of the reduced-order observer in Figure 3.4. This can be obtained by first inserting

$$\xi(s) = y_2(s) - sy_2(s) + C_2A(\Psi_2 y_2(s) + \Theta\hat{\zeta}(s)) + C_2Bu(s), \tag{3.58}$$

(see (3.56)) and

$$\hat{\zeta}(s) = (sI - F)^{-1} \begin{bmatrix} TL_1 & T(A - L_1C_1)\Psi_2 \end{bmatrix} \begin{bmatrix} y_1(s) \\ y_2(s) \end{bmatrix} + (sI - F)^{-1}TBu(s) \tag{3.59}$$

(see (3.53)) in $\eta_2(s) = y_2(s) - C_2(\Psi_2\xi(s) + \Theta\hat{\zeta}(s))$ (see Figure 3.5). Here, and in what follows, the abbreviation

$$F = T(A - L_1C_1)\Theta \tag{3.60}$$

is used. Now, $\eta_2(s)$, $\xi(s)$ and $\hat{\zeta}(s)$ can be inserted in $\varepsilon_1(s) = y_1(s) - C_1\Psi_2\eta_2(s) - C_1(\Psi_2\xi(s) + \Theta\hat{\zeta}(s))$ (see again Figure 3.5), which yields

$$\begin{aligned} \begin{bmatrix} \varepsilon_1(s) \\ \eta_2(s) \end{bmatrix} = & \begin{bmatrix} I_{m-\kappa} - C_1\Theta(sI - F)^{-1}TL_1 \\ -C_2A\Theta(sI - F)^{-1}TL_1 \end{bmatrix. \\ & \left. \begin{matrix} C_1\left\{-I - \Theta(sI - F)^{-1}T(A - L_1C_1)\right\}\Psi_2 \\ C_2\left\{sI - A - A\Theta(sI - F)^{-1}T(A - L_1C_1)\right\}\Psi_2 \end{matrix} \right] \begin{bmatrix} y_1(s) \\ y_2(s) \end{bmatrix} \\ & - \begin{bmatrix} C_1\Theta(sI - F)^{-1}TB \\ C_2\left\{I + A\Theta(sI - F)^{-1}T\right\}B \end{bmatrix} u(s). \end{aligned} \tag{3.61}$$

In the block diagram of Figure 3.7 the same transfer behaviour is characterized by

$$\begin{bmatrix} \varepsilon_1(s) \\ \eta_2(s) \end{bmatrix} = \left\{ \bar{N}_x(s)[\, L_1 \;\; \Psi_2 \,] + \bar{D}(s) \begin{bmatrix} I_{m-\kappa} & 0 \\ 0 & 0_\kappa \end{bmatrix} \right\}^{-1} \left\{ \bar{D}(s) \begin{bmatrix} y_1(s) \\ y_2(s) \end{bmatrix} - \bar{N}(s)u(s) \right\}. \tag{3.62}$$

A comparison of (3.61) and (3.62) shows that the $m \times m$ polynomial matrix

$$\tilde{\bar{D}}(s) = \bar{N}_x(s)\,[L_1 \quad \Psi_2] + \bar{D}(s) \begin{bmatrix} I_{m-\kappa} & 0 \\ 0 & 0_\kappa \end{bmatrix} \tag{3.63}$$

parameterizes the dynamics of the reduced-order observer and that

$$\det \tilde{\bar{D}}(s) = \det(sI - F) = \det(sI - T(A - L_1 C_1)\Theta). \tag{3.64}$$

Given a parameterizing polynomial matrix $\tilde{\bar{D}}(s)$ and a state-space representation (C, A, B) of the system, the time-domain parameterization of the observer *via* the matrices L_1 and Ψ_2 can be computed by solving (3.63) and *vice versa.*

By pre-multiplying (3.63) by $\bar{D}^{-1}(s)$ and using (3.26) one obtains

$$\bar{D}^{-1}(s)\tilde{\bar{D}}(s) = \begin{bmatrix} C_1 \\ C_2 \end{bmatrix} (sI - A)^{-1}[L_1 \quad \Psi_2] + \begin{bmatrix} I_{m-\kappa} & 0 \\ 0 & 0_\kappa \end{bmatrix}, \tag{3.65}$$

which is the connecting relation between the time- and the frequency-domain parameterizations of a reduced-order observer. The case of a full-order observer (*i.e.*, $n_O = n$) is characterized by $C_1 = C$, $L_1 = L$ and vanishing matrices C_2 and Ψ_2, so that (3.65) takes the form (3.33), which was derived for the full-order observer.

In the time domain the reduced-order observer is parameterized by the $m(n-\kappa)$ degrees of freedom in the two matrices L_1 and Ψ_2. In the frequency domain, the freely assignable coefficients of the $m \times m$ polynomial matrix $\tilde{\bar{D}}(s)$ parameterize the error dynamics of the observer.

In what follows the freely assignable coefficients of the parameterizing polynomial matrix $\tilde{\bar{D}}(s)$ are identified. Differing from the parameterization of state feedback in the frequency domain (see Chapter 2), the connecting relation (3.65) cannot be used directly to determine the row-degree structure of $\tilde{\bar{D}}(s)$ since

$$C_2[L_1 \quad \Psi_2] = [0 \quad I_\kappa] \tag{3.66}$$

must also be satisfied (see (3.19) and (3.20)). In order to determine the parameterizing matrix $\tilde{\bar{D}}(s)$, which takes Equation (3.66) into account, consider the transfer matrix

$$\begin{aligned} \Phi(s) &= \begin{bmatrix} I_{m-\kappa} & 0 \\ 0 & sI_\kappa \end{bmatrix} \bar{D}^{-1}(s)\tilde{\bar{D}}(s) - I \\ &= \begin{bmatrix} I_{m-\kappa} & 0 \\ 0 & sI_\kappa \end{bmatrix} \begin{bmatrix} C_1 \\ C_2 \end{bmatrix} (sI - A)^{-1}[L_1 \quad \Psi_2] - \begin{bmatrix} 0_{m-\kappa} & 0 \\ 0 & I_\kappa \end{bmatrix}, \end{aligned} \tag{3.67}$$

which is obtained by substituting $\bar{D}^{-1}(s)\tilde{\bar{D}}(s)$ by the right-hand side of (3.65). Using the expansion

$$(sI - A)^{-1} = Is^{-1} + As^{-2} + \dots \tag{3.68}$$

it is straightforward to show that

$$\begin{aligned} sC_2(sI - A)^{-1}[L_1 \quad \Psi_2] &= C_2[L_1 \quad \Psi_2] + C_2As^{-1}[L_1 \quad \Psi_2] + \dots \\ &= [0 \quad I_\kappa] + C_2A(sI - A)^{-1}[L_1 \quad \Psi_2], \end{aligned} \tag{3.69}$$

in view of (3.66), so that the transfer matrix

$$\Phi(s) = \left(\bar{D}(s)\begin{bmatrix} I_{m-\kappa} & 0 \\ 0 & s^{-1}I_\kappa \end{bmatrix}\right)^{-1} \tilde{\bar{D}}(s) - I \tag{3.70}$$

is strictly proper, which means that

$$\Pi\left\{\left(\bar{D}(s)\begin{bmatrix} I_{m-\kappa} & 0 \\ 0 & s^{-1}I_\kappa \end{bmatrix}\right)^{-1} \tilde{\bar{D}}(s)\right\} = I. \tag{3.71}$$

Consequently, the rational matrix $\bar{D}_\kappa^{-1}(s)\tilde{\bar{D}}(s) - I$ is also strictly proper, where $\bar{D}_\kappa(s)$ is defined by

$$\bar{D}_\kappa(s) = \Pi\left\{\bar{D}(s)\begin{bmatrix} I_{m-\kappa} & 0 \\ 0 & s^{-1}I_\kappa \end{bmatrix}\right\}. \tag{3.72}$$

This follows from

$$\begin{aligned} &\Pi\left\{\left(\bar{D}(s)\begin{bmatrix} I_{m-\kappa} & 0 \\ 0 & s^{-1}I_\kappa \end{bmatrix}\right)^{-1} \tilde{\bar{D}}(s)\right\} \\ &\quad = \lim_{s\to\infty}\left\{\left(\bar{D}(s)\begin{bmatrix} I_{m-\kappa} & 0 \\ 0 & s^{-1}I_\kappa \end{bmatrix}\right)^{-1} \tilde{\bar{D}}(s)\right\} \\ &\quad = \lim_{s\to\infty}\left\{\left(\Pi\left\{\bar{D}(s)\begin{bmatrix} I_{m-\kappa} & 0 \\ 0 & s^{-1}I_\kappa \end{bmatrix}\right\}\right.\right. \\ &\qquad\qquad \left.\left. + SP\left\{\bar{D}(s)\begin{bmatrix} I_{m-\kappa} & 0 \\ 0 & s^{-1}I_\kappa \end{bmatrix}\right\}\right)^{-1} \tilde{\bar{D}}(s)\right\}, \end{aligned} \tag{3.73}$$

and with

$$\lim_{s\to\infty} SP\left\{\bar{D}(s)\begin{bmatrix} I_{m-\kappa} & 0 \\ 0 & s^{-1}I_\kappa \end{bmatrix}\right\} = 0, \tag{3.74}$$

this yields

$$\lim_{s\to\infty}\left\{\left(\bar{D}(s)\begin{bmatrix} I_{m-\kappa} & 0 \\ 0 & s^{-1}I_\kappa \end{bmatrix}\right)^{-1}\tilde{\bar{D}}(s)\right\}$$

$$= \lim_{s\to\infty}\left\{\left(\Pi\left\{\bar{D}(s)\begin{bmatrix} I_{m-\kappa} & 0 \\ 0 & s^{-1}I_\kappa \end{bmatrix}\right\}\right)^{-1}\tilde{\bar{D}}(s)\right\}$$

$$= \Pi\left\{\bar{D}_\kappa^{-1}(s)\tilde{\bar{D}}(s)\right\}. \tag{3.75}$$

Thus, the polynomial matrix $\bar{D}_\kappa(s)$ satisfies

$$\Pi\left\{\bar{D}_\kappa^{-1}(s)\tilde{\bar{D}}(s)\right\} = I \tag{3.76}$$

in view of (3.71) and (3.75). The Relation (3.76) can be used to determine the row-degree structure of $\tilde{\bar{D}}(s)$ if $\bar{D}_\kappa(s)$ is row reduced. Unfortunately, not every $\bar{D}(s)$ gives rise to a row-reduced $\bar{D}_\kappa(s)$. However, in Section A.1 it is shown that for any row-reduced polynomial matrix $\bar{D}(s)$ a unimodular matrix $U_L(s)$ can always be determined, such that both the modified $\bar{D}'(s) = U_L(s)\bar{D}(s)$ and

$$\bar{D}_\kappa'(s) = \Pi\left\{\bar{D}'(s)\begin{bmatrix} I_{m-\kappa} & 0 \\ 0 & s^{-1}I_\kappa \end{bmatrix}\right\}, \quad 0 < \kappa \le m \tag{3.77}$$

are row reduced with

$$\deg\det \bar{D}_\kappa'(s) = n - \kappa. \tag{3.78}$$

Remark 3.2. If for a given $\bar{D}(s)$ the polynomial matrix $\bar{D}_\kappa(s)$ is not row reduced one can alternatively try to compute a unimodular matrix $U_L(s)$, such that $U_L(s)\bar{D}_\kappa(s)$ is row reduced. In many cases this also yields a row-reduced pair $\bar{D}'(s)$ and $\bar{D}_\kappa'(s)$. The advantage of this approach is that one can use standard software packages, such as the Polynomial Toolbox of MATLAB® (see [53]).

The observer can then be designed on the basis of the left coprime MFD $G(s) = (\bar{D}'(s))^{-1}\bar{N}'(s)$ with $\bar{N}'(s) = U_L(s)\bar{N}(s)$, which is equivalent to (3.25). Thus, without loss of generality it is assumed in the following that both $\bar{D}(s)$ and $\bar{D}_\kappa(s)$ are row reduced. In the time domain, the reduced-order observer is parameterized by the free parameters in L_1 and Ψ_2. The next theorem shows that the $m \times m$ polynomial matrix $\tilde{\bar{D}}(s)$ plays the corresponding role in the frequency domain.

Theorem 3.1 (Parameterizing polynomial matrix of the observer). *The $m \times m$ polynomial matrix $\bar{\bar{D}}(s)$ characterizing the dynamics of the reduced-order observer of the order $n_O = n - \kappa$, $0 \leq \kappa \leq m$, in the frequency domain has the properties*

$$\delta_{rj}[\bar{\bar{D}}(s)] = \delta_{rj}[\bar{D}_\kappa(s)], \quad j = 1, 2, \ldots, m, \tag{3.79}$$

and

$$\Gamma_r[\bar{\bar{D}}(s)] = \Gamma_r[\bar{D}_\kappa(s)]. \tag{3.80}$$

In (3.79) and (3.80) the polynomial matrix $\bar{D}_\kappa(s)$ is defined as

$$\bar{D}_\kappa(s) = \Pi\left\{\bar{D}(s)\begin{bmatrix} I_{m-\kappa} & 0 \\ 0 & s^{-1}I_\kappa \end{bmatrix}\right\}, \tag{3.81}$$

and it is assumed that this polynomial matrix is row reduced.

The polynomial matrix $\bar{\bar{D}}(s)$ contains exactly $m(n-\kappa)$ free parameters that also exist in the time-domain design of such observers.

Proof. Since $\bar{D}_\kappa(s)$ is row reduced (*i.e.*, $\det \Gamma_r[\bar{D}_\kappa(s)] \neq 0$), (3.76) implies $\delta_{rj}[\bar{\bar{D}}(s)] \leq \delta_{rj}[\bar{D}_\kappa(s)]$, $j = 1, 2, \ldots, m$ (see (1.71) and (1.72) in Section 1.2). Applying (1.75) one obtains

$$\Pi\{\bar{D}_\kappa^{-1}(s)\bar{\bar{D}}(s)\} = \Gamma_r^{-1}[\bar{D}_\kappa(s)]\Gamma_{\delta_r[\bar{D}_\kappa(s)]}[\bar{\bar{D}}(s)] = I. \tag{3.82}$$

Pre-multiplying this by $\Gamma_r[\bar{D}_\kappa(s)]$ leads to

$$\Gamma_{\delta_r[\bar{D}_\kappa(s)]}[\bar{\bar{D}}(s)] = \Gamma_r[\bar{D}_\kappa(s)]. \tag{3.83}$$

Since $\bar{D}_\kappa(s)$ is row reduced, also $\det \Gamma_{\delta_r[\bar{D}_\kappa(s)]}[\bar{\bar{D}}(s)] \neq 0$. On the other hand, this shows that (3.79) is satisfied, because if $\delta_{rj}[\bar{\bar{D}}(s)] < \delta_{rj}[\bar{D}_\kappa(s)]$ for any j, the corresponding row of $\Gamma_{\delta_r[\bar{D}_\kappa(s)]}[\bar{\bar{D}}(s)]$ vanishes, contradicting $\det \Gamma_{\delta_r[\bar{D}_\kappa(s)]}[\bar{\bar{D}}(s)] \neq 0$. With (3.79) satisfied, (3.80) directly follows from (3.83).

Because of (3.79) and (3.80) the number n_{fj} of free parameters in each row of $\bar{\bar{D}}(s)$ is $n_{fj} = m\delta_{rj}\left[\bar{D}_\kappa(s)\right]$. Since $\deg\det \bar{D}_\kappa(s) = n - \kappa$ (see (3.78)) and $\bar{D}_\kappa(s)$ is row reduced the row degrees of $\bar{D}_\kappa(s)$ are such that

$$\sum_{j=1}^{m} \delta_{rj}[\bar{D}_\kappa(s)] = \deg\det \bar{D}_\kappa(s) = n - \kappa. \tag{3.84}$$

Consequently, the total number n_f of free parameters in $\bar{\bar{D}}(s)$ is

$$n_f = \sum_{j=1}^{m} m\,\delta_{rj}[\bar{D}_\kappa(s)] = m(n-\kappa). \tag{3.85}$$

□

Remark 3.3. The polynomial matrix $\bar{\tilde{D}}(s)$ has the properties (3.79) and (3.80), so that it can be represented in the form

$$\bar{\tilde{D}}(s) = \operatorname{diag}\left(s^{\delta_{ri}[\bar{D}_\kappa(s)]}\right) \Gamma_r[\bar{D}_\kappa(s)] + \bar{S}(s)\bar{\tilde{D}}_c, \tag{3.86}$$

with an $m \times (n-\kappa)$ polynomial matrix $\bar{S}(s)$ and an $(n-\kappa) \times m$ constant matrix $\bar{\tilde{D}}_c$ that contains the degrees of freedom in $\bar{\tilde{D}}(s)$. If all row degrees of $\bar{D}_\kappa(s)$ are greater than zero, the polynomial matrix $\bar{S}(s)$ has the form

$$\bar{S}(s) = \operatorname{diag}\left([s^{\delta_{r1}[\bar{D}_\kappa(s)]-1} \ldots s\ 1], \ldots, [s^{\delta_{rm}[\bar{D}_\kappa(s)]-1} \ldots s\ 1]\right). \tag{3.87}$$

It has $n-\kappa$ columns, because $\sum_{j=1}^{m} \delta_{rj}[\bar{D}_\kappa(s)] = n-\kappa$ (see (3.84)). Due to the construction (3.81) the polynomial matrix $\bar{D}_\kappa(s)$ can contain rows with row degrees equal to zero. If this is the case for the ith row, a $1 \times (n-\kappa)$ zero row has to be inserted in the ith row of $\bar{S}(s)$ (see also Theorem 5.3).

Therefore, the state observer can either be parameterized in the time domain by the constant matrices L_1 and Ψ_2, or in the frequency domain by the polynomial matrix $\bar{\tilde{D}}(s)$. In both cases $m(n-\kappa)$ degrees of freedom exist.

Remark 3.4. In the frequency domain there exists a simple way to solve the eigenvalue-assignment problem by using the parameterizing matrix

$$\bar{\tilde{D}}(s) = \begin{bmatrix} \bar{\tilde{d}}_1(s) & & \\ & \cdots & \\ & & \bar{\tilde{d}}_m(s) \end{bmatrix} \Gamma_r[\bar{D}_\kappa(s)], \tag{3.88}$$

with

$$\deg \bar{\tilde{d}}_j(s) = \delta_{rj}[\bar{D}_\kappa(s)], \quad j = 1, 2, \ldots, m, \tag{3.89}$$

so that one obtains

$$\det \bar{\tilde{D}}(s) = \bar{\tilde{d}}_1(s) \cdot \ldots \cdot \bar{\tilde{d}}_m(s). \tag{3.90}$$

More degrees of freedom can possibly be used when choosing the free coefficients in $\bar{\tilde{D}}(s)$ in an appropriate way, so that (3.90) is satisfied for given polynomials $\bar{\tilde{d}}_j(s)$ (see, *e.g.*, Example 3.2).

If all degrees of freedom need to be used, the parametric approach presented in Chapter 5 can be used.

The time-domain parameterization of the reduced-order observer can be obtained from the denominator matrix $\bar{\tilde{D}}(s)$ by solving the Diophantine equation

$$\bar{N}_x(s)\bar{P}(s) + \bar{D}(s)\bar{Q}(s) = \bar{\tilde{D}}(s) \tag{3.91}$$

for the unknown polynomial matrices $\bar{P}(s)$ and $\bar{Q}(s)$. A solution of this Diophantine equation is given by the pair of constant matrices

$$\bar{P}(s) = \bar{P} = [L_1 \quad \Psi_2] \quad \text{and} \quad \bar{Q}(s) = \bar{Q} = \begin{bmatrix} I_{m-\kappa} & 0 \\ 0 & 0_\kappa \end{bmatrix}, \tag{3.92}$$

which becomes obvious by comparing (3.91) and (3.63).

Remark 3.5. If the pair $(\bar{N}_x(s), \bar{D}(s))$ is left coprime (*i.e.*, if the pair (C, A) is observable) the Diophantine equation (3.91) can be solved for arbitrary right-hand sides $\bar{\tilde{D}}(s)$ having the properties (3.79) and (3.80), *i.e.*, the dynamics of the observer can be assigned arbitrarily. If an observability defect occurs, the matrices $\bar{N}_x(s)$ and $\bar{D}(s)$ have a greatest common left divisor $L(s)$, which is no longer a unimodular matrix. Then, the Diophantine equation (3.91) obtains the form

$$L(s)\bar{N}_x^*(s)\bar{P}(s) + L(s)\bar{D}^*(s)\bar{Q}(s) = \bar{\tilde{D}}(s), \tag{3.93}$$

and this makes it obvious that a solution only exists if the polynomial matrix $\bar{\tilde{D}}(s)$ also has the form $\bar{\tilde{D}}(s) = L(s)\bar{\tilde{D}}^*(s)$. This immediately shows which part of the observer dynamics cannot be assigned arbitrarily.

Example 3.2. Frequency-domain parameterization of observers of the orders $n - m \le n_O \le n$

Considered is the system of Example 2.1, and to demonstrate the relations between the time- and the frequency parameterizations, it is assumed that there is a time-domain realization (C, A, B) of the system with

$$A = \begin{bmatrix} 0 & 1 & 0 \\ -3 & -3 & 1 \\ 1 & 0 & -3 \end{bmatrix}, \quad B = \begin{bmatrix} 0 & 0 \\ 1 & 0 \\ 0 & 1 \end{bmatrix} \quad \text{and} \quad C = \begin{bmatrix} 1 & 0 & 0 \\ 0 & 1 & 1 \\ 0 & 0 & 1 \end{bmatrix}. \tag{3.94}$$

The left coprime MFD of the transfer matrix $G(s) = C(sI - A)^{-1}B$ is

$$G(s) = \bar{D}^{-1}(s)\bar{N}(s) = \begin{bmatrix} s & -1 & 1 \\ 2 & s+3 & -1 \\ -1 & 0 & s+3 \end{bmatrix}^{-1} \begin{bmatrix} 0 & 0 \\ 1 & 1 \\ 0 & 1 \end{bmatrix}. \tag{3.95}$$

The polynomial matrix $\bar{N}_x(s)$ results as (see (3.26))

$$\bar{N}_x(s) = \begin{bmatrix} 1 & 0 & 0 \\ 0 & 1 & 1 \\ 0 & 0 & 1 \end{bmatrix}. \tag{3.96}$$

First, the design of a full-order or identity observer is discussed, *i.e.*, $\kappa = 0$ and $n_O = n = 3$. For $\kappa = 0$ (3.81) gives $\bar{D}_\kappa(s) = \bar{D}(s)$ and with $\delta_{r1}[\bar{D}(s)] = 1$, $\delta_{r2}[\bar{D}(s)] = 1$, $\delta_{r3}[\bar{D}(s)] = 1$ and $\Gamma_r[\bar{D}(s)] = I$ the parameterizing polynomial matrix of the identity observer has the general form (see (3.79) and (3.80))

$$\bar{\tilde{D}}(s) = \begin{bmatrix} s+\alpha & \beta & \gamma \\ \delta & s+\varepsilon & \varphi \\ \eta & \lambda & s+\mu \end{bmatrix}, \tag{3.97}$$

i.e., there are $mn = 9$ free parameters. Eigenvalues of the observer at $\tilde{\bar{s}}_i = -4$, $i = 1, 2, 3$, result, *e.g.*, by choosing

$$\tilde{\bar{D}}(s) = \begin{bmatrix} s & -1 & 0 \\ 16 & s+8 & 0 \\ 0 & 0 & s+4 \end{bmatrix}. \tag{3.98}$$

This $\tilde{\bar{D}}(s)$ completely parameterizes the full-order observer in the frequency domain. To obtain a time-domain realization of this observer the constant matrices

$$\bar{P} = L = \begin{bmatrix} 0 & 0 & -1 \\ 13 & 5 & 0 \\ 1 & 0 & 1 \end{bmatrix} \quad \text{and} \quad \bar{Q} = I \tag{3.99}$$

are computed from the Diophantine equation (3.91). With $T = I$, $\Theta = I$, $C_1 = C$ and vanishing C_2 and Ψ_2 the time-domain parameterization of the identity observer as shown in Figure 3.2 is complete.

Now consider the case $\kappa = 1$ or $n_O = 2$ corresponding to $C_2 = [0 \;\; 0 \;\; 1]$. For $\kappa = 1$ the polynomial matrix $\bar{D}_\kappa(s)$ takes the form (see (3.81))

$$\bar{D}_1(s) = \Pi \left\{ \begin{bmatrix} s & -1 & \frac{1}{s} \\ 2 & s+3 & -\frac{1}{s} \\ -1 & 0 & 1+\frac{3}{s} \end{bmatrix} \right\} = \begin{bmatrix} s & -1 & 0 \\ 2 & s+3 & 0 \\ -1 & 0 & 1 \end{bmatrix}, \tag{3.100}$$

which is row reduced. Therefore,

$$\tilde{\bar{D}}(s) = \begin{bmatrix} s+\alpha & \beta & \gamma \\ \delta & s+\varepsilon & \varphi \\ -1 & 0 & 1 \end{bmatrix} \tag{3.101}$$

is the parameterizing matrix containing $m(n-\kappa) = 6$ free parameters (see (3.79) and (3.80)). An observer with eigenvalues at $\tilde{\bar{s}}_i = -4$, $i = 1, 2$, results, *e.g.*, when choosing

$$\tilde{\bar{D}}(s) = \begin{bmatrix} s & -1 & 0 \\ 16 & s+8 & 0 \\ -1 & 0 & 1 \end{bmatrix}. \tag{3.102}$$

To obtain a realization of this observer, (3.102) is introduced in (3.91) giving the constant solutions

$$\bar{P} = \left[\, L_1 \,\vdots\, \Psi_2 \,\right] = \left[\begin{array}{cc:c} 0 & 0 & 0 \\ 14 & 5 & -1 \\ 0 & 0 & 1 \end{array}\right] \quad \text{and} \quad \bar{Q} = \begin{bmatrix} 1 & 0 & 0 \\ 0 & 1 & 0 \\ 0 & 0 & 0 \end{bmatrix}. \tag{3.103}$$

Using this solution in (3.21) and the resulting T in (3.22) yields

$$T = \begin{bmatrix} 0 & 1 & 1 \\ -1 & 0 & 0 \end{bmatrix} \quad \text{and} \quad \Theta = \begin{bmatrix} 0 & -1 \\ 1 & 0 \\ 0 & 0 \end{bmatrix}. \tag{3.104}$$

With (3.103) and (3.104) the reduced-order observer of the order $n_O = 2$ can be implemented as shown in Figure 3.1.

Now, consider an observer of the order $n_O = 1$ with $\kappa = 2$ and consequently

$$C_1 = \begin{bmatrix} 1 & 0 & 0 \end{bmatrix}, \tag{3.105}$$

and

$$C_2 = \begin{bmatrix} 0 & 1 & 1 \\ 0 & 0 & 1 \end{bmatrix}. \tag{3.106}$$

For $\kappa = 2$ the polynomial matrix $\bar{D}_\kappa(s)$ takes the form

$$\bar{D}_2(s) = \begin{bmatrix} s & 0 & 0 \\ 2 & 1 & 0 \\ -1 & 0 & 1 \end{bmatrix}. \tag{3.107}$$

It is also row reduced, so that the parameterizing polynomial matrix of the observer of the order $n_O = n - \kappa = 1$ has the general form

$$\bar{\bar{D}}(s) = \begin{bmatrix} s+\alpha & \beta & \gamma \\ 2 & 1 & 0 \\ -1 & 0 & 1 \end{bmatrix}. \tag{3.108}$$

Choosing $\alpha = 4$, $\beta = 0$ and $\gamma = 0$ the reduced-order observer has an eigenvalue at $s = -4$. The Diophantine equation (3.91) is solved by the constant matrices

$$L_1 = \begin{bmatrix} 4 \\ 0 \\ 0 \end{bmatrix} \quad \text{and} \quad \Psi_2 = \begin{bmatrix} 0 & 0 \\ 1 & -1 \\ 0 & 1 \end{bmatrix}, \tag{3.109}$$

and with these two solutions one finally obtains the vectors $T = \begin{bmatrix} -1 & 0 & 0 \end{bmatrix}$ and $\Theta^T = \begin{bmatrix} -1 & 0 & 0 \end{bmatrix}$.

Finally, consider the completely reduced-order observer with $\kappa = 3$, $n_O = 0$, $C_2 = C$ and vanishing C_1 and L_1. For $\kappa = 3$ the polynomial matrix $\bar{D}_\kappa(s)$ has the simple form

$$\bar{D}_3(s) = \begin{bmatrix} 1 & 0 & 0 \\ 0 & 1 & 0 \\ 0 & 0 & 1 \end{bmatrix}, \tag{3.110}$$

and this is identical with the parameterizing matrix $\bar{\bar{D}}(s)$. Solving (3.91) leads to

$$\Psi_2 = \Psi = \begin{bmatrix} 1 & 0 & 0 \\ 0 & 1 & -1 \\ 0 & 0 & 1 \end{bmatrix}. \tag{3.111}$$

Since T and Θ also vanish, the state estimate results as $\hat{x} = \Psi y$, which coincides with the obvious solution $\hat{x} = C^{-1}y$.

Example 3.3. Frequency-domain parameterization of an observer for a system, where $\bar{D}_\kappa(s)$ is not row reduced at the outset

Given is a system of the order four with two inputs and three outputs described by the state equations (3.1) and (3.2) with

$$A = \begin{bmatrix} 0 & 1 & 0 & 0 \\ -1 & -2 & 0 & 0 \\ 0 & 0 & 0 & 1 \\ -2 & -1 & -2 & -3 \end{bmatrix}, \quad B = \begin{bmatrix} 0 & 0 \\ 1 & 0 \\ 0 & 0 \\ 0 & 1 \end{bmatrix} \quad \text{and } C = \begin{bmatrix} 0 & 0 & 1 & 0 \\ 4 & 1 & 1 & 1 \\ -3 & -1 & -5 & -1 \end{bmatrix}. \tag{3.112}$$

Considered is the design of a reduced-order observer of the order $n_O = 3$, *i.e.*, $\kappa = 1$, with

$$C_1 = \begin{bmatrix} 0 & 0 & 1 & 0 \\ 4 & 1 & 1 & 1 \end{bmatrix} \quad \text{and } C_2 = \begin{bmatrix} -3 & -1 & -5 & -1 \end{bmatrix}. \tag{3.113}$$

The left coprime MFD (3.32) is then, for example, characterized by the polynomial matrices

$$\bar{D}(s) = \begin{bmatrix} s^2+2s-7 & -1 & -2 \\ 3s+31 & s+6 & 7 \\ 2s-14 & -3 & s-3 \end{bmatrix} \quad \text{and } \bar{N}(s) = \begin{bmatrix} 0 & 1 \\ 1 & 1 \\ -1 & -1 \end{bmatrix}. \tag{3.114}$$

For $\kappa = 1$ the parameterization of the reduced-order observer is based on the polynomial matrix (3.81), which here takes the form

$$\bar{D}_1(s) = \begin{bmatrix} s^2+2s-7 & -1 & 0 \\ 3s+31 & s+6 & 0 \\ 2s-14 & -3 & 1 \end{bmatrix} \quad \text{with } \Gamma_r[\bar{D}_1(s)] = \begin{bmatrix} 1 & 0 & 0 \\ 3 & 1 & 0 \\ 2 & 0 & 0 \end{bmatrix}. \tag{3.115}$$

The parameterization of the reduced-order observer, however, requires a row-reduced $\bar{D}_1(s)$. With the unimodular matrix

$$U_L(s) = \begin{bmatrix} 1 & 0 & -0.5s \\ 0 & 1 & -1.5 \\ 0 & 0 & 1 \end{bmatrix} \tag{3.116}$$

one obtains the modified factorization $G(s) = (\bar{D}'(s))^{-1}\bar{N}'(s)$ with

$$\bar{D}'(s) = U_L(s)\bar{D}(s) = \begin{bmatrix} 9s-7 & 1.5s-1 & -0.5s^2+1.5s-2 \\ 52 & s+10.5 & -1.5s+11.5 \\ 2s-14 & -3 & s-3 \end{bmatrix}, \tag{3.117}$$

and

$$\bar{N}'(s) = U_L(s)\bar{N}(s) = \begin{bmatrix} 0.5s & 0.5s+1 \\ 2.5 & 2.5 \\ -1 & -1 \end{bmatrix}. \tag{3.118}$$

Now, the matrix

$$\bar{D}_1'(s) = \begin{bmatrix} 9s-7 & 1.5s-1 & -0.5s+1.5 \\ 52 & s+10.5 & -1.5 \\ 2s-14 & -3 & 1 \end{bmatrix} \tag{3.119}$$

is row reduced and this matrix can be used to obtain the parameterizing polynomial matrix (here chosen as $\tilde{\bar{D}}(s) = \text{diag}(s+3, s+3, s+3)\Gamma_r[\bar{D}_1'(s)]$)

$$\tilde{\bar{D}}(s) = \begin{bmatrix} 9s+27 & 1.5s+4.5 & -0.5s-1.5 \\ 0 & s+3 & 0 \\ 2s+6 & 0 & 0 \end{bmatrix} \tag{3.120}$$

of a reduced-order observer having eigenvalues at $\tilde{\bar{s}}_i = -3$, $i = 1, 2, 3$.

The computation of an equivalent time-domain parameterization, which is needed for a realization of this observer, requires the MFD $C(sI-A)^{-1} = (\bar{D}'(s))^{-1}\bar{N}_x'(s)$. The polynomial matrix $\bar{N}_x'(s)$ has the form

$$\bar{N}_x'(s) = \begin{bmatrix} 1.5s & 0.5s & 2.5s+2 & 0.5s+1 \\ 8.5 & 2.5 & 8.5 & 2.5 \\ -3 & -1 & -3 & -1 \end{bmatrix}, \tag{3.121}$$

and with this matrix the Diophantine equation (3.91), which now takes the form $\bar{N}_x'(s)\bar{P}(s)+\bar{D}'(s)\bar{Q}(s) = \tilde{\bar{D}}(s)$, is solved by the pair of constant matrices

$$\bar{P} = [L_1 \quad \Psi_2] = \left[\begin{array}{cc:c} -12 & -1.5 & 0.5 \\ -28 & -5.5 & 0.5 \\ 10 & 1.5 & -0.5 \\ 14 & 2.5 & -0.5 \end{array}\right] \text{ and } \bar{Q} = \begin{bmatrix} 1 & 0 & 0 \\ 0 & 1 & 0 \\ 0 & 0 & 0 \end{bmatrix}. \tag{3.122}$$

The other matrices of the observer result as

$$T = \begin{bmatrix} -1 & 1 & 0 & 0 \\ 1 & 0 & 1 & 0 \\ 1 & 0 & 0 & 1 \end{bmatrix} \text{ and } \Theta = \begin{bmatrix} 0.5 & 2.5 & 0.5 \\ 1.5 & 2.5 & 0.5 \\ -0.5 & -1.5 & -0.5 \\ -0.5 & -2.5 & 0.5 \end{bmatrix}, \tag{3.123}$$

giving a diagonal $F = T(A - L_1C_1)\Theta$.

Note that the time-domain representations of the observer are not of interest in view of the design of observer-based compensators as described in Chapter 4, because an explicit observation of the state is not necessary in that case. The above manipulations demonstrated simply how an equivalent time-domain parameterization of the state estimator can be obtained from a given polynomial matrix $\tilde{\bar{D}}(s)$.

4

Observer-based Compensators

In the time domain the observer-based compensator is specified by the state feedback gain K and by the observer gains L_1 and Ψ_2 (see Chapters 2 and 3). This assignment defines the state-space realization of the observer-based compensator in the time domain. In the frequency domain the transfer behaviour of the observer-based compensator can be computed from the parameterizing polynomial matrices $\tilde{D}(s)$ (state feedback) and $\tilde{\bar{D}}(s)$ (observer) without recourse to a state-space realization of the system. For an implementation of the compensator, however, a realization structure is required and it can, for example, be chosen to obtain a reduced sensitivity to errors in its realization parameters (see [24]).

Due to the use of reduced-order observers, as introduced in Chapter 3, the resulting transfer functions of the observer-based compensator are strictly proper in all channels from y_1 to u and proper in the channels related to y_2. In general, the compensator that assigns the desired dynamics to the closed-loop system only needs to be proper, so that additional degrees of freedom exist, which can be parameterized explicitly when computing the transfer behaviour of the compensator.

In Section 4.1 the time-domain representation of the observer-based compensator is briefly revisited, and in Section 4.2 its frequency-domain description is introduced.

The frequency-domain design of observer-based compensators is presented in Section 4.3. It is also shown that the time- and the frequency-domain representations are equivalent and how in a reduced-order observer-based compensator the direct feedthrough between y_1 and u can be parameterized in the frequency domain. Thus, beyond the free parameters in the polynomial matrices $\tilde{D}(s)$ and $\tilde{\bar{D}}(s)$, additional degrees of freedom exist for assigning properties of the closed-loop system.

Section 4.4 contains a summary of the design steps for observer-based compensators in the frequency domain and in Section 4.5 the problems caused by restricted input signals are briefly discussed.

4.1 The Observer-based Compensator in the Time Domain

Considered is an observer-based compensator for the system

$$\dot{x}(t) = Ax(t) + Bu(t), \tag{4.1}$$

$$y(t) = Cx(t), \tag{4.2}$$

$$y_c(t) = \Xi y(t) = \Xi Cx(t) = C_c x(t), \tag{4.3}$$

where $x \in \mathbb{R}^n$ is the state, $u \in \mathbb{R}^p$ is the input, $y \in \mathbb{R}^m$ with $m \geq p$ is the measurement and $y_c \in \mathbb{R}^p$ is the controlled output (see also (2.3) and (2.4)), where the pair (A, B) is controllable and the pair (C, A) observable. The measured output is assumed to be subdivided according to

$$y(t) = \begin{bmatrix} y_1(t) \\ y_2(t) \end{bmatrix} = \begin{bmatrix} C_1 \\ C_2 \end{bmatrix} x(t), \tag{4.4}$$

where the κ measurements y_2 with $0 \leq \kappa \leq m$ are directly used in the construction of the state estimate $\hat{x}$ and the remaining $m - \kappa$ measurements are contained in y_1.

The observer-based compensator results from replacing the state x in the feedback

$$u(t) = -Kx(t) + Mr(t) \tag{4.5}$$

(see (2.5)) by the state estimate

$$\hat{x}(t) = \Theta\hat{\zeta}(t) + \Psi_2 y_2(t) \tag{4.6}$$

(see (3.16)). This estimate can be obtained from the measurement y_2 and the state $\hat{\zeta}$ of the reduced-order observer of the order $n - \kappa$ described by

$$\dot{\hat{\zeta}}(t) = T(A - L_1C_1)\Theta\hat{\zeta}(t) + [TL_1 \quad T(A - L_1C_1)\Psi_2]\begin{bmatrix} y_1(t) \\ y_2(t) \end{bmatrix} + TBu(t) \tag{4.7}$$

(see (3.15)). Thus, the state equations of the observer-based compensator of the order $n - \kappa$ have the form

$$\dot{\hat{\zeta}}(t) = T\tilde{A}\Theta\hat{\zeta}(t) + \begin{bmatrix} TL_1 & T\tilde{A}\Psi_2 \end{bmatrix}\begin{bmatrix} y_1(t) \\ y_2(t) \end{bmatrix} + TBMr(t), \tag{4.8}$$

$$u(t) = -K\Theta\hat{\zeta}(t) - K\Psi_2 y_2(t) + Mr(t), \tag{4.9}$$

where the abbreviation

$$\tilde{A} = A - L_1C_1 - BK \tag{4.10}$$

has been used.

Figure 4.1 shows a block diagram of the closed-loop system consisting of the plant (4.1)–(4.3) and the controller (4.8) and (4.9).

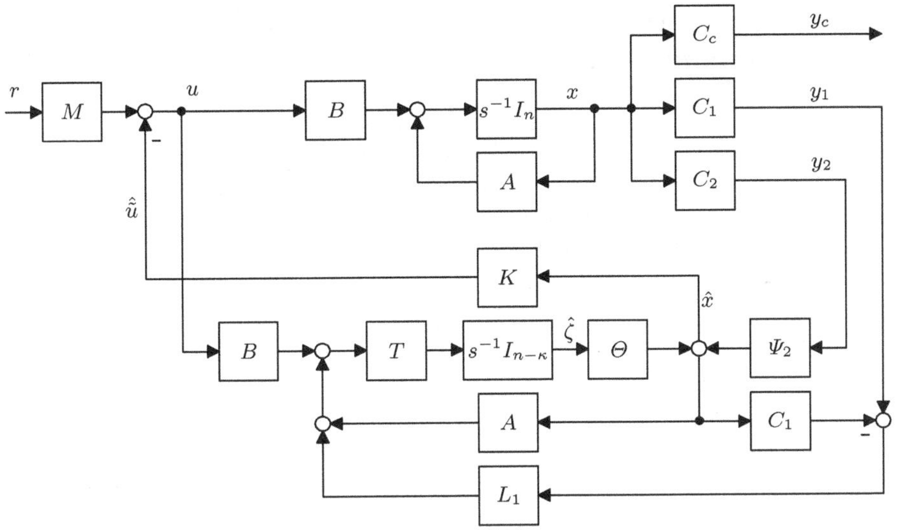

Figure 4.1. Closed-loop system with observer-based compensator in the time domain

Since the error dynamics

$$\dot{\zeta}(t) - \dot{\hat{\zeta}}(t) = T(A - L_1C_1)\Theta(\zeta(t) - \hat{\zeta}(t)) \tag{4.11}$$

(see (3.18)) do not depend on the input u, observation errors are not triggered by the reference input r to the system. Consequently, the observer exactly reproduces the states x also in the presence of reference inputs r provided that the initial observation errors have vanished. These errors decay according to the eigenvalues of the observer, *i.e.*, the zeros of $\det(sI_{n-\kappa} - T(A - L_1C_1)\Theta)$. Thus, the reference transfer behaviour of the closed-loop system in Figure 4.1 is

$$y_c(s) = C_c(sI_n - A + BK)^{-1}BMr(s), \tag{4.12}$$

where y_c is the controlled output of the system (see (4.3)).

The result of the design in the time domain are the state equations of the observer-based compensator, which define a specific realization of it. When using the frequency-domain approach one obtains the input–output behaviour of the compensator, so that a realization still needs to be chosen. At first sight this could be considered a drawback of the frequency-domain approach. Actually, it can be converted into an advantage when suitably choosing the realization structure to obtain, *e.g.*, a reduced sensitivity of the closed-loop system to inaccurate realization parameters in the compensator (see [24] or the brief discussion of this problem in [31]).

4.2 Representations of the Observer-based Compensator in the Frequency Domain

Starting from the time-domain representation in Section 4.1 two different frequency-domain representations of the observer-based compensator are considered. The first reflects the observer property of such compensators and the second is in the usual input–output structure.

The system (4.1)–(4.3) to be controlled is described by its transfer behaviour

$$y(s) = G(s)u(s) = C(sI - A)^{-1}Bu(s) = N(s)D^{-1}(s)u(s), \tag{4.13}$$

and

$$y_c(s) = G_c(s)u(s) = \Xi C(sI - A)^{-1}Bu(s) = N_c(s)D^{-1}(s)u(s), \tag{4.14}$$

where it is assumed that the pair $(N(s), D(s))$ is right coprime and the denominator matrix $D(s)$ is column reduced (see Chapter 2 and Remark 2.2). Also, a left coprime MFD

$$G(s) = \bar{D}^{-1}(s)\bar{N}(s) \tag{4.15}$$

of the transfer matrix $G(s)$ in (4.13) is considered and this MFD is assumed to be such that the matrix

$$\bar{D}_\kappa(s) = \Pi\left\{\bar{D}(s)\begin{bmatrix} I_{m-\kappa} & 0 \\ 0 & s^{-1}I_\kappa \end{bmatrix}\right\} \tag{4.16}$$

is row reduced (see (3.81)).

The frequency-domain representation of the observer-based compensator that reflects the observer property can be obtained by considering the open-loop transfer behaviour between the inputs y and u of this compensator and its output $\hat{\tilde{u}}$ (see Figure 4.1). Using the Laplace transforms

$$s\hat{\zeta}(s) = T(A - L_1C_1)\Theta\hat{\zeta}(s) + [TL_1 \quad T(A - L_1C_1)\Psi_2]\begin{bmatrix} y_1(s) \\ y_2(s) \end{bmatrix} + TBu(s) \tag{4.17}$$

of (4.7) and

$$\hat{\tilde{u}}(s) = K\Theta\hat{\zeta}(s) + K\Psi_2 y_2(s) \tag{4.18}$$

of $\hat{\tilde{u}} = K\hat{x}$ one obtains

$$\begin{aligned}\hat{\tilde{u}}(s) = \{K\Theta(sI_{n-\kappa} - F)^{-1}[TL_1 \quad T(A - L_1C_1)\Psi_2] + [0 \quad K\Psi_2]\}\begin{bmatrix} y_1(s) \\ y_2(s) \end{bmatrix} \\ + K\Theta(sI_{n-\kappa} - F)^{-1}TBu(s),\end{aligned} \tag{4.19}$$

where the abbreviation

$$F = T(A - L_1C_1)\Theta \tag{4.20}$$

has been used.

The transfer behaviour (4.19) shows that, due to the use of a reduced-order observer, there exists a direct feedthrough from the measurement y_2 to the ouput $\hat{\tilde{u}}$ of the compensator.

The input–output behaviour of the observer-based compensator between its inputs y and r and its output u can be obtained by applying the Laplace transform to the state equations (4.8) and (4.9). This yields

$$u(s) = -\left\{K\Theta(sI_{n-\kappa} - T\tilde{A}\Theta)^{-1}\begin{bmatrix}TL_1 & T\tilde{A}\Psi_2\end{bmatrix} + \begin{bmatrix}0 & K\Psi_2\end{bmatrix}\right\}\begin{bmatrix}y_1(s)\\ y_2(s)\end{bmatrix}$$
$$+\left\{-K\Theta(sI_{n-\kappa} - T\tilde{A}\Theta)^{-1}TBM + M\right\}r(s) \tag{4.21}$$

(see also (4.10)). In abbreviated notation the Equation (4.21) can also be written as

$$u(s) = -G_C(s)y(s) + G_{Cr}(s)r(s). \tag{4.22}$$

It is assumed that the parameterizations of the state feedback and the observer are such that the order of the compensator is not smaller than the order of the observer, *i.e.*, that these parameterizations do not lead to pole-zero cancellations in the transfer behaviour $u(s) = -G_C(s)y(s)$ of the observer-based compensator (see also Remark 4.1). This assumption is simply for notational convenience, because the approach presented can also be applied when such cancellations occur (see Examples 4.1 and 4.3).

When introducing the left MFDs

$$K\Theta(sI_{n-\kappa}-F)^{-1}\begin{bmatrix}TL_1 & T(A-L_1C_1)\Psi_2\end{bmatrix}+\begin{bmatrix}0 & K\Psi_2\end{bmatrix} = \Delta^{-1}(s)N_C(s), \tag{4.23}$$

and

$$K\Theta(sI_{n-\kappa} - F)^{-1}TB = \Delta^{-1}(s)N_u(s) \tag{4.24}$$

(see (4.19)) the transfer behaviour of the compensator is represented by

$$\hat{\tilde{u}}(s) = \Delta^{-1}(s)\begin{bmatrix}N_u(s) & N_C(s)\end{bmatrix}\begin{bmatrix}u(s)\\ y(s)\end{bmatrix}, \tag{4.25}$$

where $\Delta(s)$ and $[N_u(s) \quad N_C(s)]$ constitute a left coprime pair if the above-discussed pole-zero cancellations do not occur (see also Remark 4.1) and it is also assumed that the $p \times p$ polynomial matrix $\Delta(s)$ is row reduced.

With the above factorizations of the system and the compensator, the closed-loop system of Figure 4.1 can be equally represented by the schematic diagram of Figure 4.2.

Under the above assumptions the Equations (4.23) and (4.24) show that

$$\det\Delta(s) = \det(sI_{n-\kappa} - F) = \det(sI_{n-\kappa} - T(A - L_1C_1)\Theta) \tag{4.26}$$

(see (4.20)), *i.e.*, the poles of the transfer matrices (4.23) and (4.24) of the compensator are the eigenvalues of the observer. Therefore, the representation

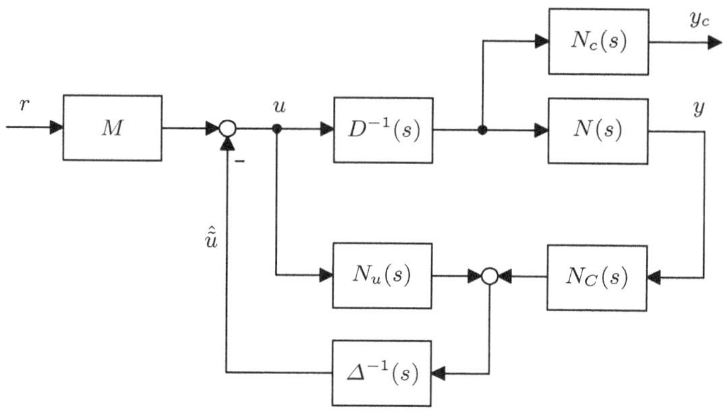

Figure 4.2. Closed-loop system in the frequency domain with compensator in the *observer structure*

of the observer-based compensator shown in Figure 4.2 is said to be in the *observer structure*. As shown in Section 4.5 this realization is of importance in view of a prevention of the undesired effects of input saturation.

By eliminating in Figure 4.2 the feedback of u onto itself *via* $\Delta^{-1}(s)N_u(s)$ and introducing the polynomial matrix

$$D_C(s) = N_u(s) + \Delta(s) \tag{4.27}$$

the closed-loop system can be represented in the form shown in Figure 4.3, where $N_C^*(s)$ and $D_C^*(s)$ are the transfer matrices

$$N_C^*(s) = \Delta^{-1}(s)N_C(s), \tag{4.28}$$

and

$$D_C^*(s) = \Delta^{-1}(s)D_C(s). \tag{4.29}$$

The reason for a representation of the compensator by the transfer matrices (4.28) and (4.29) is that they can be directly computed from the given parameterizing polynomial matrices $\tilde{D}(s)$ and $\tilde{\tilde{D}}(s)$ (see Section 4.3).

When introducing the polynomial matrix

$$N_{Cr}(s) = \Delta(s)M, \tag{4.30}$$

and observing that $D_C^{*\,-1}(s)N_C^*(s) = D_C^{-1}(s)N_C(s)$, the block diagram of Figure 4.3 can be redrawn to obtain the one in Figure 4.4, which is a form of the closed-loop system, where the observer-based compensator is represented as a dynamic output feedback controller $u(s) = -G_C(s)y(s) + G_{Cr}(s)r(s)$.

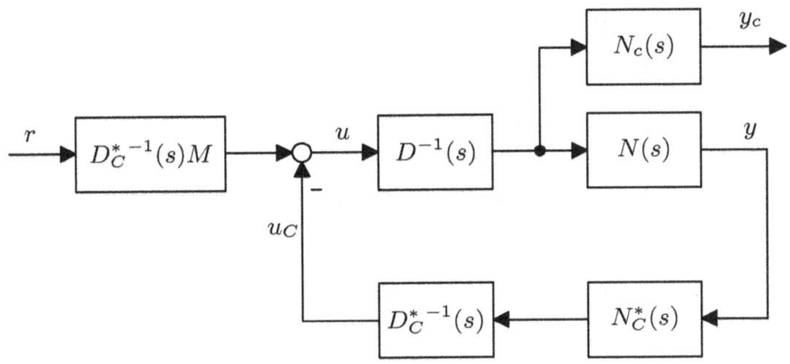

Figure 4.3. Closed-loop system with the compensator in the *output feedback structure* represented as a fraction of transfer matrices

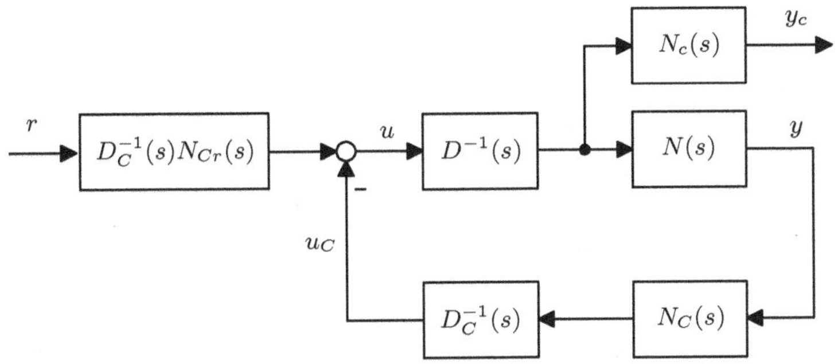

Figure 4.4. Closed-loop system in the frequency domain with compensator in the *output feedback structure*

In this structure of the closed-loop system the input–output behaviour of the plant is represented in a right coprime MFD and the transfer behaviour

$$u_C(s) = G_C(s)y(s) = D_C^{-1}(s)N_C(s)y(s) \tag{4.31}$$

of the compensator in a left coprime MFD.

Remark 4.1. Due to specific choices of the free parameters in $\tilde{D}(s)$ (state feedback) and $\tilde{\bar{D}}(s)$ (state observer) the order of the resulting observer-based compensator may be smaller than the order of the observer, *i.e.*, in the left coprime MFD (4.31) one obtains a $D_C(s)$ with $\deg\{\det(D_C(s))\} < \deg\{\det(\tilde{\bar{D}}(s))\}$. In the following development of the relations between a time- and a frequency-domain representation of observer-based compensators it is always assumed that such pole-zero cancellations do not occur. This implies that the MFD (4.25) is also left coprime, which can be shown along the following lines.

If the pair $(D_C(s), N_C(s))$ is left coprime, there exist solutions $\bar{Y}(s)$ and $\bar{X}(s)$ of the Bezout identity (see Theorem 1.2)

$$N_C(s)\bar{Y}(s) + D_C(s)\bar{X}(s) = I. \tag{4.32}$$

Since $D_C(s) = N_u(s) + \Delta(s)$ (see (4.27)) this can be rewritten as

$$[N_u(s) \quad N_C(s)] \begin{bmatrix} \bar{X}(s) \\ \bar{Y}(s) \end{bmatrix} + \Delta(s)\bar{X}(s) = I, \tag{4.33}$$

which shows that there also exist solutions $\bar{Y}'(s)$ and $\bar{X}'(s)$ of the Bezout identity (see Theorem 1.2)

$$[N_u(s) \quad N_C(s)]\bar{Y}'(s) + \Delta(s)\bar{X}'(s) = I, \tag{4.34}$$

and this Bezout identity is obviously related to the pair $(\Delta(s), [N_u(s) \quad N_C(s)])$ in (4.25). This shows that the pair $(\Delta(s), [N_u(s) \quad N_C(s)])$ is left coprime if the pair $(D_C(s), N_C(s))$ is left coprime.

The opposite, however, is not true. If the pair $(\Delta(s), [N_u(s) \quad N_C(s)])$ in (4.25) is left coprime the pair $(D_C(s), N_C(s))$ in (4.31) is not necessarily also coprime. An example that demonstrates this is characterized by the triple $\Delta(s) = s + 3$, $N_u(s) = 1$, and $N_C(s) = 4(s + 4)$, which defines a left coprime MFD (4.25). However, with $D_C(s) = N_u(s) + \Delta(s) = s + 4$ the MFD (4.31) is not left coprime.

The closed-loop system in Figure 4.4 contains the polynomial representations of the system and the compensator that are used in the design of the compensator in Section 4.3. The observer structure of Figure 4.2 can easily be obtained from this representation by use of Equation (4.27).

The reference transfer behaviour of the closed-loop system shown in Figure 4.4 is

$$y_c(s) = N_c(s) \left[N_C(s)N(s) + D_C(s)D(s)\right]^{-1} \Delta(s)Mr(s). \tag{4.35}$$

As discussed above the reference transfer behaviour of the closed-loop system in Figure 4.1 is that of constant state feedback without observer (see (4.12)). Therefore, the frequency-domain description of this behaviour can be represented by

$$y_c(s) = N_c(s)\tilde{D}^{-1}(s)Mr(s) \tag{4.36}$$

(see (2.21)). In the light of (4.36) the transfer behaviour (4.35) shows that the equation

$$N_C(s)N(s) + D_C(s)D(s) = \Delta(s)\tilde{D}(s) \tag{4.37}$$

is satisfied by an observer-based compensator.

The Diophantine equation (4.37) is the characteristic equation of the closed-loop system in polynomial matrix form and it directly demonstrates the *separation principle*, namely, that the eigenvalues of the closed-loop system with observer-based compensator are the eigenvalues assigned by the state feedback, *i.e.*, the zeros of $\det \tilde{D}(s)$, and the eigenvalues of the observer, *i.e.*, the zeros of

$$\det \Delta(s) = \det \tilde{\tilde{D}}(s) \tag{4.38}$$

(see (4.26) and (3.64)).

If the right-hand side of the Diophantine equation (4.37) is known, it can be solved for the polynomial matrices $N_C(s)$ and $D_C(s)$ of the observer-based compensator. In the frequency domain the constant state feedback control is parameterized by the $p \times p$ polynomial matrix $\tilde{D}(s)$ (see Section 2.2) and the state observer by the $m \times m$ polynomial matrix $\tilde{\tilde{D}}(s)$ (see Section 3.3). On the right-hand side of (4.37) only the parameterizing polynomial matrix $\tilde{D}(s)$ appears, whereas the observer is represented by the $p \times p$ polynomial matrix $\Delta(s)$. Though $\det \Delta(s) = \det(sI - F)$ all polynomial matrices $\Delta(s)$ having a desired characteristic polynomial do not yield a complete parameterization of the observer-based compensator.

This is already the case for single-input systems (*i.e.*, $p = 1, m \geq 1$), where the right-hand side of (4.37) is the characteristic polynomial $\Delta(s)\tilde{D}(s)$ of the closed-loop system with $\tilde{D}(s)$ a polynomial of degree n and $\Delta(s)$ a polynomial of degree $n - \kappa$ if a reduced-order observer of the order $n - \kappa$ is considered. The transfer behaviour of the system is described by the $m \times 1$ transfer vector $G(s) = N(s)\frac{1}{D(s)}$. For given polynomials $\tilde{D}(s) = \det(sI - A + bk^T)$ of the state feedback loop and $\Delta(s) = \det(sI - F)$ of the observer, the polynomials of the $1 \times m$ transfer vector $G_C(s) = \frac{1}{D_C(s)}N_C(s)$ of the compensator can be obtained by equating its coefficients in (4.37). This leads to a system of linear equations that can be solved easily. However, with the exception of the case $m = 1$, there are more free coefficients in $N_C(s)$ and $D_C(s)$ than in the characteristic polynomial $\Delta(s)\tilde{D}(s)$. For single-input systems the number of free parameters in the constant state feedback is n, and the polynomial $\tilde{D}(s)$ also contains n free coefficients. The number of coefficients in the polynomial $\Delta(s)$ is only $n - \kappa$, whereas the number of free parameters in an observer of the order $n - \kappa$ is $m(n - \kappa)$ (see Theorem 3.1). Therefore, there exist additional degrees of freedom in the corresponding linear system of equations and it is not obvious which parameters in $N_C(s)$ and $D_C(s)$ can be chosen at will, *i.e.*, which are the free and the dependent parameters. See, *e.g.*, the discussion in [31] about the choice of the free parameters. However, in the same situation the $m \times m$ polynomial matrix $\tilde{\tilde{D}}(s)$ parameterizes all degrees of freedom contained in the reduced-order observer. Thus, a design procedure is needed that, given the parameterizing polynomial matrices $\tilde{D}(s)$ and $\tilde{\tilde{D}}(s)$, yields the transfer behaviour of the observer-based compensator. This design procedure is presented in the next section.

4.3 Computation of the Observer-based Compensator in the Frequency Domain

In the frequency domain the observer-based compensator is specified by the polynomial matrices $\tilde{D}(s)$, parameterizing the state feedback and by the polynomial matrix $\tilde{\bar{D}}(s)$, parameterizing the state observer. The next theorem shows how the left coprime MFD of the compensator in the output feedback structure can be computed from these parameterizing polynomial matrices. The observer structure of the compensator can be obtained from these results by using (4.27).

Theorem 4.1 (Computation of the left MFD of the compensator). *Given are the right and left coprime MFDs*

$$G(s) = N(s)D^{-1}(s) = \bar{D}^{-1}(s)\bar{N}(s) \tag{4.39}$$

of the system, where $D(s)$ is column reduced and $\bar{D}(s)$ is such that $\bar{D}_\kappa(s)$ (see (4.16)) is row reduced. Also given are the parameterizing polynomial matrices $\tilde{D}(s)$ (state feedback as defined in Theorem 2.1) and $\tilde{\bar{D}}(s)$ (reduced-order observer as defined in Theorem 3.1).

Then, with a solution $Y(s)$ and $X(s)$ of the Bezout identity

$$Y(s)N(s) + X(s)D(s) = I_p, \tag{4.40}$$

and the polynomial matrix

$$\bar{V}(s) = \Pi\left\{\tilde{D}(s)Y(s)\bar{D}^{-1}(s)\tilde{\bar{D}}(s)\right\} \tag{4.41}$$

the transfer matrices $N_C^*(s)$ and $D_C^*(s)$ characterizing the transfer matrix $G_C(s) = D_C^{*-1}(s)N_C^*(s)$ (see also Figure 4.3) of the observer-based compensator (4.31) are given by

$$N_C^*(s) = \Delta^{-1}(s)N_C(s) = \tilde{D}(s)Y(s) - \bar{V}(s)\tilde{\bar{D}}^{-1}(s)\bar{D}(s), \tag{4.42}$$

and

$$D_C^*(s) = \Delta^{-1}(s)D_C(s) = \tilde{D}(s)X(s) + \bar{V}(s)\tilde{\bar{D}}^{-1}(s)\bar{N}(s). \tag{4.43}$$

Introducing the prime right-to-left conversion

$$\bar{V}(s)\tilde{\bar{D}}^{-1}(s) = \Delta^{-1}(s)V(s) \tag{4.44}$$

the polynomial matrices of the MFD $G_C(s) = D_C^{-1}(s)N_C(s)$, which characterizes the output feedback structure of the observer-based compensator (4.31), are given by

$$N_C(s) = \Delta(s)\tilde{D}(s)Y(s) - V(s)\bar{D}(s), \tag{4.45}$$

and

$$D_C(s) = \Delta(s)\tilde{D}(s)X(s) + V(s)\bar{N}(s). \tag{4.46}$$

The proof of this theorem can be found in Section A.2.

Remark 4.2. The above theorem draws on the results presented in [2]. In [27–29] a slightly modified version has been introduced. It yields the same results but it has the drawback that when testing various parameterizations of the compensator, it requires the solution of a Diophantine equation for each new parameterizing matrix $\tilde{D}(s)$.

Both design procedures presented in [2] and [29] yield a factorization of the compensator in terms of rational matrices (see Figure 4.3). This necessitates an additional right-to-left factorization to obtain the polynomial matrices of the compensators MFD. Theorem 4.1, however, directly yields these polynomial matrices.

Remark 4.3. There exists a dual representation of the closed-loop system, where the system to be controlled is represented in a left coprime MFD and the compensator in a right coprime MFD

$$G_C(s) = \bar{N}_C(s)\bar{D}_C^{-1}(s). \tag{4.47}$$

The polynomial matrices $\bar{N}_C(s)$ and $\bar{D}_C(s)$ result from a dual version of Theorem 4.1.

The right coprime MFD of the compensator can also be obtained from the above results by the left-to-right conversion

$$D_C^{-1}(s)N_C(s) = \bar{N}_C(s)\bar{D}_C^{-1}(s). \tag{4.48}$$

This leads to the dual version

$$\bar{N}(s)\bar{N}_C(s) + \bar{D}(s)\bar{D}_C(s) = \bar{\tilde{D}}(s)\bar{\Delta}(s) \tag{4.49}$$

of the characteristic equation (4.37) of the closed-loop system in polynomial matrix form, where $\bar{\tilde{D}}(s)$ is the parameterizing polynomial matrix of the observer and

$$\det \bar{\Delta}(s) = \det \tilde{D}(s). \tag{4.50}$$

For the details of this dual representation the reader is referred to [2,25,27,28].

Because only y_2 is used in the reconstruction (3.16) of the state of the system, all transfer functions of the compensator between y_1 and u are strictly proper, whereas the transfer functions between y_2 and u are proper. When, for some reason (see, for example, the design of the compensator in Section 11.5), the transfer functions of the compensator between y_1 and u do not also have to be strictly proper, a compensator containing $p(m-\kappa)$ additional degrees of freedom can be designed that assures the same eigenvalues of the closed-loop system. The parameterization of such compensators is formulated in the following theorem.

Theorem 4.2 (Parameterization of an additional feedthrough in the compensator). *Given is an observer-based compensator (4.31) of the order $n-\kappa$, $0 \leq \kappa \leq m$. Applying the design described in Theorem 4.1 one obtains a compensator whose transfer functions between y_1 and u are strictly proper.*

A feedthrough F_1 in the transfer functions of the compensator between y_1 and u results when substituting $\bar{V}(s)$ in Theorem 4.1 by

$$\bar{V}_F(s) = \bar{V}(s) - [F_1 \quad 0_{p\times\kappa}], \tag{4.51}$$

where F_1 is a $p \times (m-\kappa)$ constant matrix containing additional $p(m-\kappa)$ degrees of freedom. This compensator also assigns the dynamics of the closed-loop system specified by the polynomial matrices $\tilde{D}(s)$ and $\bar{\tilde{D}}(s)$.

Proof. Pre-multiplying (4.27) by $\Delta^{-1}(s)$ and substituting (4.24) yields

$$\Delta^{-1}(s)D_C(s) = I_p + K\Theta(sI-F)^{-1}TB, \tag{4.52}$$

which demonstrates that the polynomial part of $\Delta^{-1}(s)D_C(s)$ is the identity matrix.

As $\Delta(s)$ is assumed to be row reduced (*i.e.*, $\det \Gamma_r[\Delta(s)] \neq 0$) this implies $\delta_{rj}[D_C(s)] \leq \delta_{rj}[\Delta(s)]$, $j = 1, 2, \ldots, p$ (see (1.71) and (1.72) in Section 1.2). Applying (1.75) one obtains

$$\Pi\{\Delta^{-1}(s)D_C(s)\} = \Gamma_r^{-1}[\Delta(s)]\Gamma_{\delta_r[\Delta(s)]}[D_C(s)] = I_p. \tag{4.53}$$

Pre-multiplying this by $\Gamma_r[\Delta(s)]$ leads to

$$\Gamma_{\delta_r[\Delta(s)]}[D_C(s)] = \Gamma_r[\Delta(s)]. \tag{4.54}$$

Since $\Delta(s)$ is row reduced, also $\det \Gamma_{\delta_r[\Delta(s)]}[D_C(s)] \neq 0$. This, on the other hand, shows that

$$\delta_{rj}[D_C(s)] = \delta_{rj}[\Delta(s)], \quad j = 1, 2, \ldots, p \tag{4.55}$$

is satisfied, because if $\delta_{rj}[D_C(s)] < \delta_{rj}[\Delta(s)]$ for any j, the corresponding row of $\Gamma_{\delta_r[\Delta(s)]}[D_C(s)]$ would vanish, contradicting $\det \Gamma_{\delta_r[\Delta(s)]}[D_C(s)] \neq 0$. With (4.55) satisfied

$$\Gamma_r[D_C(s)] = \Gamma_r[\Delta(s)] \tag{4.56}$$

directly follows from (4.54).

Since the row degrees and the highest row-degree-coefficient matrices of $\Delta(s)$ and $D_C(s)$ coincide, the polynomial parts (or the feedthroughs) of both transfer matrices $G_C(s) = D_C^{-1}(s)N_C(s)$ and $N_C^*(s) = \Delta^{-1}(s)N_C(s)$ (see (1.75)) also coincide. Obviously (4.42) can be written as

$$N_C^*(s) = [\tilde{D}(s)Y(s)\bar{D}^{-1}(s)\bar{\tilde{D}}(s) - \bar{V}(s)]\bar{\tilde{D}}^{-1}(s)\bar{D}(s). \tag{4.57}$$

By construction (see (4.41)) the polynomial matrix $\bar{V}(s)$ assures that the transfer matrix in square brackets in (4.57) is strictly proper because it can be written as

$$\begin{aligned}\tilde{D}(s)Y(s)\bar{D}^{-1}(s)\bar{\tilde{D}}(s) - \bar{V}(s) \\ &= \tilde{D}(s)Y(s)\bar{D}^{-1}(s)\bar{\tilde{D}}(s) - \Pi\{\tilde{D}(s)Y(s)\bar{D}^{-1}(s)\bar{\tilde{D}}(s)\} \\ &= SP\{\tilde{D}(s)Y(s)\bar{D}^{-1}(s)\bar{\tilde{D}}(s)\}. \end{aligned} \tag{4.58}$$

Using $C_2\Psi_2 = I_\kappa$ (see (3.6)) the polynomial part of the transfer matrix $\bar{\tilde{D}}^{-1}(s)\bar{D}(s)$ (see (A.57)) takes the form

$$\Pi\{\bar{\tilde{D}}^{-1}(s)\bar{D}(s)\} = \begin{bmatrix} I_{m-\kappa} & -C_1\Psi_2 \\ 0 & sI_\kappa - C_2A\Psi_2 \end{bmatrix}, \tag{4.59}$$

so that the first $m-\kappa$ columns of $N_C^*(s)$ are strictly proper. Due to the improper element sI_κ in (4.59) the pole-zero difference in the last κ columns is reduced by one, so that the transfer functions of the compensator between y_2 and u are only proper. When replacing $\bar{V}(s)$ in (4.57) by $\bar{V}_F(s) = \bar{V}(s) - [F_1 \quad 0_{p\times\kappa}]$ the polynomial part of the first $m-\kappa$ columns of $[\tilde{D}(s)Y(s)\bar{D}^{-1}(s)\bar{\tilde{D}}(s) - \bar{V}_F(s)]$ is F_1. In view of (4.59) the multiplication with $\bar{\tilde{D}}^{-1}(s)\bar{D}(s)$ does not change this polynomial part, so that F_1 parameterizes the feedthrough in the first $m-\kappa$ columns of the transfer matrix $G_C(s)$.

That this additional feedthrough does not influence the characteristic polynomial of the closed-loop system can be shown by an investigation of

$$\begin{aligned} N_C^*(s)N(s) + D_C^*(s)D(s) \\ &= [\tilde{D}(s)Y(s) - \bar{V}_F(s)\bar{\tilde{D}}^{-1}(s)\bar{D}(s)]N(s) \\ &\quad + [\tilde{D}(s)X(s) + \bar{V}_F(s)\bar{\tilde{D}}^{-1}(s)\bar{N}(s)]D(s) \\ &= \tilde{D}(s)Y(s)N(s) + \tilde{D}(s)X(s)D(s) \\ &\quad + \bar{V}_F(s)\bar{\tilde{D}}^{-1}(s)[\bar{N}(s)D(s) - \bar{D}(s)N(s)] \end{aligned} \tag{4.60}$$

(see (4.42) and (4.43)) that, because of $N(s)D^{-1}(s) = \bar{D}^{-1}(s)\bar{N}(s)$, gives

$$N_C^*(s)N(s) + D_C^*(s)D(s) = \tilde{D}(s)Y(s)N(s) + \tilde{D}(s)X(s)D(s) = \tilde{D}(s), \tag{4.61}$$

(see (4.40)) and this is independent of the polynomial matrix $\bar{V}_F(s)$. In view of (4.42) and (4.43) a pre-multiplication of (4.61) by $\Delta(s)$ yields the Diophantine equation (4.37). This shows that the dynamics specified by $\Delta(s)$ and $\tilde{D}(s)$ are assigned to the closed-loop system. This completes the proof. □

Remark 4.4. The additional feedthrough F_1 in Theorem 4.2 yields the feedback

$$\begin{aligned} u(t) &= -K\Theta\hat{\zeta}(t) - K\Psi_2 y_2(t) - KF_1(y_1(t) - \hat{y}_1(t)) + TBMr(t) \\ &= -K(\Theta - F_1C_1\Theta)\hat{\zeta}(t) - K(\Psi_2 - F_1C_1\Psi_2)y_2(t) - KF_1y_1(t) + TBMr(t) \end{aligned} \tag{4.62}$$

in view of (3.8) and (4.9) in the time domain. Obviously, the additional feedthrough F_1 does not change the dynamics of the closed-loop system since $x - \hat{x} \to 0$ for $t \to \infty$ and thus $y_1 - \hat{y}_1 \to 0$ for $t \to \infty$ independently of u. This verifies that F_1 contains additional degrees of freedom after the assignment of the closed-loop dynamics.

4.4 Summary of the Steps for the Design of Observer-based Compensators in the Frequency Domain

The design of an observer-based compensator in the frequency domain starts from the $m \times p$ transfer matrix

$$G(s) = N(s)D^{-1}(s) = \bar{D}^{-1}(s)\bar{N}(s) \tag{4.63}$$

of the system, where $N(s)$ and $D(s)$ constitute a right coprime and $\bar{N}(s)$ and $\bar{D}(s)$ a left coprime pair of polynomial matrices. The order of the system is assumed to be n.

1. State Feedback Design

Provided that the $p \times p$ polynomial matrix $D(s)$ is column reduced, the state feedback control is parameterized in the frequency domain by the $p \times p$ polynomial matrix $\tilde{D}(s)$ that contains exactly the same number pn of free parameters as the state feedback gain K in the time domain when it has been chosen such that

$$\delta_{ci}[\tilde{D}(s)] = \delta_{ci}[D(s)], \quad i = 1, 2, \ldots, p, \tag{4.64}$$

and

$$\Gamma_c[\tilde{D}(s)] = \Gamma_c[D(s)]. \tag{4.65}$$

The eigenvalues of the closed-loop system are the roots of $\det \tilde{D}(s)$.

2. Observer Design

The state observer of the order $n_O = n - \kappa$ with $0 \leq \kappa \leq m$, is parameterized in the frequency domain by the $m \times m$ polynomial matrix $\tilde{\bar{D}}(s)$ that contains exactly the $m(n - \kappa)$ free parameters that influence the transfer behaviour of the observer-based compensator, provided that the $m \times m$ polynomial matrix

$$\bar{D}_\kappa(s) = \Pi\left\{\bar{D}(s)\begin{bmatrix} I_{m-\kappa} & 0 \\ 0 & s^{-1}I_\kappa \end{bmatrix}\right\} \tag{4.66}$$

is row reduced, and $\tilde{\bar{D}}(s)$ satisfies the restrictions

$$\delta_{rj}[\tilde{\bar{D}}(s)] = \delta_{rj}[\bar{D}_\kappa(s)], \quad j = 1, 2, \ldots, m, \tag{4.67}$$

and

$$\Gamma_r[\tilde{\bar{D}}(s)] = \Gamma_r[\bar{D}_\kappa(s)]. \tag{4.68}$$

The eigenvalues of the observer are the roots of $\det \tilde{\bar{D}}(s)$.

This formulation contains all state observers as special cases, namely the so-called identity observer of the order $n_O = n$ (for $\kappa = 0$) and the completely reduced-order observer of the order $n_O = n - m$ (for $\kappa = m$) and all observers of intermediate orders $n - m \leq n_O \leq n$.

3. Left Coprime MFD of the Observer-based Compensator

The left MFD of the compensator's transfer matrix can be obtained along the lines of Theorem 4.1. First, solve the Bezout identity

$$Y(s)N(s) + X(s)D(s) = I_p \tag{4.69}$$

for the polynomial matrices $Y(s)$ and $X(s)$. Then, compute

$$\bar{V}(s) = \Pi\{\tilde{D}(s)Y(s)\bar{D}^{-1}(s)\tilde{\bar{D}}(s)\}, \tag{4.70}$$

and carry out the prime right-to-left conversion

$$\bar{V}(s)\tilde{\bar{D}}^{-1}(s) = \Delta^{-1}(s)V(s), \tag{4.71}$$

such that $\Delta(s)$ is row reduced. The polynomial matrices of the left MFD $G_C(s) = D_C^{-1}(s)N_C(s)$ of the observer-based compensator are then given by

$$N_C(s) = \Delta(s)\tilde{D}(s)Y(s) - V(s)\bar{D}(s), \tag{4.72}$$

$$D_C(s) = \Delta(s)\tilde{D}(s)X(s) + V(s)\bar{N}(s). \tag{4.73}$$

The transfer functions of this compensator are strictly proper in the channels between y_1 and u, and proper in the channels between y_2 and u. An additional feedback F_1 in the channels between y_1 and u can be parameterized when substituting the polynomial matrix $\bar{V}(s)$ that results from (4.70) by

$$\bar{V}_F(s) = \bar{V}(s) - [F_1 \quad 0_{p\times\kappa}]. \tag{4.74}$$

This shows that the complete design of an observer-based compensator can be carried out in the frequency domain without recurrence to time-domain quantities.

Example 4.1. Frequency-domain design of a SISO compensator with pole-zero cancellation in $G_C(s)$

To demonstrate the design steps of Theorem 4.1 a very simple SISO example is investigated. The plant considered is of the order $n = 2$ with one input u and one output y and its transfer behaviour $y(s) = G(s)u(s)$ with $G(s) = N(s)D^{-1}(s) = \bar{D}^{-1}(s)\bar{N}(s)$ is characterized by the polynomials

$$N(s) = \bar{N}(s) = s + 2, \tag{4.75}$$

and

$$D(s) = \bar{D}(s) = s^2 + 2s + 1. \tag{4.76}$$

By constant state feedback the eigenvalues of the plant are shifted to $\tilde{s}_1 = -3$ and $\tilde{s}_2 = -4$, so that the parameterizing polynomial $\tilde{D}(s)$ (see (2.28) and (2.29)) has the form

$$\tilde{D}(s) = s^2 + 7s + 12. \tag{4.77}$$

Since the order of the system is $n = 2$ and only $m = 1$ measurement exists, one must use a state observer. The minimal order of this observer is $n-m = 1$ and it is assumed to have an eigenvalue at $\tilde{\bar{s}}_1 = -3$. With $\bar{D}_1(s) = s+2$ (see (3.81)) its parameterizing polynomial $\tilde{\bar{D}}(s)$ is consequently given by

$$\tilde{\bar{D}}(s) = s + 3 \tag{4.78}$$

(see (3.79) and (3.80)). Following the design steps of Theorem 4.1 the solutions $Y(s) = -s$ and $X(s) = 1$ of the Bezout identity (4.40) are computed first.

The rational quantity $\tilde{D}(s)Y(s)\bar{D}^{-1}(s)\tilde{\bar{D}}(s)$ has the form

$$\frac{\tilde{D}(s)Y(s)\tilde{\bar{D}}(s)}{\bar{D}(s)} = \frac{-s^4 - 10s^3 - 33s^2 - 36s}{s^2 + 2s + 1}, \tag{4.79}$$

and with

$$\Pi\left\{\frac{\tilde{D}(s)Y(s)\tilde{\bar{D}}(s)}{\bar{D}(s)}\right\} = -s^2 - 8s - 16, \tag{4.80}$$

and

$$SP\left\{\frac{\tilde{D}(s)Y(s)\tilde{\bar{D}}(s)}{\bar{D}(s)}\right\} = \frac{4s + 16}{s^2 + 2s + 1} \tag{4.81}$$

the polynomial (4.41) is

$$\bar{V}(s) = \Pi\left\{\frac{\tilde{D}(s)Y(s)\tilde{\bar{D}}(s)}{\bar{D}(s)}\right\} = -s^2 - 8s - 16. \tag{4.82}$$

The prime right-to-left conversion (4.44)

$$\bar{V}(s)\tilde{\bar{D}}^{-1}(s) = \Delta^{-1}(s)V(s) \tag{4.83}$$

yields $\Delta(s) = \tilde{\bar{D}}(s)$ and $V(s) = \bar{V}(s)$ because the two polynomials do not have a common zero.

With this result the numerator polynomial (4.45) of the compensator takes the form

$$N_C(s) = \Delta(s)\tilde{D}(s)Y(s) - V(s)\bar{D}(s) = 4s + 16, \tag{4.84}$$

and the denominator polynomial (4.46) is

$$D_C(s) = \Delta(s)\tilde{D}(s)X(s) + V(s)\bar{N}(s) = s + 4. \tag{4.85}$$

As the two polynomials $N_C(s)$ and $D_C(s)$ have a common zero, the feedback controller $u(s) = -G_C(s)y(s)$ is a mere proportional controller. Therefore, if only the problem of stabilization by dynamic output feedback is considered (*i.e.*, $r \equiv 0$ in Figure 4.4) this result shows that the order of the compensator can be reduced to zero, because the eigenvalue of the closed-loop system at $\tilde{s}_2 = -4$ cancels in the transfer function $G_C(s)$. With this proportional controller the two eigenvalues of the closed-loop system are the eigenvalue at $\tilde{\tilde{s}}_1 = -3$ of the observer and the remaining eigenvalue at $\tilde{s}_1 = -3$ assigned by $\tilde{D}(s)$.

However, by adding the reference channel $G_{Cr}(s) = D_C^{-1}(s)N_{Cr}(s)$ with $N_{Cr}(s) = \Delta(s)M$ (see Figure 4.4) or by using the observer structure of Figure 4.2, the observer-based compensator is a system of the order one and the eigenvalues of the closed-loop system are located at $\tilde{s}_1 = -3$, $\tilde{s}_2 = -4$ and $\tilde{\tilde{s}}_1 = -3$.

Example 4.2. Design of an observer-based compensator with identity observer in the frequency domain

In Example 2.2 a state feedback control was considered for the system whose transfer matrix is characterized by the right coprime MFD

$$G(s) = N(s)D^{-1}(s) = \begin{bmatrix} 1 & 0 \\ s & 1 \\ 0 & 1 \end{bmatrix} \begin{bmatrix} s^2+3s+3 & -1 \\ -1 & s+3 \end{bmatrix}^{-1}. \tag{4.86}$$

A parameterizing polynomial matrix that assigns all three eigenvalues $\tilde{s}_i$ of the closed-loop system at $s = -5$ has, *e.g.*, the form

$$\tilde{D}(s) = \begin{bmatrix} s^2+10s+25 & 0 \\ 0 & s+5 \end{bmatrix}. \tag{4.87}$$

Based on the left coprime MFD

$$G(s) = \bar{D}^{-1}(s)\bar{N}(s) = \begin{bmatrix} s & -1 & 1 \\ 2 & s+3 & -1 \\ -1 & 0 & s+3 \end{bmatrix}^{-1} \begin{bmatrix} 0 & 0 \\ 1 & 1 \\ 0 & 1 \end{bmatrix} \tag{4.88}$$

of this system, a full-order observer with eigenvalues $\tilde{\tilde{s}}_i = -4$, $i = 1, 2, 3$ was designed in Example 3.2. Its parameterizing polynomial matrix is

$$\tilde{\bar{D}}(s) = \begin{bmatrix} s & -1 & 0 \\ 16 & s+8 & 0 \\ 0 & 0 & s+4 \end{bmatrix}. \tag{4.89}$$

Given the right coprime MFD (4.86), the Bezout identity (4.69) is solved by the pair

$$Y(s) = \begin{bmatrix} 1 & 0 & 0 \\ 0 & 0 & 1 \end{bmatrix} \quad \text{and} \quad X(s) = \begin{bmatrix} 0 & 0 \\ 0 & 0 \end{bmatrix}. \tag{4.90}$$

With this solution, (4.70) yields

$$\bar{V}(s) = \begin{bmatrix} s^2+10s+38 & 5 & -s-10 \\ 1 & 0 & s+6 \end{bmatrix} \tag{4.91}$$

and by the right-to-left conversion (4.71) one obtains

$$V(s) = \begin{bmatrix} s^2+14s+68 & s+11 & 5s+26 \\ s+8 & 1 & s^2+10s+24 \end{bmatrix}, \tag{4.92}$$

and

$$\Delta(s) = \begin{bmatrix} s+4 & 6s+24 \\ 0 & s^2+8s+16 \end{bmatrix}. \tag{4.93}$$

Inserting these results in (4.72) and (4.73) finally gives

$$N_C(s) = \begin{bmatrix} 104 & 35 & -15 \\ 2s+22 & 5 & s+1 \end{bmatrix}, \tag{4.94}$$

and

$$D_C(s) = \begin{bmatrix} s+11 & 6s+37 \\ 1 & s^2+10s+25 \end{bmatrix}. \tag{4.95}$$

The row degrees of $N_C(s)$ are smaller than the row degrees of $D_C(s)$, so that the matrix $\Gamma_{\delta_r[D_C(s)]}[N_C(s)]$ vanishes. Therefore, this compensator is strictly proper (see (1.75)). It assigns the three eigenvalues $\tilde{s}_i$ of the controlled system at $s=-5$ and the three eigenvalues $\tilde{\tilde{s}}_i$ of the observer at $s=-4$.

Example 4.3. Design of an observer-based compensator with reduced-order observer, where the order of the resulting compensator is smaller than the order of the observer

In Example 3.3 a reduced-order observer of the order 3 is parameterized in the frequency domain on the basis of the left coprime MFD

$$\begin{aligned} G(s) &= \bar{D}^{-1}(s)\bar{N}(s) \\ &= \begin{bmatrix} 9s-7 & 1.5s-1 & -0.5s^2+1.5s-2 \\ 52 & s+10.5 & -1.5s+11.5 \\ 2s-14 & -3 & s-3 \end{bmatrix}^{-1} \begin{bmatrix} 0.5s & 0.5s+1 \\ 2.5 & 2.5 \\ -1 & -1 \end{bmatrix} \end{aligned} \tag{4.96}$$

of a system of the order 4 and the parameterizing polynomial matrix of this observer has the form

$$\tilde{\bar{D}}(s) = \begin{bmatrix} 9s+27 & 1.5s+4.5 & -0.5s-1.5 \\ 0 & s+3 & 0 \\ 2s+6 & 0 & 0 \end{bmatrix} \tag{4.97}$$

(see (3.120)).

A right coprime MFD of the system is, *e.g.*,

$$G(s) = N(s)D^{-1}(s) = \begin{bmatrix} 0 & 1 \\ s+4 & s+1 \\ -s-3 & -s-5 \end{bmatrix} \begin{bmatrix} s^2+2s+1 & 0 \\ s+2 & s^2+3s+2 \end{bmatrix}^{-1}. \tag{4.98}$$

Assume that the state feedback for this system is parameterized by the polynomial matrix

$$\tilde{D}(s) = \begin{bmatrix} s^2+10s+25 & 0 \\ s+5 & s^2+10s+25 \end{bmatrix}, \tag{4.99}$$

i.e., all closed-loop poles are placed at $s = -5$. Then, the Bezout identity (4.69) is solved by the pair

$$Y(s) = \begin{bmatrix} 4 & 1 & 1 \\ 1 & 0 & 0 \end{bmatrix} \quad \text{and} \quad X(s) = \begin{bmatrix} 0 & 0 \\ 0 & 0 \end{bmatrix}. \tag{4.100}$$

With this solution, (4.70) yields

$$\bar{V}(s) = \begin{bmatrix} 4s^2+28s-48 & s^2+8.5s+4.5 & 0.5s+5.5 \\ s^2+24s+147 & 2.5s+21 & -0.5s-5 \end{bmatrix}, \tag{4.101}$$

and the right-to-left conversion (4.71) leads to

$$V(s) = \begin{bmatrix} -s-11 & s^2+10s+21 & 2s^2+18.5s+25.5 \\ s+10 & s+6 & 0.5s^2+7.5s+28.5 \end{bmatrix}, \tag{4.102}$$

and

$$\Delta(s) = \begin{bmatrix} s+3 & 0 \\ 0 & s+3 \end{bmatrix}. \tag{4.103}$$

Though the parameterizing matrix (4.97) of the observer has three zeros at $s = -3$, the polynomial matrix $\Delta(s)$ has only two zeros at $s = -3$. This is a consequence of the fact that the order of the resulting observer-based compensator is only two and not three, because it contains a pole-zero cancellation at $s = -3$. This effect also occurs in the time-domain approach where, however, it becomes not as obvious as in the frequency domain.

Inserting $V(s)$ and $\Delta(s)$ in (4.72) and (4.73) finally yields

$$N_C(s) = \begin{bmatrix} -512 & -80 & 16s-112 \\ 292 & 47.5 & -13.5s+51.5 \end{bmatrix}, \tag{4.104}$$

and

$$D_C(s) = \begin{bmatrix} s+27 & 16 \\ -13.5 & s-3.5 \end{bmatrix}. \tag{4.105}$$

This compensator assigns the eigenvalues of the closed-loop system at $\tilde{s}_i = -5$, $i = 1, 2, 3$ and $\tilde{\bar{s}}_j = -3$, $j = 1, 2$ and it has a direct feedthrough

$$\begin{aligned} G_C(\infty) &= \Gamma_r^{-1}\left[D_C(s)\right]\Gamma_{\delta_r[D_C(s)]}\left[N_C(s)\right] = \Gamma_{\delta_r[D_C(s)]}\left[N_C(s)\right] \\ &= \begin{bmatrix} 0 & 0 & 16 \\ 0 & 0 & -13.5 \end{bmatrix} \end{aligned} \tag{4.106}$$

in the transfer functions between y_3 and u.

4.5 Prevention of Problems Caused by Input-signal Restrictions

If, as assumed above, the system to be controlled is linear the behaviour of the closed-loop system is the same whether the compensator has been realized in the dynamic output feedback structure, as in Figure 4.4, or in the observer structure, as in Figure 4.2.

Due to technological restrictions, limited energy, or safety requirements the output signal

$$u(s) = -D_C^{-1}(s)N_C(s)y(s) + D_C^{-1}(s)\Delta(s)Mr(s) \tag{4.107}$$

of the compensator (see Figure 4.4) can only be transferred to the linear system within a limited amplitude range. The most commonly used non-linear element describing such a behaviour is the saturation non-linearity $u_s = \text{sat}_{u_0}(u)$, whose p components are defined by

$$\text{sat}_{u_0}(u_i) = \begin{cases} u_{0i} \text{ if } u_i > u_{0i} \\ u_i \text{ if } -u_{0i} \leq u_i \leq u_{0i} \ ; \ u_{0i} > 0; \ i = 1, 2, \ldots, p \\ -u_{0i} \text{ if } u_i < -u_{0i}. \end{cases} \tag{4.108}$$

Figure 4.5 shows a block diagram of the closed loop, where the compensator is realized as in Figure 4.4 and an input-amplitude restriction $u_s = \text{sat}_{u_0}(u)$ is at the input of the system.

If input saturation becomes active, the feedback is interrupted, because variations of the output signal of the plant have no influence on the input of the plant during saturation. If the compensator contains badly damped or unstable modes (*e.g.*, integral action), these modes can develop freely in this open-loop situation, causing large and badly decaying overshoots in the reference transients, the so-called *controller windup* (see [31]).

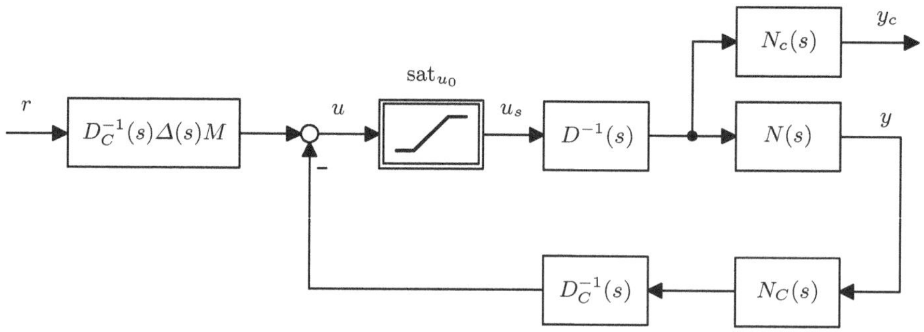

Figure 4.5. Control loop with input saturation

When computing the polynomial matrix (see also (4.27))

$$N_u(s) = D_C(s) - \Delta(s) \tag{4.109}$$

one can implement the compensator in the observer structure (see Figure 4.2) and add a model of the input saturation at the output of the compensator as shown in Figure 4.6. This is also called the *observer technique* for the prevention of controller windup.

The insertion of a model of the saturation non-linearity at the output of the controller has two advantages. First, the non-linearities in the actual actuator never become active, so that they can be neglected in the investigation of the properties of the closed-loop system. Second, the actual input to the linear system, namely the saturated input signal u_s, also drives the observer, so that observation errors are not triggered by an input signal u whose components u_i, $i = 1, 2, \ldots, p$ pass the saturation limits u_{0i}.

This systematically prevents the controller windup, because now the reference behaviour of the loop is the same as if only constant feedback of measured states of the system had been applied.

Also, the above open-loop argument can be used to explain why controller windup does not occur in the loop of Figure 4.6. When input saturation takes place in the scheme of Figure 4.6, all open-loop transfer functions of the compensator characterize stable systems. The poles of the transfer matrices between u_s and $\hat{\tilde{u}}$ and y and $\hat{\tilde{u}}$ are the stable eigenvalues of the observer, so that the possibly existing badly damped or unstable eigenvalues of the controller are not effective during saturation.

However, also after controller windup has been removed the closed-loop system remains non-linear and it can be destabilized by the saturating element. Here, and in what follows, the system is supposed to be asymptotically

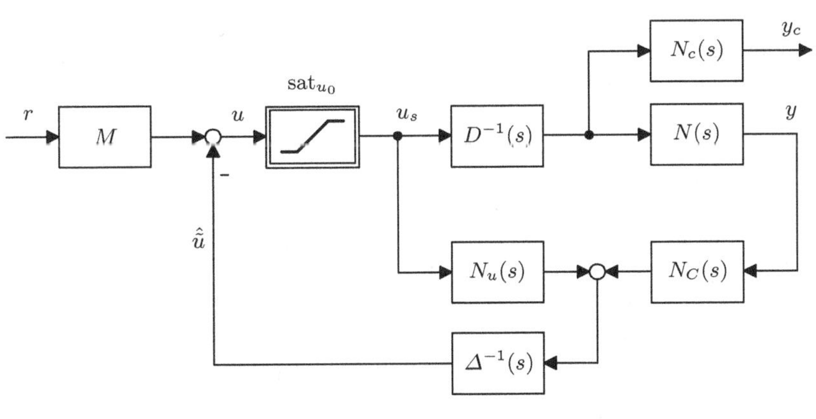

Figure 4.6. Control loop with prevention of controller windup

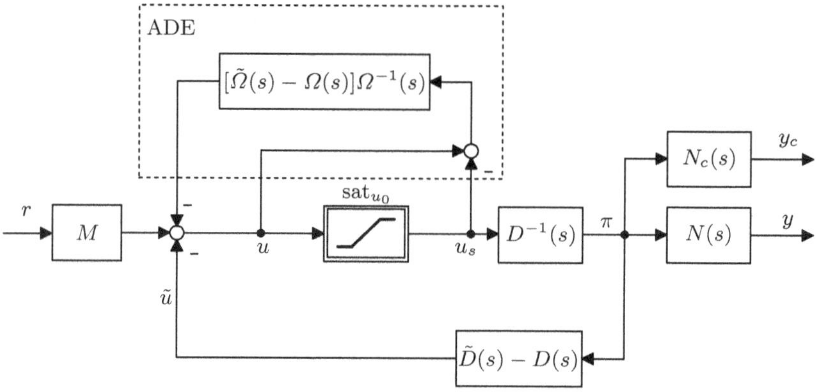

Figure 4.7. Prevention of plant windup in the frequency domain

stable. There are various criteria that provide sufficient conditions for the absolute stability of a non-linear closed loop, consisting of a linear stable system and a non-linearity of the sector type (see [63]). These criteria assure that there exists a globally asymptotically stable equilibrium point of the closed-loop system for any non-linearity in the sector.

The transfer matrix $G_L(s)$ of the linear part $u(s) = -G_L(s)u_s(s)$ of the loop in Figure 4.6 has the form

$$G_L(s) = \Delta^{-1}(s)(N_C(s)N(s)D^{-1}(s) + N_u(s)). \tag{4.110}$$

By using (4.27) and (4.37) this becomes

$$G_L(s) = \tilde{D}(s)D^{-1}(s) - I_p. \tag{4.111}$$

If $G_L(s)$ violates one of the criteria for absolute stability, the reference transients of the loop can exhibit undesired overshoots or limit cycles when input saturation becomes active. This effect is the so-called *plant windup*, which is caused by roots of $\tilde{D}(s)$ that are located too far in the left-half complex plane in view of the existing ampitude limitations (see [31]).

This plant windup can be prevented by an *additional dynamic element* (ADE), which is described in the frequency domain by the block diagram shown in Figure 4.7. The controlled system is represented by a constant state feedback loop (see Figure 2.2 and (2.18)), because it is assumed that the above-discussed prevention of controller windup has been applied.

The transfer matrix $u(s) = -G_{L\Omega}(s)u_s(s)$ of the open loop now has the form

$$G_{L\Omega}(s) = \Omega(s)\tilde{\Omega}^{-1}(s)\tilde{D}(s)D^{-1}(s) - I_p. \tag{4.112}$$

This form makes it obvious how the danger of plant windup can be prevented. First, choose

$$\tilde{\Omega}(s) = \tilde{D}(s) \tag{4.113}$$

to compensate the "fast" $\tilde{D}(s)$, which causes the plant windup and then assign a $p \times p$ polynomial matrix $\Omega(s)$, such that the resulting transfer matrix

$$G_{L\Omega}(s) = \Omega(s)D^{-1}(s) - I_p \tag{4.114}$$

satisfies the chosen stability criterion. When the input saturation is no longer active, *i.e.*, when $u - u_s$ vanishes, the roots of $\det \Omega(s)$ (see Figure 4.7) characterize the influence of the decaying transients of the ADE on the behaviour of the closed-loop system. However, after the transients of the ADE have settled, the dynamics of the closed-loop system are those assigned by the "fast" $\tilde{D}(s)$.

In order to obtain an ADE that does not lead to an algebraic loop, the polynomial matrix $\Omega(s)$ must satisfy the restrictions

$$\delta_{ci}[\Omega(s)] = \delta_{ci}[D(s)], \quad i = 1, 2, \ldots, p, \tag{4.115}$$

and

$$\Gamma_c[\Omega(s)] = \Gamma_c[D(s)]. \tag{4.116}$$

This is true because $\tilde{\Omega}(s) = \tilde{D}(s)$, so that $\tilde{\Omega}(s)$ satisfies the restrictions (2.28) and (2.29) with $\tilde{D}(s)$ replaced by $\tilde{\Omega}(s)$. Then, with $D(s)$ column reduced (4.116) assures that $\Pi\left\{\tilde{\Omega}(s)\Omega^{-1}(s)\right\} = \Gamma_{\delta_c[\Omega(s)]}[\tilde{\Omega}(s)]\Gamma_c^{-1}[\Omega(s)] = I_p$ (see (1.22)).

The time-domain equivalent of this ADE for MIMO systems was presented in [31].

A polynomial matrix $\Omega(s)$ that assures stability of the non-linear closed-loop system consisting of the saturation non-linearity and the linear system can, for example, be obtained *via* the Kalman–Yacubovich lemma (see [63]). A frequency-domain version of this lemma was derived in [65].

Theorem 4.3 (Computation of a stabilizing polynomial matrix for the ADE). *Given the $p \times p$ polynomial matrix $D(s)$ of the right coprime MFD (4.13) of the system to be controlled with* $\det D(s)$ *a Hurwitz polynomial, an arbitrarily chosen $p \times p$ polynomial matrix $N_Q(s)$ having the property*

$$\delta_{ci}[N_Q(s)] < \delta_{ci}[D(s)], \quad i = 1, 2, \ldots, p, \tag{4.117}$$

and a constant matrix W satisfying

$$W^T W = 2I_p. \tag{4.118}$$

Then, the closed-loop system in Figure 4.7 does not have an algebraic loop and is asymptotically stable for $r = 0$ if $\Omega(s)$ is a solution of the bilateral symmetric polynomial matrix equation

$$\begin{aligned} D^T(-s)\Omega(s) + \Omega^T(-s)D(s) = 2D^T(-s)D(s) + N_Q^T(-s)N_Q(s) \\ + D^T(-s)W^T N_Q(s) + N_Q^T(-s)WD(s), \end{aligned} \tag{4.119}$$

and $[\Omega(s) - D(s)]D^{-1}(s)$ is right coprime.

Proof. Consider the controllable and observable state-space representation

$$\dot{x}(t) = Ax(t) + Bu(t), \tag{4.120}$$

$$y(t) = Cx(t) \tag{4.121}$$

of the system (4.13) and the right coprime MFD

$$(sI - A)^{-1}B = N_x(s)D^{-1}(s). \tag{4.122}$$

Since the column degrees and the highest column-degree-coefficient matrices of $\Omega(s)$ and $D(s)$ must coincide $\Omega(s)$ can be regarded as a polynomial matrix that parameterizes a state feedback $u = -Kx$ of the system (4.120) in the frequency domain. Then, the transfer matrix (4.114) can be represented by

$$G_{L\Omega}(s) = K(sI - A)^{-1}B = \Omega(s)D^{-1}(s) - I_p, \tag{4.123}$$

where $\det \Omega(s) = c\det(sI - A + BK), c \neq 0$, is satisfied.

Next, it is shown that the equilibrium point $x = 0$ of the non-linear closed-loop system consisting of a linear system with transfer behaviour $u(s) = G_{L\Omega}(s)\bar{u}(s)$ with $\bar{u} = -u_s$ and a saturation non-linearity $u_s = \Psi(u) = \text{sat}_{u_0}(u)$ satisfying the sector condition $\Psi \in [0, I_p]$, *i.e.*, $\Psi^T(u)(\Psi(u) - u) \leq 0$ (see (4.108)) is globally asymptotically stable if $\tilde{G}_{L\Omega}(s) = I + K(sI - A)^{-1}B$ is positive real. To this end, a loop transformation is applied to the closed-loop system by adding a unity feedforward to the linear system yielding the modified transfer behaviour

$$\begin{aligned}\tilde{u}(s) &= \tilde{G}_{L\Omega}(s)\tilde{\bar{u}}(s) = (I_p + G_{L\Omega}(s))\tilde{\bar{u}}(s)\\ &= (I_p + K(sI - A)^{-1}B)\tilde{\bar{u}}(s).\end{aligned} \tag{4.124}$$

The additional feedforward is compensated in the closed loop by a unity feedback of the non-linearity Ψ giving the modified non-linearity $\tilde{u}_s = \tilde{\Psi}(\tilde{u})$ with $\tilde{\bar{u}} = -\tilde{u}_s$ that satisfies the sector condition $\tilde{\Psi} \in (0, \infty)$, *i.e.*, $\tilde{u}^T\tilde{\Psi}(\tilde{u}) > 0$, $\tilde{u} \neq 0$ and $\tilde{\Psi}(0) = 0$ (see [38]).

The controllable and observable state-space representation

$$\dot{x}(t) = Ax(t) + B\tilde{\bar{u}}(t), \tag{4.125}$$

$$\tilde{u}(t) = Kx(t) + \tilde{\bar{u}}(t) \tag{4.126}$$

of $\tilde{G}_{L\Omega}(s)$ (see (4.120) and (4.124)) is positive real iff there exist matrices $P = P^T > 0$, Q, and W solving

$$A^TP + PA = -Q^TQ, \tag{4.127}$$

$$B^TP + W^TQ = K, \tag{4.128}$$

$$W^TW = 2I_p \tag{4.129}$$

by the positive real lemma (see Lemma 6.2 in [38]). Note that (4.127) requires that A has no eigenvalues in $\text{Re}(s) > 0$, which is satisfied because $\det D(s) =$

$c\det(sI-A), c \neq 0$ is a Hurwitz polynomial by assumption. By considering the Lyapunov function candidate $V(s) = \frac{1}{2}x^T P x$ and by using the same reasoning as in the proof of Theorem 7.1 in [38] it is straightforward to show that

$$\dot{V}(x) = -\frac{1}{2}\|Qx + W\tilde{\tilde{u}}\|^2 - \tilde{u}^T\tilde{\Psi}(\tilde{u}) \leq 0, \tag{4.130}$$

since $\tilde{u}^T\tilde{\Psi}(\tilde{u}) > 0$ for $\tilde{u} \neq 0$. A pre-requisite for $\dot{V}(x)$ to vanish is $\tilde{u} \equiv 0$, which implies $\tilde{\tilde{u}} = -\tilde{u}_s = -\tilde{\Psi}(0) \equiv 0$. Consequently, $\dot{V}(x) \equiv 0$ only holds if $x \equiv 0$ since (K, A) is observable. This shows that $x = 0$ is a globally asymptotically stable equilibrium point of the closed-loop system.

The design equation (4.119) for $\Omega(s)$, such that $\tilde{G}_{L\Omega}(s)$ is positive real can be derived by multiplying (4.127) by -1 and adding $sP - sP$. This yields

$$(-sI - A^T)P + P(sI - A) = Q^T Q. \tag{4.131}$$

Pre-multiplying this by $B^T(-sI - A^T)^{-1}$ and postmultiplying by $(sI - A)^{-1}B$ and substituting $B^T P = K - W^T Q$ from (4.128) leads to

$$\begin{aligned} K(sI - A)^{-1}B + B^T(-sI - A^T)^{-1}K^T & \\ - W^T Q(sI - A)^{-1}B - B^T(-sI - A^T)^{-1}Q^T W & \\ = B^T(-sI - A^T)^{-1}Q^T Q(sI - A)^{-1}B. & \end{aligned} \tag{4.132}$$

Using (4.123) and (4.122) and introducing

$$N_Q(s) = QN_x(s) \tag{4.133}$$

(which obviously has column degrees less than $D(s)$) this takes the form

$$\begin{aligned} \{\Omega(s)D^{-1}(s) - I_p\} + \{(D^T(-s))^{-1}\Omega^T(-s) - I_p\} & \\ - W^T N_Q(s)D^{-1}(s) - (D^T(-s))^{-1}N_Q^T(-s)W & \\ = (D^T(-s))^{-1}N_Q^T(-s)N_Q(s)D^{-1}(s). & \end{aligned} \tag{4.134}$$

Pre-multiplying this by $D^T(-s)$ and postmultiplying by $D(s)$ leads to the design equation (4.119). The assumption that $[\Omega(s) - D(s)]D^{-1}(s)$ is right coprime implies that (K, A) is observable and (A, B) is controllable, which is a usual assumption. □

The above solution for the prevention of plant windup can only be applied in the presence of stable systems. In [30] a solution has been presented that can also be applied in the presence of unstable systems. A discussion of this solution, however, is beyond the scope of this book on observer-based compensators. For a more comprehensive discussion of the problems caused by input-signal restrictions see, *e.g.*, [31].

Figure 4.8 shows the structure of the closed-loop system when the measures for the prevention of controller windup and plant windup have been applied.

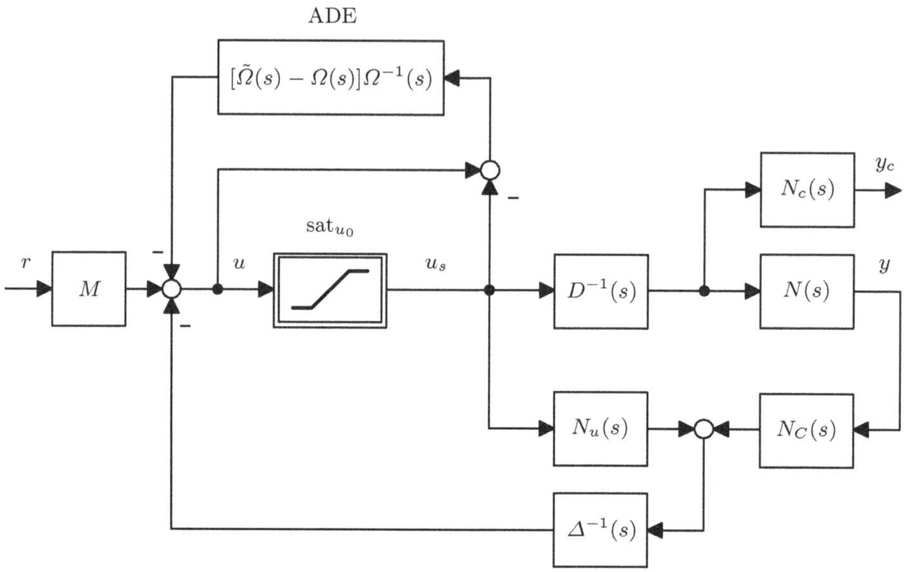

Figure 4.8. Closed-loop system with prevention of controller and plant windup

Example 4.4. Observer-based compensator with full-order observer and feed-through, such that integral controller action results. Discussion of windup effects

Considered is a stable system of the order four with two inputs and two outputs, where $y_c = y$ and therefore $\Xi = I$ and $N_c(s) = N(s)$. Its transfer behaviour is described either by its right coprime MFD

$$G(s) = N(s)D^{-1}(s) \tag{4.135}$$
$$= \begin{bmatrix} s+4 & 2 \\ -4 & 2s+4 \end{bmatrix} \begin{bmatrix} s^2+5s+4 & s+1 \\ -s-4 & s^2+5s+5 \end{bmatrix}^{-1},$$

or by its left coprime MFD

$$G(s) = \bar{D}^{-1}(s)\bar{N}(s) \tag{4.136}$$
$$= \begin{bmatrix} s^2+4.25s+3 & 0.375s+1.5 \\ -1.5s-2 & s^2+5.75s+7 \end{bmatrix}^{-1} \begin{bmatrix} s+3.25 & 1.75 \\ -3.5 & 2s+5.5 \end{bmatrix}.$$

The system can be decoupled in a stable scheme. A decoupled reference transfer behaviour, where the two freely assignable eigenvalues are located at $\tilde{s}_i = -5$, $i = 1, 2$, results in

$$\tilde{D}(s) = \begin{bmatrix} s^2+9s+20 & 2s+10 \\ -2s-10 & s^2+7s+10 \end{bmatrix}, \tag{4.137}$$

and

$$M = \begin{bmatrix} 5 & 0 \\ 0 & 2.5 \end{bmatrix} \tag{4.138}$$

(for details, see Chapter 6).

A full-order or identity observer for this system that has eigenvalues at $\tilde{\tilde{s}}_j = -5$, $j = 1, 2, 3, 4$, is, *e.g.*, parameterized by the polynomial matrix

$$\tilde{\tilde{D}}(s) = \begin{bmatrix} s^2+10s+25 & 0 \\ 0 & s^2+10s+25 \end{bmatrix}. \tag{4.139}$$

If the design steps (4.69)–(4.73) are applied all transfer functions of the observer-based compensator between y and u are strictly proper.

The feedthrough in the four compensator channels can be considered a design parameter to obtain desired effects in the closed-loop system. A rejection of constantly acting disturbances results, for example, when the feedthrough of the compensator is parameterized by

$$F_1 = \begin{bmatrix} 10.75 & -0.375 \\ 0.75 & 4.625 \end{bmatrix}. \tag{4.140}$$

This F_1 was chosen by trial and error to obtain an integral controller action. The transfer matrix of the compensator exhibits this feedthrough if the polynomial matrix $\bar{V}(s)$ resulting from (4.70) is replaced by $\bar{V}_F(s) = \bar{V}(s) - F_1$ (see (4.74)).

Following the design steps of Section 4.4 one obtains the pair

$$Y(s) = \begin{bmatrix} 0.375s+0.5 & 0.0625s \\ -0.25s & 0.125s+0.5 \end{bmatrix}, \tag{4.141}$$

and

$$X(s) = \begin{bmatrix} -0.375 & -0.125 \\ 0.25 & -0.25 \end{bmatrix} \tag{4.142}$$

as a solution of the Bezout identity (4.69). Substituting this in (4.70) yields

$$\bar{V}(s) = \begin{bmatrix} 0.375s^3+5.625s^2+29.125s+57.625 & 0.0625s^3+0.9375s^2+4.6875s+7.4375 \\ -0.25s^3-3.75s^2-18.75s-30.5 & 0.125s^3+1.875s^2+9.875s+20.25 \end{bmatrix}, \tag{4.143}$$

and with the above F_1 the Relation (4.74) yields

$$\bar{V}_F(s) = \bar{V}(s) - F_1 = \begin{bmatrix} 0.375s^3 + 5.625s^2 + 29.125s + 46.875 & 0.0625s^3 + 0.9375s^2 + 4.6875s + 7.8125 \\ -0.25s^3 - 3.75s^2 - 18.75s - 31.25 & 0.125s^3 + 1.875s^2 + 9.875s + 15.625 \end{bmatrix}. \tag{4.144}$$

Carrying out the right-to-left conversion (4.71), which now takes the form $\bar{V}_F(s)\tilde{\bar{D}}^{-1}(s) = \Delta^{-1}(s)V_F(s)$, yields

$$V_F(s) = \begin{bmatrix} 0.375s^3 + 5.625s^2 + 29.125s + 46.875 & 0.0625s^3 + 0.9375s^2 + 4.6875s + 7.8125 \\ -0.25s^3 - 3.75s^2 - 18.75s - 31.25 & 0.125s^3 + 1.875s^2 + 9.875s + 15.625 \end{bmatrix}, \tag{4.145}$$

and

$$\Delta(s) = \begin{bmatrix} s^2+10s+25 & 0 \\ 0 & s^2+10s+25 \end{bmatrix}. \tag{4.146}$$

This finally leads to (see (4.72))

$$\begin{aligned} N_C(s) &= \Delta(s)\tilde{D}(s)Y(s) - V_F(s)\bar{D}(s) \\ &= \begin{bmatrix} 10.75s^2+72s+125 & -0.375s^2 - 1.5s \\ 0.75s^2 + s & 4.625s^2+34s+62.5 \end{bmatrix}, \end{aligned} \tag{4.147}$$

and (see (4.73))

$$\begin{aligned} D_C(s) &= \Delta(s)\tilde{D}(s)X(s) + V_F(s)\bar{N}(s) \\ &= \begin{bmatrix} s^2+3.25s & 1.75s \\ -1.75s & s^2+2.75s \end{bmatrix}. \end{aligned} \tag{4.148}$$

The feedthrough of the transfer matrix $G_C(s) = D_C^{-1}(s)N_C(s)$ of this compensator is indeed as parameterized by F_1, because $\Gamma_{\delta_r[D_C(s)]}[N_C(s)] = F_1$ and $\Gamma_r[D_C(s)] = I$ (see (1.75)). An inspection of (4.148) shows that this compensator contains integral action.

With the matrix M defined in (4.138) the numerator of the reference transfer matrix of the compensator is

$$N_{Cr}(s) = \Delta(s)M = \begin{bmatrix} 5s^2 + 50s + 125 & 0 \\ 0 & 2.5s^2+25s+62.5 \end{bmatrix}. \tag{4.149}$$

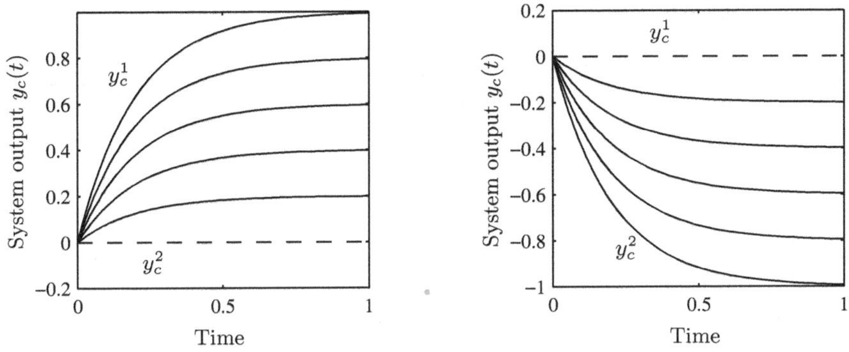

(a) Output $y_c(t)$ of the system

(b) Output $y_c(t)$ of the system

Figure 4.9. Reference behaviour of the closed-loop system

Figure 4.9 shows the transients of the closed-loop system for reference input signals

$$r_i(t) = r_{Si}1(t), \quad i = 1, 2, \tag{4.150}$$

where $1(t)$ is the unit step function. Figure 4.9a depicts the case $0 < r_{S1} \leq 1$, $r_{S2} = 0$ and Figure 4.9b the case $r_{S1} = 0$, $-1 \leq r_{S2} < 0$. This clearly demonstrates the decoupled behaviour.

For linear unconstrained input signals u, the behaviour of the loops in Figures 4.5 and 4.6 is identical. If, however, input saturation becomes active, badly damped or unstable modes of the compensator can give rise to controller windup in the loop of Figure 4.5, whereas this is not the case in the loop of Figure 4.6. This can be demonstrated by the control system investigated here, because two zeros of $\det D_C(s)$ are located on the imaginary axis of the complex plane.

Assume that an input saturation (4.108) with $u_{01} = 1.1$ and $u_{02} = 1.4$ exists, and that the compensator is realized as shown in Figure 4.5. The solid lines in Figure 4.10 show the transients caused by a reference input (4.150) with $r_{S1} = 0.8$, $r_{S2} = -0.6$.

The applied reference signal amplitudes lead to an input saturation and this amplitude restriction causes transients with a considerable overshoot behaviour. If, however,

$$N_u(s) = D_C(s) - \Delta(s) = \begin{bmatrix} -6.75s - 25 & 1.75s \\ -1.75s & -7.25s - 25 \end{bmatrix} \tag{4.151}$$

is used to obtain a realization of the compensator according to Figure 4.6, the same input signals give rise to the transients in dashed lines in Figure 4.10.

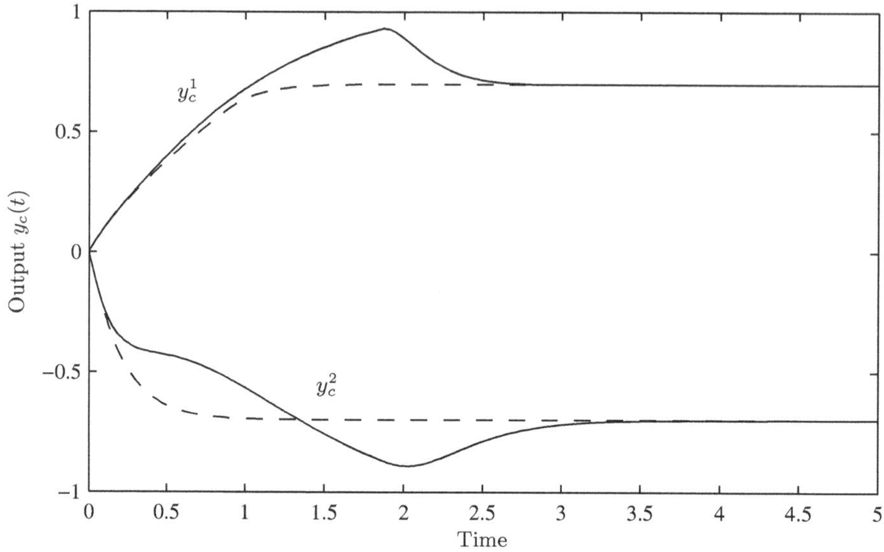

Figure 4.10. Reactions of the closed-loop systems in Figures 4.5 and 4.6 in the presence of input saturation

They are no longer impaired by overshoots, because the observer structure of the compensator in Figure 4.6 assures that the unstable modes of the compensator are not effective in case of input saturation. If saturation is active the open loop is a stable system having the eigenvalues of the system and the observer. Thus, the so-called controller windup is prevented in a systematic manner (see [31]).

It can be shown that, given the above matrices $D(s)$ and $\tilde{D}(s)$, the transfer matrix $G_L(s) = \tilde{D}(s)D^{-1}(s) - I$ satisfies the circle criterion (see, *e.g.*, [59]). Therefore, no danger of plant windup exists in the closed-loop system according to Figure 4.6.

Since, in general, it cannot be ruled out that the compensator has poorly damped or unstable modes a realization according to Figure 4.6 is always advisable.

5

Parametric Compensator Design

In Chapters 2 and 3 it is shown that the assignment of poles to the state feedback loop and to the observer does not uniquely define the observer-based compensator in general. There exist additional degrees of freedom in the design that can be exploited to meet further requirements for the closed-loop system. In order to allow a systematical use of this design freedom the so-called parametric approach was developed in the time domain. This design method was originally proposed independently in [54] and in [16]. Developments of the parametric state feedback design are reported in [52, 55], while a comprehensive review of the different time-domain approaches to the eigenstructure-assignment problem is provided in [43, 64]. This method uses the set of closed-loop eigenvalues and the corresponding invariant parameter vectors as design parameters. The parameter vectors can be chosen to accommodate various design specifications without changing the desired eigenvalue assignment. For a given set of closed-loop eigenvalues and parameter vectors parametric expressions can be derived for computing the corresponding state feedback and observer gains. This leads to an explicit parametric solution of the underlying non-linear eigenvalue-placement problem.

In this chapter it is shown that the parametric approach can also be formulated in the frequency domain. By introducing the set of closed-loop poles and the set of so-called pole directions as free design parameters, an explicit parametric expression is obtained for the polynomial matrix $\tilde{D}(s)$ parameterizing the state feedback in the frequency domain. The pole directions are closely related with the invariant parameter vectors used in the time-domain approach, so that connecting relations can be formulated. These connecting relations allow conversion of the time-domain results into equivalent frequency-domain results and *vice versa.* The frequency-domain parameterization is then extended to the parametric design of reduced-order observers in the frequency domain. To this end, an explicit expression is derived for the polynomial matrix $\tilde{\bar{D}}(s)$ that characterizes the dynamics of the reduced-order observer in the frequency domain. The two polynomial matrices $\tilde{D}(s)$ and $\tilde{\bar{D}}(s)$ completely parameterize the reduced-order observer-based compensator (see Chapter 4), so that the presented results yield a parameterization of the observer-based compensator in the frequency domain.

Another result of this chapter is the parameterization of the reduced-order observer proposed by Uttam and O'Halloran (see Chapter 3). This observer has the advantage that the matrices that parameterize the $m(n-\kappa)$ influential parameters of the observer can be directly identified. So far, the eigenvalue-assignment problem for this kind of observer was not solved in explicit form. By using the connecting relation between the design parameters in the time and in the frequency domain an explicit expression for the parameterizing matrices is derived such that a solution of the eigenvalue-assignment problem for the observer of Uttam and O'Halloran is achieved.

The next section briefly reviews the parametric approach to the design of state feedback controllers in the time domain. Then, the parametric approach to the frequency-domain design of state feedback is presented in Section 5.2 and the relationship with the corresponding time-domain parameterization is investigated. Section 5.3 shows that the state feedback controller can be computed without solving a Diophantine equation if the closed-loop dynamics are specified by the closed-loop poles and their pole directions. The frequency-domain parameterization of reduced-order observers is derived in Section 5.4. A time-domain parameterization of reduced-order observers is presented in Section 5.5 that can be directly obtained from the corresponding frequency-domain result.

5.1 Parametric Design of State Feedback in the Time Domain

Consider a linear time-invariant system described by the state equations

$$\dot{x}(t) = Ax(t) + Bu(t), \tag{5.1}$$

with state $x \in \mathbb{R}^n$ and input $u \in \mathbb{R}^p$. In the following, it is assumed that the matrix B in (5.1) has full column rank and that the pair (A, B) is controllable.

In the time domain the state feedback

$$u(t) = -Kx(t) \tag{5.2}$$

is parameterized by the constant $p \times n$ feedback gain K that assigns the eigenvalues $\tilde{s}_\nu$, $\nu = 1, 2, \ldots, n$, of the closed-loop system if

$$\det(sI - A + BK) = \prod_{\nu=1}^{n}(s - \tilde{s}_\nu). \tag{5.3}$$

In the multivariable case (*i.e.*, $p > 1$) (5.3) yields a set of n non-linear equations for the np elements of K. Consequently, there exist $n(p-1)$ degrees of freedom to influence additional properties of the closed-loop system (see also [52]). In the following, the parametric approach according to [54, 56] is applied to determine an explicit solution of (5.3). The advantage of this approach is that the degrees of freedom remaining after the eigenvalue assignment can be used to assign further properties of the closed-loop system without changing the eigenvalue placement.

This approach uses two sets of free parameters, namely the set of closed-loop eigenvalues $\tilde{s}_\nu$, $\nu = 1, 2, \ldots, n$, and the set of invariant parameter vectors p_ν. The *invariant parameter vectors* p_ν of the controlled plant are defined by

$$p_\nu = K\tilde{v}_\nu, \quad \nu = 1, 2, \ldots, n, \tag{5.4}$$

where $\tilde{v}_\nu$ are the closed-loop eigenvectors. They can be obtained from

$$(\tilde{s}_\nu I - A + BK)\tilde{v}_\nu = 0, \quad \nu = 1, 2, \ldots, n, \tag{5.5}$$

provided that $A+BK$ has n linearly independent eigenvectors. Equation (5.4) shows that for each conjugate complex pair of closed-loop eigenvalues the corresponding parameter vectors are also conjugate complex, and for each real closed-loop eigenvalue the corresponding parameter vector is a real-valued vector. This property has to be taken into account when assigning the invariant parameter vectors to obtain a real-valued K. By substituting (5.4) in (5.5) one can derive an explicit expression for the closed-loop eigenvectors

$$\tilde{v}_\nu = (A - \tilde{s}_\nu I)^{-1} B p_\nu, \quad \nu = 1, 2, \ldots, n. \tag{5.6}$$

The inverse matrix in (5.6) exists if the set of closed-loop eigenvalues does not include an open-loop eigenvalue. After choosing the closed-loop eigenvalues $\tilde{s}_\nu$ and the invariant parameter vectors p_ν the state feedback gain K assigning the closed-loop properties is given in the next theorem.

Theorem 5.1 (Parametric expression for the state feedback gain). *Given the system (5.1) and a set of numbers* $\tilde{s}_1, \tilde{s}_2, \ldots, \tilde{s}_n$ *not including any open-loop eigenvalue together with the corresponding constant* $p \times 1$ *vectors* $p_1, p_2, \ldots, p_n$. *Further assume that the* $\tilde{s}_\nu$ *and the* p_ν *have been chosen such that the vectors (5.6) are linearly independent. Then, the feedback gain*

$$K = \begin{bmatrix} p_1 \ldots p_n \end{bmatrix} \begin{bmatrix} (A - \tilde{s}_1 I)^{-1} B p_1 \ldots (A - \tilde{s}_n I)^{-1} B p_n \end{bmatrix}^{-1} \tag{5.7}$$

assigns the eigenvalues $\tilde{s}_1, \tilde{s}_2, \ldots, \tilde{s}_n$ *and the invariant parameter vectors* $p_1, p_2, \ldots, p_n$ *to the closed-loop system. Conversely, each feedback gain* K *can be expressed by (5.7) if it assigns the closed-loop eigenvalues* $\tilde{s}_1, \tilde{s}_2, \ldots, \tilde{s}_n$ *and corresponding invariant parameter vectors* $p_1, p_2, \ldots, p_n$, *such that the closed-loop eigenvectors are linearly independent.*

Proof. For a proof of Theorem 5.1 see [54,56]. □

Remark 5.1. The parametric expression (5.7) only holds for non-defective matrices $A-BK$ where the closed-loop eigenvectors $\tilde{v}_\nu$ in (5.6) are linearly independent. This restriction can always be satisfied by an appropriate choice of the invariant parameter vectors if the algebraic multiplicity of the closed-loop eigenvalues is less than or equal to p. In view of the assumption $\operatorname{rank} B = p$ this property follows from the fact that for every multiple eigenvalue only up to p linearly independent eigenvectors (5.6) exist.

Remark 5.2. The restriction that the closed-loop eigenvalues do not coincide with any open-loop eigenvalue can be removed (for details see [56]).

The n invariant parameter vectors p_ν contain $n(p-1)$ degrees of freedom, though the n vectors p_ν have np elements. This is due to the fact that the multiplication of p_ν by a non-zero constant does not change the result for (5.7) since these constants cancel in (5.7). Thus, the set of closed-loop eigenvalues and the corresponding invariant parameter vectors completely parameterize the degrees of freedom in K. The advantage of the explicit characterization of the feedback gain in (5.7) lies in the ability to choose a stabilizing set of closed-loop eigenvalues and to satisfy additional design criteria by choosing the invariant parameter vectors without affecting the closed-loop stability.

5.2 Parametric Design of State Feedback in the Frequency Domain

In the frequency domain the state feedback control is characterized by the polynomial matrix $\tilde{D}(s)$ (see Chapter 2). Thus, a frequency-domain formulation of the parametric approach must yield a $\tilde{D}(s)$, so that after choosing the n closed-loop poles the remaining $n(p-1)$ degrees of freedom are arbitrarily assignable. This becomes possible by introducing the pole directions.

5.2.1 Definition of the Pole Directions

In the following, it is assumed that the algebraic multiplicity μ_ν of each closed-loop pole $\tilde{s}_\nu$ is less than or equal to p and coincides with the corresponding geometric multiplicity, *i.e.*, all pole directions q_ν corresponding to a multiple pole $\tilde{s}_\nu$ (see (5.9)) are linearly independent (for further details see, *e.g.*, [13]). Since the closed-loop poles $\tilde{s}_\nu$ are a solution of

$$\det \tilde{D}(s) = 0 \tag{5.8}$$

(see (2.20)) the constant $p \times p$ matrix $\tilde{D}(\tilde{s}_\nu)$ is rank deficient with $\operatorname{rank} \tilde{D}(\tilde{s}_\nu) = p - \mu_\nu$ (see the Smith–Form of $\tilde{D}(s)$, *e.g.*, in [36]). Therefore, there exists a solution $q_\nu \neq 0$ of

$$\tilde{D}(\tilde{s}_\nu)q_\nu = 0, \quad \nu = 1, 2, \ldots, n. \tag{5.9}$$

For distinct poles (*i.e.*, $\mu_\nu = 1$) the vector q_ν in (5.9) is unique up to a multiplicative constant and for multiple poles there exist μ_ν linearly independent pole directions. Consequently, every closed-loop pole can be associated with a vector q_ν that is called the *pole direction* q_ν of the closed-loop system.

Remark 5.3. For real closed-loop poles the pole directions are real-valued vectors, and for conjugate complex closed-loop poles the pole directions are also conjugate complex.

5.2.2 Parametric Expression of the State Feedback

In the following, an explicit parametric expression for the closed-loop denominator matrix $\tilde{D}(s)$ is derived such that it assigns the poles $\tilde{s}_1, \tilde{s}_2, \ldots, \tilde{s}_n$, *i.e.*,

$$\det \tilde{D}(\tilde{s}_\nu) = 0, \quad \nu = 1, 2, \ldots, n, \tag{5.10}$$

and the pole directions $q_1, q_2, \ldots, q_n$, *i.e.*,

$$\tilde{D}(\tilde{s}_\nu) q_\nu = 0, \quad \nu = 1, 2, \ldots, n \tag{5.11}$$

to the closed-loop system. In view of Remark 5.3 the assigned pole directions q_ν for real closed-loop poles have to be real-valued vectors, and for conjugate complex closed-loop poles the pole directions q_ν have to be conjugate complex.

Theorem 5.2 (Parametric expression for the polynomial matrix of the state feedback). *Consider the transfer behaviour of the system (5.1) in a right coprime MFD*

$$x(s) = N_x(s) D^{-1}(s) u(s), \tag{5.12}$$

with $D(s)$ column reduced and a set of numbers $\tilde{s}_1, \tilde{s}_2, \ldots, \tilde{s}_n$ together with the corresponding constant $p \times 1$ vectors $q_1, q_2, \ldots, q_n$. Further, assume that given the $n \times p$ polynomial matrix[1]

$$S(s) = \operatorname{diag}\left(\left[s^{\delta_{c1}[D(s)]-1} \ \ldots \ s \ 1\right]^T, \ldots, \left[s^{\delta_{cp}[D(s)]-1} \ \ldots \ s \ 1\right]^T\right) \tag{5.13}$$

the $\tilde{s}_\nu$ and the q_ν have been chosen such that the vectors $S(\tilde{s}_\nu) q_\nu$, $\nu = 1, 2, \ldots, n$ are linearly independent. Then, the denominator matrix

$$\tilde{D}(s) = D(s) - \left[D(\tilde{s}_1) q_1 \ \ldots \ D(\tilde{s}_n) q_n\right] \left[S(\tilde{s}_1) q_1 \ \ldots \ S(\tilde{s}_n) q_n\right]^{-1} S(s) \tag{5.14}$$

assigns the poles $\tilde{s}_1, \tilde{s}_2, \ldots, \tilde{s}_n$ and the pole directions $q_1, q_2, \ldots, q_n$ to the closed-loop system. Conversely, each denominator matrix $\tilde{D}(s)$ can be expressed by (5.14) if it assigns the closed-loop poles $\tilde{s}_1, \tilde{s}_2, \ldots, \tilde{s}_n$ and corresponding pole directions $q_1, q_2, \ldots, q_n$, such that the vectors $S(\tilde{s}_\nu) q_\nu$, $\nu = 1, 2, \ldots, n$ are linearly independent.

[1] The matrix $S(s)$ has n rows since $\sum_{i=1}^{p} \delta_{ci}[D(s)] = n$ (see (1.14)).

Proof. First, it is shown that any polynomial matrix $\tilde{D}(s)$ can be parameterized by (5.14) if the assumptions of Theorem 5.2 hold. The denominator matrix $\tilde{D}(s)$ can be represented by

$$\tilde{D}(s) = \Gamma_c[D(s)]\left(\operatorname{diag}(s^{\delta_{ci}[D(s)]}) + \tilde{D}_c S(s)\right), \tag{5.15}$$

where (2.28) and (2.29) have been used. In (5.15) the polynomial matrix $S(s)$ is defined by (5.13) and the $p \times n$ constant matrix $\tilde{D}_c$ contains the freely assignable coefficients of $\tilde{D}(s)$. Inserting the closed-loop poles $\tilde{s}_\nu$ in (5.15) and postmultiplying it by the pole directions q_ν yields

$$\Gamma_c[D(s)]\left(\operatorname{diag}(\tilde{s}_\nu^{\delta_{ci}[D(s)]})q_\nu + \tilde{D}_c S(\tilde{s}_\nu)q_\nu\right) = 0, \tag{5.16}$$

which has to vanish in view of (5.9). Since the denominator matrix $D(s)$ of the system is column reduced (*i.e.*, $\Gamma_c[D(s)]$ is non-singular (see Section 1.1)) the expression in the brackets has to vanish, giving

$$\tilde{D}_c S(\tilde{s}_\nu)q_\nu = -\operatorname{diag}(\tilde{s}_\nu^{\delta_{ci}[D(s)]})q_\nu, \quad \nu = 1, 2, \ldots, n. \tag{5.17}$$

The matrix $\tilde{D}_c$ of coefficients in (5.17) can be represented by

$$\tilde{D}_c = D_c + \Gamma_c^{-1}[D(s)]\tilde{M}, \tag{5.18}$$

in which $\tilde{M}$ is a suitable $p \times n$ matrix. In (5.18) D_c is the matrix of coefficients in

$$D(s) = \Gamma_c[D(s)]\left(\operatorname{diag}(s^{\delta_{ci}[D(s)]}) + D_c S(s)\right). \tag{5.19}$$

Substituting (5.18) in (5.17) yields

$$D_c S(\tilde{s}_\nu)q_\nu + \Gamma_c^{-1}[D(s)]\tilde{M}S(\tilde{s}_\nu)q_\nu = -\operatorname{diag}(\tilde{s}_\nu^{\delta_{ci}[D(s)]})q_\nu, \tag{5.20}$$

which gives, with (5.19),

$$\begin{aligned}\tilde{M}S(\tilde{s}_\nu)q_\nu &= -\Gamma_c[D(s)]\left(\operatorname{diag}(\tilde{s}_\nu^{\delta_{ci}[D(s)]}) + D_c S(\tilde{s}_\nu)\right)q_\nu \\ &= -D(\tilde{s}_\nu)q_\nu, \quad \nu = 1, 2, \ldots, n\end{aligned} \tag{5.21}$$

after a simple rearrangement. The n Relations (5.21) can be written in the matrix form

$$\tilde{M}\left[S(\tilde{s}_1)q_1 \ldots S(\tilde{s}_n)q_n\right] = -\left[D(\tilde{s}_1)q_1 \ldots D(\tilde{s}_n)q_n\right]. \tag{5.22}$$

Since by assumption the vectors $S(\tilde{s}_\nu)q_\nu$, $\nu = 1, 2, \ldots, n$, are linearly independent, (5.22) is solvable for $\tilde{M}$, giving

$$\tilde{M} = -\left[D(\tilde{s}_1)q_1 \ldots D(\tilde{s}_n)q_n\right]\left[S(\tilde{s}_1)q_1 \ldots S(\tilde{s}_n)q_n\right]^{-1}. \tag{5.23}$$

By substituting (5.18) in (5.15) the result

$$\tilde{D}(s) = D(s) + \tilde{M}S(s) \tag{5.24}$$

is obtained in view of (5.19). Then, the parametric expression (5.14) follows from inserting (5.23) in (5.24).

If, on the other hand, $\tilde{D}(s)$ according to (5.14) is computed from the numbers $\tilde{s}_\nu$ with appropriate vectors q_ν (see Remark 5.3) the closed-loop poles $\tilde{s}_\nu$ with corresponding pole directions q_ν result. To show this, insert $\tilde{s}_\nu$ in (5.14) and postmultiply the result by q_ν, yielding

$$\tilde{D}(\tilde{s}_\nu)q_\nu = D(\tilde{s}_\nu)q_\nu - \left[D(\tilde{s}_1)q_1 \ldots D(\tilde{s}_n)q_n\right]\left[S(\tilde{s}_1)q_1 \ldots S(\tilde{s}_n)q_n\right]^{-1} S(\tilde{s}_\nu)q_\nu. \tag{5.25}$$

By using

$$S(\tilde{s}_\nu)q_\nu = \left[S(\tilde{s}_1)q_1 \ldots S(\tilde{s}_n)q_n\right] e_\nu, \tag{5.26}$$

and

$$\left[D(\tilde{s}_1)q_1 \ldots D(\tilde{s}_n)q_n\right] e_\nu = D(\tilde{s}_\nu)q_\nu, \tag{5.27}$$

where e_ν is the νth unit vector, the right-hand side of (5.25) vanishes, giving

$$\tilde{D}(\tilde{s}_\nu)q_\nu = 0, \quad \nu = 1, 2, \ldots, n. \tag{5.28}$$

Thus, the numbers $\tilde{s}_\nu$ and the vectors q_ν are the closed-loop poles and the corresponding pole directions (see (5.10) and (5.11)) since (5.28) only holds for a non-trivial vector q_ν if $\det \tilde{D}(\tilde{s}_\nu) = 0$. □

Remark 5.4. It is well known that $n(p-1)$ degrees of freedom remain in the state feedback problem after the n closed-loop poles have been specified. The n pole directions, however, consist of pn free elements. This is no contradiction since every pole direction only contains $p-1$ influential parameters. A multiplication of each q_ν by a non-zero constant does not change the result for $\tilde{D}(s)$ since these constants cancel in (5.14). Consequently, only the direction of each vector q_ν is of interest, but not its length.

Remark 5.5. Different from the parametric approach in the time domain (see Section 5.1) the parametric expression (5.14) is also valid if closed-loop poles coincide with open-loop poles.

Remark 5.6. In general, the parametric expression (5.14) can be used to assign closed-loop poles with algebraic multiplicity up to p, where p is the number of inputs. This is due to the fact that the polynomial matrix $S(s)$ in (5.13) has rank p for all s. Consequently, only a maximum number of p linearly independent vectors $S(\tilde{s}_\nu)q_\nu$ exist for a multiple closed-loop pole. Closed-loop poles with algebraic multiplicity greater than p can be assigned by introducing the *generalized pole directions* (for details see [13]).

Example 5.1. Parametric design of state feedback in the frequency domain
Consider the system used in Example 2.2 whose input–output behaviour $y(s) = G(s)u(s)$ is characterized by the transfer matrix

$$G(s) = N(s)D^{-1}(s) = \begin{bmatrix} 1 & 0 \\ s & 1 \\ 0 & 1 \end{bmatrix} \begin{bmatrix} s^2+3s+3 & -1 \\ -1 & s+3 \end{bmatrix}^{-1} \tag{5.29}$$

in a right coprime MFD. In the frequency domain, the state feedback control is characterized by the denominator matrix

$$\tilde{D}(s) = \begin{bmatrix} s^2+\alpha s+\beta & \gamma \\ \delta s+\varepsilon & s+\varphi \end{bmatrix}, \tag{5.30}$$

where the degrees of freedom are contained in the $pn = 6$ parameters α, β, γ, δ, ε and φ (see Example 2.2). In order to assign the poles $\tilde{s}_1$, $\tilde{s}_2$ and $\tilde{s}_3$ to the closed-loop system the polynomal matrix $\tilde{D}(s)$ has to satisfy

$$\det \tilde{D}(s) = (s - \tilde{s}_1)(s - \tilde{s}_2)(s - \tilde{s}_3). \tag{5.31}$$

Consequently, there exist $n(p-1) = 3$ additional degrees of freedom since (5.31) only yields 3 conditions for the 6 freely assignable parameters in $\tilde{D}(s)$. One possibility to use this freedom is to determine 3 parameters such that $\tilde{D}(s)$ becomes diagonal. Then, the remaining free parameters can easily be chosen to satisfy (5.31) (see Example 2.2). However, these 3 free parameters can also be used to accommodate further design specification. This is facilitated by the parametric expression (5.14). It uses the matrix $S(s)$ that here has the form

$$S(s) = \begin{bmatrix} s & 0 \\ 1 & 0 \\ 0 & 1 \end{bmatrix}, \tag{5.32}$$

in view of $\delta_{c1}[D(s)] = 2$ and $\delta_{c2}[D(s)] = 1$ (see (5.13)). With this result a parametric expression for the denominator matrix assigning the closed-loop poles $\tilde{s}_1$, $\tilde{s}_2$ and $\tilde{s}_3$ is given by

$$\tilde{D}(s) = D(s) - \begin{bmatrix} D(\tilde{s}_1)q_1 & D(\tilde{s}_2)q_2 & D(\tilde{s}_3)q_3 \end{bmatrix} \begin{bmatrix} S(\tilde{s}_1)q_1 & S(\tilde{s}_2)q_2 & S(\tilde{s}_3)q_3 \end{bmatrix}^{-1} S(s), \tag{5.33}$$

in which the directions of $q_i = \begin{bmatrix} q_{i1} & q_{i2} \end{bmatrix}^T$, $i = 1, 2, 3$, parameterize the 3 remaining degrees of freedom. By an appropriate choice of the closed-loop poles and their pole directions q_i desired properties can be assigned to the closed-loop system. An example of such properties is the complete decoupling of the reference behaviour, where the corresponding reference transfer matrix is diagonal. This will be discussed in detail in Chapter 6.

5.2.3 Relation Between the Pole Directions and the Closed-loop Eigenvectors

A relation between the pole directions and the closed-loop eigenvectors can be obtained by using the connecting relation

$$\tilde{D}(s) = D(s) + KN_x(s) \tag{5.34}$$

between the parametrization of state feedback control in the time and in the frequency domain (see (2.18)). Substituting s by the closed-loop poles $\tilde{s}_\nu$ in (5.34) and multiplying the result from the right by q_ν gives

$$\tilde{D}(\tilde{s}_\nu)q_\nu = D(\tilde{s}_\nu)q_\nu + KN_x(\tilde{s}_\nu)q_\nu, \quad \nu = 1, 2, \ldots, n, \tag{5.35}$$

which implies

$$KN_x(\tilde{s}_\nu)q_\nu = -D(\tilde{s}_\nu)q_\nu, \tag{5.36}$$

in view of (5.11). Now, consider the expression

$$(\tilde{s}_\nu I - A + BK)N_x(\tilde{s}_\nu)q_\nu = (\tilde{s}_\nu I - A)N_x(\tilde{s}_\nu)q_\nu + BKN_x(\tilde{s}_\nu)q_\nu. \tag{5.37}$$

Substituting (5.36) in (5.37) and observing that the right coprime MFD

$$(sI - A)^{-1}B = N_x(s)D^{-1}(s) \tag{5.38}$$

implies

$$(sI - A)N_x(s) = BD(s), \tag{5.39}$$

the expression (5.37) can be rewritten as

$$(\tilde{s}_\nu I - A + BK)N_x(\tilde{s}_\nu)q_\nu = 0. \tag{5.40}$$

Comparing this with the eigenvector equation

$$(\tilde{s}_\nu I - A + BK)\tilde{v}_\nu = 0 \tag{5.41}$$

of the closed-loop system (see (5.5)), the relation between the pole directions q_ν and the closed-loop eigenvectors $\tilde{v}_\nu$ is obviously given by

$$\tilde{v}_\nu = N_x(\tilde{s}_\nu)q_\nu, \quad \nu = 1, 2, \ldots, n. \tag{5.42}$$

A pre-requisite for this is that all closed-loop poles are distinct, because then each closed-loop eigenvalue $\tilde{s}_\nu$ is accompanied by only one linearly independent eigenvector $\tilde{v}_\nu$.

5.2.4 Relation Between the Pole Directions and the Invariant Parameter Vectors

For given closed-loop pole locations $\tilde{s}_\nu$ one can immediately substitute (5.42) and (5.36) in (5.4) to obtain

$$p_\nu = K\tilde{v}_\nu = KN_x(\tilde{s}_\nu)q_\nu = -D(\tilde{s}_\nu)q_\nu, \tag{5.43}$$

so that the invariant parameter vectors p_ν and the pole directions q_ν are related by

$$p_\nu = -D(\tilde{s}_\nu)q_\nu, \quad \nu = 1, 2, \ldots, n. \tag{5.44}$$

This constitutes the connecting relation between the parametric approaches to the design of state feedback in the time and in the frequency domain. Therefore, one can use existing results for the parametric approach in the time domain (see, *e.g.*, [43, 57]) to solve the same problems in the frequency domain.

5.3 Parameterization of the State Feedback Gain Using the Pole Directions

After specifying the closed-loop dynamics by the closed-loop denominator matrix $\tilde{D}(s)$ the feedback gain K in (5.2) is obtained from the constant solution $P(s) = K$ and $Q(s) = I$ of the Diophantine equation

$$P(s)N_x(s) + Q(s)D(s) = \tilde{D}(s) \tag{5.45}$$

(see (2.39)). However, when using the closed-loop poles and the corresponding pole directions as design parameters, the feedback gain K can be directly computed by means of the explicit expression

$$K = -\left[D(\tilde{s}_1)q_1 \ \ldots \ D(\tilde{s}_n)q_n\right]\left[N_x(\tilde{s}_1)q_1 \ \ldots \ N_x(\tilde{s}_n)q_n\right]^{-1}. \tag{5.46}$$

This can be derived when writing the Equations (5.36) in the matrix form

$$\left[D(\tilde{s}_1)q_1 \ \ldots \ D(\tilde{s}_n)q_n\right] = -K\left[N_x(\tilde{s}_1)q_1 \ \ldots \ N_x(\tilde{s}_n)q_n\right], \tag{5.47}$$

and when assuming that the poles $\tilde{s}_\nu$ and the pole directions q_ν have been chosen such that the vectors $N_x(\tilde{s}_\nu)q_\nu$ are linearly independent. Thus, the state feedback gain K can be parameterized by the closed-loop poles and the corresponding pole directions according to (5.46). By using the same reasoning as in the proof of Theorem 5.2 it is straightforward to verify that (5.46) assigns the desired poles and pole directions to the closed-loop system.

If n linearly independent outputs are available for measurement the controller assigning all closed-loop poles is a static output feedback

$$u(s) = -Py(s), \tag{5.48}$$

with the constant $p \times n$ feedback gain P. If the transfer behaviour between u and y of the system (5.1) is given by

$$y(s) = C(sI - A)^{-1}Bu(s) = N(s)D^{-1}(s)u(s) \tag{5.49}$$

the feedback gain P can be obtained from the constant solution $P(s) = P$ and $Q(s) = I$ of the Diophantine equation

$$P(s)N(s) + Q(s)D(s) = \tilde{D}(s) \tag{5.50}$$

that results when substituting $N_x(s)$ in (5.45) by $N(s)$. Thus, the output feedback controller assigning the characteristic polynomial $\det \tilde{D}(s)$ to the closed-loop system is attainable without recurrence to time-domain results. Furthermore, the constant solution P of the Diophantine equation (5.50) can be parameterized by

$$P = -\begin{bmatrix} D(\tilde{s}_1)q_1 & \dots & D(\tilde{s}_n)q_n \end{bmatrix} \begin{bmatrix} N(\tilde{s}_1)q_1 & \dots & N(\tilde{s}_n)q_n \end{bmatrix}^{-1}, \tag{5.51}$$

in view of (5.46).

Example 5.2. Parametric design of static output feedback in the frequency domain

Consider the system already investigated in Example 5.1 with the right coprime MFD

$$G(s) = N(s)D^{-1}(s) = \begin{bmatrix} 1 & 0 \\ s & 1 \\ 0 & 1 \end{bmatrix} \begin{bmatrix} s^2 + 3s + 3 & -1 \\ -1 & s+3 \end{bmatrix}^{-1}. \tag{5.52}$$

The characteristic polynomial of the system is given by $\det D(s) = (s+2)^3$, *i.e.*, the system is of the order three. Since the system has the same number of linearly independent measurable outputs all states can be determined directly from the measurements (see also Example 2.2). Consequently, the state feedback controller is an output feedback controller

$$u(s) = -Py(s). \tag{5.53}$$

This output feedback controller, which is parameterized by the closed-loop poles and the corresponding pole directions, can be computed by using the results of Section 5.3. After assigning three closed-loop poles $\tilde{s}_1$, $\tilde{s}_2$, $\tilde{s}_3$ to stabilize the system, the corresponding pole directions q_1, q_2, q_3 can be chosen to satisfy further design specifications without affecting the closed-loop poles. The resulting output feedback gain P is

$$P = -\begin{bmatrix} D(\tilde{s}_1)q_1 & D(\tilde{s}_2)q_2 & D(\tilde{s}_3)q_3 \end{bmatrix} \begin{bmatrix} N(\tilde{s}_1)q_1 & N(\tilde{s}_2)q_2 & N(\tilde{s}_3)q_3 \end{bmatrix}^{-1}. \tag{5.54}$$

5.4 Parametric Design of Reduced-order Observers in the Frequency Domain

Considered is a system of the order n represented by its transfer behaviour $y(s) = G(s)u(s)$ with the transfer matrix $G(s)$ given by the left coprime MFD

$$G(s) = \bar{D}^{-1}(s)\bar{N}(s). \tag{5.55}$$

Further assume that the m outputs y with $m < n$ are measurable and that $\bar{D}_\kappa(s)$ is row reduced (see (3.81)). In the frequency domain an observer of the order $n - \kappa$, $0 \leq \kappa \leq m$, can be designed for this system by specifying the denominator matrix $\tilde{\bar{D}}(s)$ (see Chapter 3). This polynomial matrix assigns the poles $\tilde{\bar{s}}_1, \tilde{\bar{s}}_2, \ldots, \tilde{\bar{s}}_{n-\kappa}$ to the observer if

$$\det \tilde{\bar{D}}(s) = \prod_{i=1}^{n-\kappa} (s - \tilde{\bar{s}}_i). \tag{5.56}$$

By computing the determinant on the left-hand side of (5.56) one obtains $n-\kappa$ non-linear equations for the $m(n-\kappa)$ freely assignable elements of $\tilde{\bar{D}}(s)$. Thus, if the system has more than one output (*i.e.*, $m > 1$) there exist additional $(m-1)(n-\kappa)$ degrees of freedom in the design. In the following, a parametric approach is presented that yields an explicit solution of the pole-placement problem (5.56) and also parameterizes the additional degrees of freedom of the observer independently from the pole assignment.

5.4.1 Definition of the Observer Pole Directions

The $(m-1)(n-\kappa)$ degrees of freedom remaining in the observer after the pole assignment can be parameterized by the pole directions. In the following, it is assumed that the algebraic multiplicity μ_ν of each observer pole $\tilde{\bar{s}}_\nu$ is less than or equal to m and coincides with the corresponding geometric multiplicity, *i.e.*, all pole directions $\bar{q}_\nu$ corresponding to a multiple pole $\tilde{\bar{s}}_\nu$ (see (5.58)) are linearly independent (for further details see, *e.g.*, [13]). Since the observer poles $\tilde{\bar{s}}_\nu$ are a solution of

$$\det \tilde{\bar{D}}(s) = 0 \tag{5.57}$$

the constant $m \times m$ matrix $\tilde{\bar{D}}(\tilde{\bar{s}}_\nu)$ is rank deficient with $\operatorname{rank} \tilde{\bar{D}}(\tilde{\bar{s}}_\nu) = m - \mu_\nu$ (see the Smith–Form of $\tilde{\bar{D}}(s)$ in, *e.g.*, [36]). Therefore, there exists a solution $\bar{q}_\nu \neq 0$ of

$$\bar{q}_\nu^T \tilde{\bar{D}}(\tilde{\bar{s}}_\nu) = 0^T, \quad \nu = 1, 2, \ldots, n - \kappa. \tag{5.58}$$

For distinct observer poles (*i.e.*, $\mu_\nu = 1$) the vector $\bar{q}_\nu$ in (5.58) is unique up to a multiplicative constant and for multiple observer poles there exist μ_ν linearly independent pole directions. Consequently, every observer pole can be associated with a vector $\bar{q}_\nu$, which is called the *pole direction* $\bar{q}_\nu$ *of the observer*.

Remark 5.7. For real observer poles the pole directions are real-valued vectors, and for conjugate complex observer poles the pole directions are also conjugate complex.

5.4.2 Parametric Expression for the Observer Design

Using the pole directions defined in Section 5.4.1 the pole-placement problem (5.56) can be reformulated in the following way. Determine a denominator matrix $\tilde{\bar{D}}(s)$ that assigns the observer poles $\tilde{\bar{s}}_1, \tilde{\bar{s}}_2, \ldots, \tilde{\bar{s}}_{n-\kappa}$, *i.e.*,

$$\det \tilde{\bar{D}}(\tilde{\bar{s}}_\nu) = 0, \quad \nu = 1, 2, \ldots, n - \kappa, \tag{5.59}$$

and the pole directions $\bar{q}_1, \bar{q}_2, \ldots, \bar{q}_{n-\kappa}$, *i.e.*,

$$\bar{q}_\nu^T \tilde{\bar{D}}(\tilde{\bar{s}}_\nu) = 0^T, \quad \nu = 1, 2, \ldots, n - \kappa \tag{5.60}$$

to the observer. In view of Remark 5.7 the assigned pole directions $\bar{q}_\nu$ for real observer poles have to be real-valued vectors, and for conjugate complex observer poles the pole directions $\bar{q}_\nu$ have to be chosen conjugate complex. The next theorem shows that the problem of assigning the observer poles and their pole directions can be solved explicitly.

Theorem 5.3 (Parametric expression for the polynomial matrix of the observer). *Consider a system of the order n with m outputs y represented by the transfer behaviour*

$$y(s) = \bar{D}^{-1}(s)\bar{N}(s)u(s), \tag{5.61}$$

in which $\bar{N}(s)$ and $\bar{D}(s)$ are left coprime polynomial matrices with $\bar{D}(s)$ such that

$$\bar{D}_\kappa(s) = \Pi\left\{\bar{D}(s)\begin{bmatrix} I_{m-\kappa} & 0 \\ 0 & s^{-1}I_\kappa \end{bmatrix}\right\} \tag{5.62}$$

is row reduced. Further assume that the numbers $\tilde{\bar{s}}_1, \tilde{\bar{s}}_2, \ldots, \tilde{\bar{s}}_{n-\kappa}$ and the corresponding $m \times 1$ vectors $\bar{q}_1, \bar{q}_2, \ldots, \bar{q}_{n-\kappa}$ are given, such that the vectors $\bar{S}^T(\tilde{\bar{s}}_\nu)\bar{q}_\nu$, $\nu = 1, 2, \ldots, n - \kappa$, are linearly independent, where the $(n - \kappa) \times m$ polynomial matrix[2]

$$\bar{S}^T(s) = \begin{bmatrix} \bar{\sigma}_1(s) & 0 & \ldots & 0 \\ 0 & \bar{\sigma}_2(s) & \ldots & 0 \\ 0 & 0 & \ldots & 0 \\ \vdots & \vdots & \vdots & \vdots \\ 0 & 0 & 0 & 0 \\ 0 & 0 & 0 & \bar{\sigma}_m(s) \end{bmatrix} \tag{5.63}$$

[2] The row reducedness of the polynomial matrix $\bar{D}_\kappa(s)$ implies $\sum_{j=1}^m \delta_{rj}[\bar{D}_\kappa(s)] = n - \kappa$ (see proof of Theorem 3.1), such that the polynomial matrix $\bar{S}^T(s)$ has exactly $n - \kappa$ rows.

contains the vectors

$$\bar{\sigma}_j(s) = \left[s^{\delta_{rj}[\bar{D}_\kappa(s)]-1} \ldots s\ 1\right]^T, \quad j = 1, 2, \ldots, m, \tag{5.64}$$

whenever $\delta_{rj}[\bar{D}_\kappa(s)] \geq 1$ *is satisfied. Otherwise, if the row degree* $\delta_{rj}[\bar{D}_\kappa(s)]$ *is zero, an* $(n-\kappa) \times 1$ *zero column has to be inserted in the* j*th column of* $\bar{S}^T(s)$, i.e.,

$$\bar{S}^T(s) = \begin{bmatrix} \bar{\sigma}_1(s) & 0 & \ldots & 0 & 0 \\ 0 & \bar{\sigma}_2(s) & \ldots & 0 & 0 \\ 0 & 0 & \ldots & 0 & 0 \\ \vdots & \vdots & \vdots & \vdots & \vdots \\ 0 & 0 & 0 & 0 & 0 \\ 0 & 0 & 0 & \underbrace{0}_{j\text{th column}} & \bar{\sigma}_m(s) \end{bmatrix}. \tag{5.65}$$

Then, the denominator matrix of the observer

$$\begin{aligned} \tilde{\bar{D}}^T(s) &= \bar{D}_\kappa^T(s) \\ &- \left[\bar{D}_\kappa^T(\tilde{\bar{s}}_1)\bar{q}_1 \ldots \bar{D}_\kappa^T(\tilde{\bar{s}}_{n-\kappa})\bar{q}_{n-\kappa}\right]\left[\bar{S}^T(\tilde{\bar{s}}_1)\bar{q}_1 \ldots \bar{S}^T(\tilde{\bar{s}}_{n-\kappa})\bar{q}_{n-\kappa}\right]^{-1}\bar{S}^T(s) \end{aligned} \tag{5.66}$$

assigns the poles $\tilde{\bar{s}}_1, \tilde{\bar{s}}_2, \ldots, \tilde{\bar{s}}_{n-\kappa}$ *and the corresponding pole directions* $\bar{q}_1, \bar{q}_2, \ldots, \bar{q}_{n-\kappa}$ *to the observer. Conversely, each denominator matrix* $\tilde{\bar{D}}(s)$ *of the observer can be expressed by (5.66) if it assigns the observer poles* $\tilde{\bar{s}}_1, \tilde{\bar{s}}_2, \ldots, \tilde{\bar{s}}_{n-\kappa}$ *and the corresponding pole directions* $\bar{q}_1, \bar{q}_2, \ldots, \bar{q}_{n-\kappa}$*, such that the vectors* $\bar{S}^T(\tilde{\bar{s}}_\mu)\bar{q}_\mu$*,* $\mu = 1, 2, \ldots, n-\kappa$*, are linearly independent.*

This theorem can be proven as Theorem 5.2 by applying the duality relations

$$D(s) \to \bar{D}_\kappa^T(s), \quad \tilde{D}(s) \to \tilde{\bar{D}}^T(s), \quad \tilde{D}_C \to \tilde{\bar{D}}_C^T, \quad \tilde{s}_\nu \to \tilde{\bar{s}}_\nu, \quad q_\nu \to \bar{q}_\nu, \tag{5.67}$$

in which the $(n-\kappa) \times m$ constant matrix $\tilde{\bar{D}}_C$ is the coefficient matrix in the representation

$$\tilde{\bar{D}}(s) = \left(\bar{S}(s)\tilde{\bar{D}}_C + \text{diag}(s^{\delta_{rj}[\bar{D}_\kappa(s)]})\right)\Gamma_r[\bar{D}_\kappa(s)] \tag{5.68}$$

of $\tilde{\bar{D}}(s)$ (see (5.15)).

Remark 5.8. The set of $n-\kappa$ numbers $\tilde{\bar{s}}_1, \tilde{\bar{s}}_2, \ldots, \tilde{\bar{s}}_{n-\kappa}$ with the corresponding $n-\kappa$ constant $m \times 1$ vectors $\bar{q}_1, \bar{q}_2, \ldots, \bar{q}_{n-\kappa}$ contain $m(n-\kappa)$ degrees of freedom, since each pole direction only contributes $m-1$ influential design parameters. This result can be verified by similar arguments as in Remark 5.4.

Remark 5.9. In the case of full-order observers (*i.e.*, $\kappa = 0$) the parametric expression (5.66) for the polynomial matrix $\tilde{\bar{D}}(s)$ can only be used to assign observer poles with algebraic multiplicity not larger than m. This is due to the fact that at most m linearly independent vectors $\bar{S}^T(\tilde{\bar{s}}_\nu)\bar{q}_\nu$ exist for the same observer pole $\tilde{\bar{s}}_\nu$, since rank $\bar{S}^T(s) = m$ for all s. For reduced-order observers (*i.e.*, $\kappa > 0$) the algebraic multiplicity of the assignable observer poles is further reduced if $\delta_{rj}[\bar{D}_\kappa(s)] = 0$ for some j. By Theorem 5.3 this fact leads to zero columns in the polynomial matrix $\bar{S}^T(s)$, such that its rank is reduced for all s. Consequently, the number of linearly independent vectors $\bar{S}^T(\tilde{\bar{s}}_\nu)\bar{q}_\nu$ for the same observer pole $\tilde{\bar{s}}_\nu$ is also decreased. Observer poles with algebraic multiplicity larger than m can be assigned by introducing the *generalized observer pole directions*. For deriving the corresponding parametric expression for $\tilde{\bar{D}}(s)$ the results in [13] can be used.

Example 5.3. Parametric design of different observers in the frequency domain

Consider the system of Example 3.2 with its transfer matrix $G(s)$ represented by the left coprime MFD

$$G(s) = \bar{D}^{-1}(s)\bar{N}(s) = \begin{bmatrix} s & -1 & 1 \\ 2 & s+3 & -1 \\ -1 & 0 & s+3 \end{bmatrix}^{-1} \begin{bmatrix} 0 & 0 \\ 1 & 1 \\ 0 & 1 \end{bmatrix}. \tag{5.69}$$

In Example 3.2 it is shown that a full-order observer (*i.e.*, $\kappa = 0$) is parameterized by the polynomial matrix

$$\tilde{\bar{D}}(s) = \begin{bmatrix} s+\alpha & \beta & \gamma \\ \delta & s+\varepsilon & \varphi \\ \eta & \lambda & s+\mu \end{bmatrix}. \tag{5.70}$$

If this polynomial matrix satisfies

$$\det \tilde{\bar{D}}(s) = (s - \tilde{\bar{s}}_1)(s - \tilde{\bar{s}}_2)(s - \tilde{\bar{s}}_3) \tag{5.71}$$

the poles $\tilde{\bar{s}}_1, \tilde{\bar{s}}_2, \tilde{\bar{s}}_3$ are assigned to the observer. This yields $n = 3$ non-linear equations for determining the $nm = 9$ free parameters in (5.70). Consequently, there remain $(m-1)n = 6$ additional degrees of freedom in the observer design. These degrees of freedom can be parameterized by the observer pole directions $\bar{q}_1, \bar{q}_2, \bar{q}_3$ without affecting the pole placement. In order to determine the polynomial matrix $\tilde{\bar{D}}(s)$ such that the observer has the desired observer poles and the corresponding pole directions the polynomial matrix $\bar{S}^T(s)$ in (5.63) is needed. In view of $\bar{D}_0(s) = \bar{D}(s)$ and $\delta_{ri}[\bar{D}(s)] = 1$, $i = 1, 2, 3$, this matrix is $\bar{S}^T(s) = I$. Then, the polynomial matrix

$$\tilde{\bar{D}}^T(s) = \bar{D}^T(s) - \left[\bar{D}^T(\tilde{\bar{s}}_1)\bar{q}_1 \; \bar{D}^T(\tilde{\bar{s}}_2)\bar{q}_2 \; \bar{D}^T(\tilde{\bar{s}}_3)\bar{q}_3\right] \left[\bar{q}_1 \; \bar{q}_2 \; \bar{q}_3\right]^{-1} \tag{5.72}$$

assigns the observer poles and pole directions (see (5.66)). Obviously, the pole directions have to be chosen linearly independent, such that the inverse matrix in (5.72) exists.

If the last output of (5.69) is used to reconstruct the state directly (*i.e.*, $\kappa = 1$) a reduced-order observer of the order $n - \kappa = 2$ can be designed. For $\kappa = 1$ the polynomial matrix (5.62) has the form

$$\bar{D}_1(s) = \Pi\left\{\begin{bmatrix} s & -1 & \frac{1}{s} \\ 2 & s+3 & -\frac{1}{s} \\ -1 & 0 & 1+\frac{3}{s} \end{bmatrix}\right\} = \begin{bmatrix} s & -1 & 0 \\ 2 & s+3 & 0 \\ -1 & 0 & 1 \end{bmatrix} \tag{5.73}$$

(see Example 3.2). Since $\bar{D}_1(s)$ is row reduced the left coprime MFD (5.69) is a valid polynomial description of the plant for the observer design (see Chapter 3). On the basis of $\bar{D}_1(s)$ the general form of the polynomial matrix characterizing the dynamics of the reduced-order observer is

$$\tilde{\bar{D}}(s) = \begin{bmatrix} s+\alpha & \beta & \gamma \\ \delta & s+\varepsilon & \varphi \\ -1 & 0 & 1 \end{bmatrix} \tag{5.74}$$

(see also Example 3.2). This matrix assigns the observer poles $\tilde{\bar{s}}_1$ and $\tilde{\bar{s}}_2$ if

$$\det \tilde{\bar{D}}(s) = (s - \tilde{\bar{s}}_1)(s - \tilde{\bar{s}}_2) \tag{5.75}$$

is satisfied. This leads to $n - \kappa = 2$ non-linear equations for the $m(n-\kappa) = 6$ free coefficients of the polynomial matrix $\tilde{\bar{D}}(s)$ in (5.74). The remaining $(m-1)(n-\kappa) = 4$ degrees of freedom can be parameterized by the pole directions. With $\delta_{r1}[\bar{D}_1(s)] = \delta_{r2}[\bar{D}_1(s)] = 1$ and $\delta_{r3}[\bar{D}_1(s)] = 0$ the matrix $\bar{S}^T(s)$ in (5.66) takes the form

$$\bar{S}^T(s) = \begin{bmatrix} 1 & 0 & 0 \\ 0 & 1 & 0 \end{bmatrix}, \tag{5.76}$$

where a 2×1 zero vector is inserted in the last column of $\bar{S}^T(s)$, because $\delta_{r3}[\bar{D}_1(s)] = 0$ (see Theorem 5.3). By using this result the parametric expression for $\tilde{\bar{D}}(s)$ is given by

$$\tilde{\bar{D}}^T(s) = \bar{D}_1^T(s) - \begin{bmatrix} \bar{D}_1^T(\tilde{\bar{s}}_1)\bar{q}_1 & \bar{D}_1^T(\tilde{\bar{s}}_2)\bar{q}_2 \end{bmatrix} \begin{bmatrix} \bar{q}_{11} & \bar{q}_{21} \\ \bar{q}_{12} & \bar{q}_{22} \end{bmatrix}^{-1} \begin{bmatrix} 1 & 0 & 0 \\ 0 & 1 & 0 \end{bmatrix}. \tag{5.77}$$

The polynomial matrix $\tilde{\bar{D}}(s)$ in (5.77) assigns the observer poles $\tilde{\bar{s}}_1$ and $\tilde{\bar{s}}_2$ and the corresponding pole directions $\bar{q}_i = \begin{bmatrix} \bar{q}_{i1} & \bar{q}_{i2} & \bar{q}_{i3} \end{bmatrix}^T$, $i = 1, 2$. A prerequisite for the parametric expression (5.77) is that the inverse matrix on the right-hand side exists. This restricts the choice of the pole directions.

5.4.3 Relation Between the Observer Pole Directions and the Left Eigenvectors of the Observer

In the following, a relationship between the observer pole directions and the left eigenvectors of the observer is established under the assumption that the observer eigenvalues are distinct. In the next section this result will be used to parameterize the gain in the time-domain representation of the observer by the observer poles and their pole directions. Consider a minimal state-space realization (C, A, B) of the transfer matrix $G(s)$ in (5.55). In Chapter 3 it was shown that in the time domain an observer is represented by

$$\dot{\hat{\zeta}}(t) = T(A - L_1C_1)\Theta\hat{\zeta}(t) + \begin{bmatrix} TL_1 & T(A - L_1C_1)\Psi_2 \end{bmatrix}\begin{bmatrix} y_1(t) \\ y_2(t) \end{bmatrix} + TBu(t), \tag{5.78}$$

$$\hat{x}(t) = \Theta\hat{\zeta}(t) + \Psi_2 y_2(t), \tag{5.79}$$

(see (3.15) and (3.16)) where the partition

$$C = \begin{bmatrix} C_1 \\ C_2 \end{bmatrix} \tag{5.80}$$

is used (see (3.2) and (3.3)). The left eigenvectors $\tilde{\bar{w}}_\nu$ of the observer (5.78) are the non-vanishing solutions of

$$\tilde{\bar{w}}_\nu^T(\tilde{\bar{s}}_\nu I - T(A - L_1C_1)\Theta) = 0^T, \quad \nu = 1, 2, \ldots, n - \kappa, \tag{5.81}$$

where it is assumed that the observer eigenvalues $\tilde{\bar{s}}_\nu$ are distinct. Consider the connecting relation

$$\tilde{\bar{D}}(s) = \bar{D}(s)\begin{bmatrix} I_{m-\kappa} & 0 \\ 0 & 0_\kappa \end{bmatrix} + \bar{N}_x(s)\begin{bmatrix} L_1 & \Psi_2 \end{bmatrix}, \tag{5.82}$$

(see (3.63)) in which the polynomial matrix $\bar{N}_x(s)$ is obtained from (C, A, B) by

$$C(sI - A)^{-1} = \bar{D}^{-1}(s)\bar{N}_x(s). \tag{5.83}$$

Replacing s by $\tilde{\bar{s}}_\nu$ in (5.82) and pre-multiplying the result by $\bar{q}_\nu^T$ gives

$$\bar{q}_\nu^T\bar{D}(\tilde{\bar{s}}_\nu)\begin{bmatrix} I_{m-\kappa} & 0 \\ 0 & 0_\kappa \end{bmatrix} = -\bar{q}_\nu^T\bar{N}_x(\tilde{\bar{s}}_\nu)\begin{bmatrix} L_1 & \Psi_2 \end{bmatrix}, \quad \nu = 1, 2, \ldots, n - \kappa, \tag{5.84}$$

in view of (5.58). From (5.84) one obtains the two relations

$$\bar{q}_\nu^T\bar{D}(\tilde{\bar{s}}_\nu)\begin{bmatrix} I_{m-\kappa} \\ 0 \end{bmatrix} = -\bar{q}_\nu^T\bar{N}_x(\tilde{\bar{s}}_\nu)L_1, \tag{5.85}$$

and

$$\bar{q}_\nu^T\bar{N}_x(\tilde{\bar{s}}_\nu)\Psi_2 = 0^T. \tag{5.86}$$

Now consider the equation

$$\bar{q}_\nu^T \bar{N}_x(\tilde{\bar{s}}_\nu)(\tilde{\bar{s}}_\nu I - A + L_1 C_1)\Theta = \bar{q}_\nu^T \bar{N}_x(\tilde{\bar{s}}_\nu)(\tilde{\bar{s}}_\nu I - A)\Theta + \bar{q}_\nu^T \bar{N}_x(\tilde{\bar{s}}_\nu) L_1 C_1 \Theta. \tag{5.87}$$

Rewriting (5.83) with (5.80) in the form

$$\bar{N}_x(s)(sI - A) = \bar{D}(s) \begin{bmatrix} C_1 \\ C_2 \end{bmatrix}, \tag{5.88}$$

and using (5.85), the right-hand side of (5.87) becomes

$$\bar{q}_\nu^T \bar{N}_x(\tilde{\bar{s}}_\nu)(\tilde{\bar{s}}_\nu I - A)\Theta + \bar{q}_\nu^T \bar{N}_x(\tilde{\bar{s}}_\nu) L_1 C_1 \Theta = \bar{q}_\nu^T \bar{D}(\tilde{\bar{s}}_\nu) \begin{bmatrix} C_1\Theta \\ C_2\Theta \end{bmatrix} - \bar{q}_\nu^T \bar{D}(\tilde{\bar{s}}_\nu) \begin{bmatrix} C_1\Theta \\ 0 \end{bmatrix}, \tag{5.89}$$

and since (3.6) implies $C_2\Theta = 0$, the right-hand side of (5.89) vanishes, so that (5.87) yields

$$\bar{q}_\nu^T \bar{N}_x(\tilde{\bar{s}}_\nu)(\tilde{\bar{s}}_\nu I - A + L_1 C_1)\Theta = 0^T. \tag{5.90}$$

Combining (5.86) and (5.90) in one equation leads to

$$\bar{q}_\nu^T \bar{N}_x(\tilde{\bar{s}}_\nu) \begin{bmatrix} (\tilde{\bar{s}}_\nu I - A + L_1 C_1)\Theta & \Psi_2 \end{bmatrix} = 0^T. \tag{5.91}$$

In view of $T\Theta = I_{n-\kappa}$ (see (3.6)) the eigenvector equation (5.81) for the left eigenvectors $\tilde{\bar{w}}_\nu$ can be written as

$$\tilde{\bar{w}}_\nu^T T(\tilde{\bar{s}}_\nu I - A + L_1 C_1)\Theta = 0^T, \tag{5.92}$$

and with $T\Psi_2 = 0$ (see (3.6)) this gives

$$\tilde{\bar{w}}_\nu^T T \begin{bmatrix} (\tilde{\bar{s}}_\nu I - A + L_1 C_1)\Theta & \Psi_2 \end{bmatrix} = 0^T. \tag{5.93}$$

Thus, by comparing (5.91) with (5.93) the pole directions $\bar{q}_\nu$ and the left eigenvectors $\tilde{\bar{w}}_\nu$ of the observer are obviously related by

$$\tilde{\bar{w}}_\nu^T T = \bar{q}_\nu^T \bar{N}_x(\tilde{\bar{s}}_\nu), \quad \nu = 1, 2, \ldots, n - \kappa. \tag{5.94}$$

A pre-requiste for this is that the $n \times n$ matrix $\begin{bmatrix} (\tilde{\bar{s}}_\nu I - A + L_1 C_1)\Theta & \Psi_2 \end{bmatrix}$ has rank $n - 1$. In order to prove this claim consider the solution ξ of

$$\xi^T \begin{bmatrix} (\tilde{\bar{s}}_\nu I - A + L_1 C_1)\Theta & \Psi_2 \end{bmatrix} = 0^T. \tag{5.95}$$

The general solution of

$$\xi^T \Psi_2 = 0^T \tag{5.96}$$

is

$$\xi^T = v^T T, \tag{5.97}$$

where v^T can be an arbitrary vector since the rows of T form a basis of the left kernel of the $n \times \kappa$ matrix Ψ_2. This is due to the fact that $T\Psi_2 = 0$ holds with $\operatorname{rank} T = n - \kappa$ and $\operatorname{rank} \Psi_2 = \kappa$ (see (3.5)). Hence,

$$v^T T \left[(\tilde{\bar{s}}_\nu I - A + L_1 C_1)\Theta \quad \Psi_2 \right] = 0^T \tag{5.98}$$

in view of (5.95) and (5.97). Since for each distinct observer eigenvalue $\tilde{\bar{s}}_\nu$ only one linearly independent vector $\tilde{\bar{w}}_\nu$ satisfies the eigenvector equation (5.92) of the observer there is only one linearly independent vector $\xi^T = v^T T$ that solves (5.95). Consequently, the $n \times n$ matrix $\left[T(\tilde{\bar{s}}_\nu I - A + L_1 C_1)\Theta \quad \Psi_2 \right]$ has rank $n - 1$.

5.4.4 Parameterization of Observers in the Time Domain Using the Pole Directions

In some applications one is interested in observing the unmeasurable states of the system for diagnostics. Then, a parameterization of the state-space representation (5.78) and (5.79) of the observer is required. In what follows it will be shown that if the observer poles and the corresponding pole directions of $\tilde{\bar{D}}(s)$ are known, then the matrices L_1 and Ψ_2 characterizing the observer dynamics can be determined without solving the Diophantine equation (3.91). In order to derive a parametric expression for L_1 and Ψ_2 consider

$$C_2 L_1 = 0_{\kappa, m-\kappa} \quad \text{and} \quad C_2 \Psi_2 = I_\kappa, \tag{5.99}$$

which have to be satisfied by any observer matrices L_1 and Ψ_2 (see (3.19) and (3.20)). This result can be compactly written as

$$C_2 \left[L_1 \quad \Psi_2 \right] = \left[0_{\kappa, m-\kappa} \quad I_\kappa \right], \tag{5.100}$$

and it yields κm conditions for the elements in both matrices. The remaining $m(n-\kappa)$ degrees of freedom are parameterized by the observer poles and their pole directions if, additionally,

$$\bar{q}_\nu^T \bar{N}_x(\tilde{\bar{s}}_\nu) \left[L_1 \quad \Psi_2 \right] = -\bar{q}_\nu^T \bar{D}(\tilde{\bar{s}}_\nu) \begin{bmatrix} I_{m-\kappa} & 0 \\ 0 & 0_\kappa \end{bmatrix}, \quad \nu = 1, 2, \ldots, n - \kappa \tag{5.101}$$

holds (see (5.84)). Thus, the matrices L_1 and Ψ_2 must satisfy

$$\begin{bmatrix} \bar{q}_1^T \bar{N}_x(\tilde{\bar{s}}_1) \\ \vdots \\ \bar{q}_{n-\kappa}^T \bar{N}_x(\tilde{\bar{s}}_{n-\kappa}) \\ C_2 \end{bmatrix} \left[L_1 \quad \Psi_2 \right] = \begin{bmatrix} -\begin{bmatrix} \bar{q}_1^T \bar{D}(\tilde{\bar{s}}_1) \\ \vdots \\ \bar{q}_{n-\kappa}^T \bar{D}(\tilde{\bar{s}}_{n-\kappa}) \end{bmatrix} \begin{bmatrix} I_{m-\kappa} & 0 \\ 0 & 0_\kappa \end{bmatrix} \\ \left[0_{\kappa, m-\kappa} \quad I_\kappa \right] \end{bmatrix}, \tag{5.102}$$

in view of (5.100) and (5.101). If the observer poles and the corresponding pole directions are chosen such that the matrix pre-multiplying $\left[L_1 \quad \Psi_2 \right]$ in (5.102) is non-singular, (5.102) can be solved for $\left[L_1 \quad \Psi_2 \right]$, yielding the parametric expression

$$[L_1 \quad \Psi_2] = \begin{bmatrix} \bar{q}_1^T \bar{N}_x(\tilde{\bar{s}}_1) \\ \vdots \\ \bar{q}_{n-\kappa}^T \bar{N}_x(\tilde{\bar{s}}_{n-\kappa}) \\ C_2 \end{bmatrix}^{-1} \begin{bmatrix} -\begin{bmatrix} \bar{q}_1^T \bar{D}(\tilde{\bar{s}}_1) \\ \vdots \\ \bar{q}_{n-\kappa}^T \bar{D}(\tilde{\bar{s}}_{n-\kappa}) \end{bmatrix} \begin{bmatrix} I_{m-\kappa} & 0 \\ 0 & 0_\kappa \end{bmatrix} \\ \begin{bmatrix} 0_{\kappa,m-\kappa} & I_\kappa \end{bmatrix} \end{bmatrix}. \tag{5.103}$$

To show that (5.103) assigns the observer poles $\tilde{\bar{s}}_1, \ldots, \tilde{\bar{s}}_{n-\kappa}$ and the corresponding pole directions $\bar{q}_1, \ldots, \bar{q}_{n-\kappa}$ insert (5.103) in the connecting relation (5.82), substitute s by $\tilde{\bar{s}}_\nu$ and pre-multiply by $\bar{q}_\nu^T$, which gives

$$\bar{q}_\nu^T \tilde{\bar{D}}(\tilde{\bar{s}}_\nu) = \bar{q}_\nu^T \bar{D}(\tilde{\bar{s}}_\nu) \begin{bmatrix} I_{m-\kappa} & 0 \\ 0 & 0_\kappa \end{bmatrix} \tag{5.104}$$

$$+ \bar{q}_\nu^T \bar{N}_x(\tilde{\bar{s}}_\nu) \begin{bmatrix} \bar{q}_1^T \bar{N}_x(\tilde{\bar{s}}_1) \\ \vdots \\ \bar{q}_{n-\kappa}^T \bar{N}_x(\tilde{\bar{s}}_{n-\kappa}) \\ C_2 \end{bmatrix}^{-1} \begin{bmatrix} -\begin{bmatrix} \bar{q}_1^T \bar{D}(\tilde{\bar{s}}_1) \\ \vdots \\ \bar{q}_{n-\kappa}^T \bar{D}(\tilde{\bar{s}}_{n-\kappa}) \end{bmatrix} \begin{bmatrix} I_{m-\kappa} & 0 \\ 0 & 0_\kappa \end{bmatrix} \\ \begin{bmatrix} 0_{\kappa,m-\kappa} & I_\kappa \end{bmatrix} \end{bmatrix}.$$

By substituting

$$e_\nu^T \begin{bmatrix} \bar{q}_1^T \bar{N}_x(\tilde{\bar{s}}_1) \\ \vdots \\ \bar{q}_{n-\kappa}^T \bar{N}_x(\tilde{\bar{s}}_{n-\kappa}) \\ C_2 \end{bmatrix} = \bar{q}_\nu^T \bar{N}_x(\tilde{\bar{s}}_\nu), \quad \nu = 1, 2, \ldots, n-\kappa \tag{5.105}$$

in

$$e_\nu^T \begin{bmatrix} \bar{q}_1^T \bar{N}_x(\tilde{\bar{s}}_1) \\ \vdots \\ \bar{q}_{n-\kappa}^T \bar{N}_x(\tilde{\bar{s}}_{n-\kappa}) \\ C_2 \end{bmatrix} \begin{bmatrix} \bar{q}_1^T \bar{N}_x(\tilde{\bar{s}}_1) \\ \vdots \\ \bar{q}_{n-\kappa}^T \bar{N}_x(\tilde{\bar{s}}_{n-\kappa}) \\ C_2 \end{bmatrix}^{-1} = e_\nu^T, \tag{5.106}$$

the relation

$$\bar{q}_\nu^T \bar{N}_x(\tilde{\bar{s}}_\nu) \begin{bmatrix} \bar{q}_1^T \bar{N}_x(\tilde{\bar{s}}_1) \\ \vdots \\ \bar{q}_{n-\kappa}^T \bar{N}_x(\tilde{\bar{s}}_{n-\kappa}) \\ C_2 \end{bmatrix}^{-1} = e_\nu^T \tag{5.107}$$

holds, so that (5.104) simplifies to

$$\bar{q}_\nu^T \tilde{\bar{D}}(\tilde{\bar{s}}_\nu) = \bar{q}_\nu^T \bar{D}(\tilde{\bar{s}}_\nu) \begin{bmatrix} I_{m-\kappa} & 0 \\ 0 & 0_\kappa \end{bmatrix} - \bar{q}_\nu^T \bar{D}(\tilde{\bar{s}}_\nu) \begin{bmatrix} I_{m-\kappa} & 0 \\ 0 & 0_\kappa \end{bmatrix} = 0^T, \quad \nu = 1, 2, \ldots, n-\kappa. \tag{5.108}$$

This shows that $\tilde{\bar{s}}_\nu$ is an observer pole and that $\bar{q}_\nu$ is the corresponding pole direction, because (5.108) is only satisfied for any $\bar{q}_\nu \neq 0$ if $\det \tilde{\bar{D}}(\tilde{\bar{s}}_\nu) = 0$. Inserting

$$\begin{bmatrix} 0_{\kappa,m-\kappa} & I_\kappa \end{bmatrix} \begin{bmatrix} \bar{q}_1^T \bar{N}_x(\tilde{\bar{s}}_1) \\ \vdots \\ \bar{q}_{n-\kappa}^T \bar{N}_x(\tilde{\bar{s}}_{n-\kappa}) \\ C_2 \end{bmatrix} = C_2 \tag{5.109}$$

in

$$\begin{bmatrix} 0_{\kappa,m-\kappa} & I_\kappa \end{bmatrix} \begin{bmatrix} \bar{q}_1^T \bar{N}_x(\tilde{\bar{s}}_1) \\ \vdots \\ \bar{q}_{n-\kappa}^T \bar{N}_x(\tilde{\bar{s}}_{n-\kappa}) \\ C_2 \end{bmatrix} \begin{bmatrix} \bar{q}_1^T \bar{N}_x(\tilde{\bar{s}}_1) \\ \vdots \\ \bar{q}_{n-\kappa}^T \bar{N}_x(\tilde{\bar{s}}_{n-\kappa}) \\ C_2 \end{bmatrix}^{-1} = \begin{bmatrix} 0_{\kappa,m-\kappa} & I_\kappa \end{bmatrix}, \tag{5.110}$$

yields

$$C_2 \begin{bmatrix} \bar{q}_1^T \bar{N}_x(\tilde{\bar{s}}_1) \\ \vdots \\ \bar{q}_{n-\kappa}^T \bar{N}_x(\tilde{\bar{s}}_{n-\kappa}) \\ C_2 \end{bmatrix}^{-1} = \begin{bmatrix} 0_{\kappa,m-\kappa} & I_\kappa \end{bmatrix}, \tag{5.111}$$

so that it is straightforward to show that for (5.103) the Condition (5.100) holds. The remaining matrices T and Θ of the observer can be computed from Ψ_2 by using the Relations (3.21) and (3.22).

Example 5.4. Time-domain representation of an observer designed by the parametric approach in the frequency domain

If one is interested in a realization of the observers designed in Example 5.3 the parameterizing matrices in the time domain have to be determined on the basis of the state-space model characterized by the matrices A, B and C in (3.94). In Example 3.2 this is accomplished by solving the Diophantine equation (3.91). In the following, a parametric class of matrices assigning the observer poles and the corresponding pole directions is derived. A full-order observer for (3.94) (*i.e.*, $\kappa = 0$) is given by

$$\dot{\hat{\zeta}}(t) = (A - LC)\hat{\zeta}(t) + Ly(t) + Bu(t), \tag{5.112}$$

$$\hat{x}(t) = \hat{\zeta}(t), \tag{5.113}$$

resulting from the reduced-order observer (5.78) and (5.79) by setting $L = L_1$, $T = \Theta = I$ and $C_1 = C$, whereas C_2 and Ψ_2 are vanishing (see also (3.23)). Then, the parametric expression (5.103) becomes

$$L = \begin{bmatrix} \bar{q}_1^T \bar{N}_x(\tilde{\bar{s}}_1) \\ \bar{q}_2^T \bar{N}_x(\tilde{\bar{s}}_2) \\ \bar{q}_3^T \bar{N}_x(\tilde{\bar{s}}_3) \end{bmatrix}^{-1} \begin{bmatrix} -\bar{q}_1^T \bar{D}(\tilde{\bar{s}}_1) \\ -\bar{q}_2^T \bar{D}(\tilde{\bar{s}}_2) \\ -\bar{q}_3^T \bar{D}(\tilde{\bar{s}}_3) \end{bmatrix}, \tag{5.114}$$

which is a solution of the Diophantine equation (3.91) with $\tilde{\bar{D}}(s)$ given by (5.70). The polynomial matrix $\bar{N}_x(s)$ needed in (5.114) can be obtained by the right-to-left matrix fraction conversion (5.83) yielding

$$\bar{N}_x(s) = \bar{M} = \begin{bmatrix} 1 & 0 & 0 \\ 0 & 1 & 1 \\ 0 & 0 & 1 \end{bmatrix} \tag{5.115}$$

(see (3.96)). Then, (5.114) can be written as

$$L = \bar{M}^{-1} \left(\begin{bmatrix} \bar{q}_1 \ \bar{q}_2 \ \bar{q}_3 \end{bmatrix}^{-1} \right)^T \begin{bmatrix} -\bar{q}_1^T \bar{D}(\tilde{\bar{s}}_1) \\ -\bar{q}_2^T \bar{D}(\tilde{\bar{s}}_2) \\ -\bar{q}_3^T \bar{D}(\tilde{\bar{s}}_3) \end{bmatrix}. \tag{5.116}$$

This shows that the pole directions of the observer have to satisfy the same conditions as in the frequency domain (see (5.72)). When using the output y_2 defined by $C_2 = \begin{bmatrix} 0 & 0 & 1 \end{bmatrix}$ (see (3.94)) for directly reconstructing the state (*i.e.*, $\kappa = 1$) a reduced-order observer (5.78) and (5.79) of the order $n - \kappa = 2$ results. In the frequency domain this observer is parameterized by the polynomial matrix $\tilde{\bar{D}}(s)$ in (5.74). The corresponding time-domain parameterization (5.103) has the form

$$\begin{aligned} \begin{bmatrix} L_1 & \Psi_2 \end{bmatrix} &= \begin{bmatrix} \bar{q}_1^T \bar{N}_x(\tilde{\bar{s}}_1) \\ \bar{q}_2^T \bar{N}_x(\tilde{\bar{s}}_2) \\ \begin{bmatrix} 0 \ 0 \ 1 \end{bmatrix} \end{bmatrix}^{-1} \begin{bmatrix} -\begin{bmatrix} \bar{q}_1^T \bar{D}(\tilde{\bar{s}}_1) \\ \bar{q}_2^T \bar{D}(\tilde{\bar{s}}_2) \end{bmatrix} \begin{bmatrix} 1 \ 0 \ 0 \\ 0 \ 1 \ 0 \\ 0 \ 0 \ 0 \end{bmatrix} \\ \begin{bmatrix} 0 & 0 & 1 \end{bmatrix} \end{bmatrix} \\ &= \begin{bmatrix} \bar{q}_{11} \ \bar{q}_{12} \ \bar{q}_{12} + \bar{q}_{13} \\ \bar{q}_{21} \ \bar{q}_{22} \ \bar{q}_{22} + \bar{q}_{23} \\ 0 \quad 0 \quad 1 \end{bmatrix}^{-1} \begin{bmatrix} -\begin{bmatrix} \bar{q}_1^T \bar{D}(\tilde{\bar{s}}_1) \\ \bar{q}_2^T \bar{D}(\tilde{\bar{s}}_2) \end{bmatrix} \begin{bmatrix} 1 \ 0 \ 0 \\ 0 \ 1 \ 0 \\ 0 \ 0 \ 0 \end{bmatrix} \\ \begin{bmatrix} 0 & 0 & 1 \end{bmatrix} \end{bmatrix}, \end{aligned} \tag{5.117}$$

where $\bar{q}_i = \begin{bmatrix} \bar{q}_{i1} \ \bar{q}_{i2} \ \bar{q}_{i3} \end{bmatrix}^T$, $i = 1, 2$, are the observer pole directions. Since the inverse matrix in (5.117) only exists if the inverse matrix in (5.77) exists, the parametric expressions in the time and in the frequency domain describe the same parametric class of observers. The remaining matrices T and Θ can readily be obtained by solving (3.21) and (3.22), such that the designed reduced-order observer can be implemented in the form (5.78) and (5.79).

5.5 Parametric Design of Reduced-order Observers in the Time Domain

Consider the observer

$$\dot{\hat{\zeta}}(t) = T(A-L_1C_1)\Theta\hat{\zeta}(t) + [TL_1\ T(A-L_1C_1)\Psi_2]\begin{bmatrix} y_1(t) \\ y_2(t) \end{bmatrix} + TBu(t), \quad (5.118)$$

$$\hat{x}(t) = \Theta\hat{\zeta}(t) + \Psi_2 y_2(t), \quad (5.119)$$

with $\hat{\zeta} \in \mathbb{R}^{n-\kappa}$, $0 \leq \kappa \leq m$, (see Section 3.1) for the controllable and observable system

$$\dot{x}(t) = Ax(t) + Bu(t), \quad (5.120)$$

$$y_1(t) = C_1x(t), \quad (5.121)$$

$$y_2(t) = C_2x(t). \quad (5.122)$$

In order to assign the eigenvalues $\tilde{\bar{s}}_1, \tilde{\bar{s}}_2, \ldots, \tilde{\bar{s}}_{n-\kappa}$ to the error dynamics

$$\dot{\zeta}(t) - \dot{\hat{\zeta}}(t) = T(A - L_1C_1)\Theta(\zeta(t) - \hat{\zeta}(t)) \quad (5.123)$$

of the observer (see (3.18)) a solution of

$$\det(sI - T(A - L_1C_1)\Theta) = \prod_{\nu=1}^{n-\kappa}(s - \tilde{\bar{s}}_\nu) \quad (5.124)$$

has to be determined. This problem cannot be solved directly. On the one hand, the matrices T and Θ depend on Ψ_2 in view of (3.21) and (3.22) and, on the other hand, the matrix L_1 has to satisfy (5.99). By using the results of the last section a solution of this problem is attainable in explicit form by deriving a parametric expression for L_1 and Ψ_2. The next theorem contains the corresponding parameterization of these matrices and it shows that the degrees of freedom that remain after the eigenvalue assignment can be parameterized by introducing parameter vectors for the observer.

Theorem 5.4 (Parametric expression for the observer in the time domain). *Given a set of different numbers $\tilde{\bar{s}}_1$, $\tilde{\bar{s}}_2$, ..., $\tilde{\bar{s}}_{n-\kappa}$ that do not coincide with any eigenvalue of the system (5.120) and a set of corresponding $m \times 1$ vectors $\bar{p}_1$, $\bar{p}_2$, ..., $\bar{p}_{n-\kappa}$. Assume that the $\tilde{\bar{s}}_\nu$ and the $\bar{p}_\nu$ have been chosen such that the row vectors $\bar{p}_\nu^T C(A - \tilde{\bar{s}}_\nu I)^{-1}$, $\nu = 1, 2, \ldots, n-\kappa$, and the rows of C_2 are linearly independent. Then,*

$$[L_1\ \ \Psi_2] = \begin{bmatrix} \bar{p}_1^T C(A - \tilde{\bar{s}}_1 I)^{-1} \\ \vdots \\ \bar{p}_{n-\kappa}^T C(A - \tilde{\bar{s}}_{n-\kappa} I)^{-1} \\ C_2 \end{bmatrix}^{-1} \begin{bmatrix} \begin{bmatrix} \bar{p}_1^T \\ \vdots \\ \bar{p}_{n-\kappa}^T \end{bmatrix} \begin{bmatrix} I_{m-\kappa} & 0 \\ 0 & 0_\kappa \end{bmatrix} \\ [0_{\kappa,m-\kappa}\ \ I_\kappa] \end{bmatrix} \quad (5.125)$$

satisfies (5.100) and assigns the eigenvalues $\tilde{\bar{s}}_1, \tilde{\bar{s}}_2, \ldots, \tilde{\bar{s}}_{n-\kappa}$ *along with the vectors* $\tilde{\bar{w}}_1^T T, \tilde{\bar{w}}_2^T T, \ldots, \tilde{\bar{w}}_{n-\kappa}^T T$ *to the observer (5.118) and (5.119). The vectors* $\bar{p}_1, \bar{p}_2, \ldots, \bar{p}_n$ *are the invariant parameter vectors of the observer satisfying*

$$\tilde{\bar{w}}_\nu^T T \begin{bmatrix} L_1 & K\Psi_2 \end{bmatrix} = \bar{p}_\nu^T \begin{bmatrix} I_{m-\kappa} & 0 \\ 0 & 0_\kappa \end{bmatrix}, \quad \nu = 1, 2, \ldots, n-\kappa, \tag{5.126}$$

and

$$\tilde{\bar{w}}_\nu^T T = \bar{p}_\nu^T C(A - \tilde{\bar{s}}_\nu I)^{-1}, \quad \nu = 1, 2, \ldots, n-\kappa. \tag{5.127}$$

The relation between the pole directions $\bar{q}_\nu$ *(frequency domain) and the parameter vectors* $\bar{p}_\nu$ *(time domain) of the observer is*

$$\bar{p}_\nu^T = -\bar{q}_\nu^T \bar{D}(\tilde{\bar{s}}_\nu), \quad \nu = 1, 2, \ldots, n-\kappa \tag{5.128}$$

for distinct eigenvalues $\tilde{\bar{s}}_\nu$*, where* $\bar{D}(s)$ *is the denominator matrix in the left MFD (5.55) such that* $\bar{D}_\kappa(s)$ *(see (5.62)) is row reduced. Conversely, every matrix pair* L_1 *and* Ψ_2 *can be expressed by (5.125) if the observer eigenvalues are different from the eigenvalues of the system (5.120) and if the row vectors* $\bar{p}_\nu^T C(A - \tilde{\bar{s}}_\nu I)^{-1}$*,* $\nu = 1, 2, \ldots, n-\kappa$*, and the rows of* C_2 *are linearly independent.*

Proof. For the proof of the first part of Theorem 5.4 pre-multiply (5.82) by $-\bar{p}_\nu^T \bar{D}^{-1}(\tilde{\bar{s}}_\nu)$ (the inverse matrix exists since the observer eigenvalues do not coincide with any eigenvalues of the system (5.120) by assumption), substitute $s = \tilde{\bar{s}}_\nu$ and use (5.83) to obtain

$$\begin{aligned} -\bar{p}_\nu^T \bar{D}^{-1}(\tilde{\bar{s}}_\nu)\tilde{\bar{D}}(\tilde{\bar{s}}_\nu) &= -\bar{p}_\nu^T \begin{bmatrix} I_{m-\kappa} & 0 \\ 0 & 0_\kappa \end{bmatrix} - \bar{p}_\nu^T \bar{D}^{-1}(\tilde{\bar{s}}_\nu)\bar{N}_x(\tilde{\bar{s}}_\nu) \begin{bmatrix} L_1 & \Psi_2 \end{bmatrix} \\ &= -\bar{p}_\nu^T \begin{bmatrix} I_{m-\kappa} & 0 \\ 0 & 0_\kappa \end{bmatrix} + \bar{p}_\nu^T C(A - \tilde{\bar{s}}_\nu I)^{-1} \begin{bmatrix} L_1 & \Psi_2 \end{bmatrix}. \end{aligned} \tag{5.129}$$

By substituting

$$e_\nu^T \begin{bmatrix} \bar{p}_1^T C(A - \tilde{\bar{s}}_1 I)^{-1} \\ \vdots \\ \bar{p}_{n-\kappa}^T C(A - \tilde{\bar{s}}_{n-\kappa} I)^{-1} \\ C_2 \end{bmatrix} = \bar{p}_\nu^T C(A - \tilde{\bar{s}}_\nu I)^{-1}, \quad \nu = 1, 2, \ldots, n-\kappa \tag{5.130}$$

in

$$e_\nu^T \begin{bmatrix} \bar{p}_1^T C(A - \tilde{\bar{s}}_1 I)^{-1} \\ \vdots \\ \bar{p}_{n-\kappa}^T C(A - \tilde{\bar{s}}_{n-\kappa} I)^{-1} \\ C_2 \end{bmatrix} \begin{bmatrix} \bar{p}_1^T C(A - \tilde{\bar{s}}_1 I)^{-1} \\ \vdots \\ \bar{p}_{n-\kappa}^T C(A - \tilde{\bar{s}}_{n-\kappa} I)^{-1} \\ C_2 \end{bmatrix}^{-1} = e_\nu^T \tag{5.131}$$

the result

$$\bar{p}_\nu^T C(A-\tilde{\bar{s}}_\nu I)^{-1} \begin{bmatrix} \bar{p}_1^T C(A-\tilde{\bar{s}}_1 I)^{-1} \\ \vdots \\ \bar{p}_{n-\kappa}^T C(A-\tilde{\bar{s}}_{n-\kappa} I)^{-1} \\ C_2 \end{bmatrix}^{-1} = e_\nu^T \tag{5.132}$$

is obtained. Thus, by inserting (5.125) in (5.129) and observing (5.132) yields

$$-\bar{p}_\nu^T \bar{D}^{-1}(\tilde{\bar{s}}_\nu)\tilde{\bar{D}}(\tilde{\bar{s}}_\nu) = 0^T, \quad \nu = 1,2,\ldots,n-\kappa. \tag{5.133}$$

The parametric expression (5.125) assigns the observer eigenvalues $\tilde{\bar{s}}_\nu$, since $-\bar{p}_\nu^T \bar{D}^{-1}(\tilde{\bar{s}}_\nu)$ is a non-zero vector, such that the matrix $\tilde{\bar{D}}(\tilde{\bar{s}}_\nu)$ is rank deficient and $\tilde{\bar{s}}_\nu$ is a solution of

$$\det \tilde{\bar{D}}(s) = \det(sI_{n-\kappa} - T(A - L_1 C_1)\Theta) = 0 \tag{5.134}$$

(see (3.64)). Thus, (5.128) holds in view of (5.58) and $\operatorname{rank} \tilde{\bar{D}}(\tilde{\bar{s}}_\nu) = m-1$ (see the Smith-Form of $\tilde{\bar{D}}(s)$ for distinct poles in, *e.g.*, [36]). Equation (5.127) follows directly from (5.94) when taking (5.83) and (5.128) into account. By pre-multiplying (5.125) with $\left[(\tilde{\bar{w}}_1^T T)^T \; (\tilde{\bar{w}}_2^T T)^T \; \ldots \; (\tilde{\bar{w}}_{n-\kappa}^T T)^T \; C_2^T\right]^T$ and using (5.127) it becomes obvious that both the Relation (5.126) for the parameter vectors $\bar{p}_\nu$ and the Restriction (5.100) are satisfied.

In order to show that (5.127) is satisfied consider

$$C(sI-A)^{-1} = \bar{D}^{-1}(s)\bar{N}_x(s), \tag{5.135}$$

(see (5.83)) which can be rewritten as

$$\bar{N}_x(s)(sI-A) = \bar{D}(s)C. \tag{5.136}$$

Substituting the observer poles $\tilde{\bar{s}}_\nu$ in (5.136) and pre-multiplying with $\bar{q}_\nu^T$ gives

$$\bar{q}_\nu^T \bar{N}_x(\tilde{\bar{s}}_\nu)(\tilde{\bar{s}}_\nu I - A) = \bar{q}_\nu^T \bar{D}(\tilde{\bar{s}}_\nu)C. \tag{5.137}$$

By inserting (5.94) and (5.128) one obtains

$$\tilde{\bar{w}}_\nu^T T(\tilde{\bar{s}}_\nu I - A) = -\bar{p}_\nu^T C. \tag{5.138}$$

Since $\tilde{\bar{s}}_\nu I - A$ is invertible by assumption, the Equation (5.138) can be solved for $\tilde{\bar{w}}_\nu^T T$, yielding (5.127). Substituting (5.94), (5.127) and (5.128) in (5.102) gives

$$\begin{bmatrix} \bar{p}_1^T C(A-\tilde{\bar{s}}_1 I)^{-1} \\ \vdots \\ \bar{p}_{n-\kappa}^T C(A-\tilde{\bar{s}}_{n-\kappa} I)^{-1} \\ C_2 \end{bmatrix} \begin{bmatrix} L_1 & \Psi_2 \end{bmatrix} = \begin{bmatrix} \begin{bmatrix} \bar{p}_1^T \\ \vdots \\ \bar{p}_{n-\kappa}^T \end{bmatrix} \begin{bmatrix} I_{m-\kappa} & 0 \\ 0 & 0_\kappa \end{bmatrix} \\ \begin{bmatrix} 0_{\kappa,m-\kappa} & I_\kappa \end{bmatrix} \end{bmatrix}. \tag{5.139}$$

As the row vectors in the leftmost matrix in (5.139) are assumed to be linearly independent, (5.125) can be obtained from (5.139). □

Remark 5.10. For a real observer eigenvalue $\tilde{\bar{s}}_\nu$ the corresponding parameter vector $\bar{p}_\nu$ has to be a real-valued vector and for each conjugate complex pair of observer eigenvalues the parameter vectors have to be chosen conjugate complex.

Remark 5.11. The set of $n-\kappa$ numbers $\tilde{\bar{s}}_1, \tilde{\bar{s}}_2, \ldots, \tilde{\bar{s}}_{n-\kappa}$ with the corresponding $n-\kappa$ constant $m \times 1$ vectors $\bar{p}_1, \bar{p}_2, \ldots, \bar{p}_{n-\kappa}$ contain $m(n-\kappa)$ degrees of freedom, since each parameter vector $\bar{p}_\nu$ only contributes $m-1$ influential design parameters. This is due to the fact that a multiplication of each parameter vector by a non-zero constant does not change the result for (5.125).

Remark 5.12. In the case of a full-order observer

$$\dot{\hat{\zeta}}(t) = (A - LC)\hat{\zeta}(t) + Ly(t) + Bu(t), \tag{5.140}$$

$$\hat{x}(t) = \hat{\zeta}(t), \tag{5.141}$$

i.e., $\kappa = 0$, $C_1 = C$, $C_2 = 0$, $T = \Theta = I$, $L_1 = L$ and $\Psi_2 = 0$ the parametric expression (5.125) coincides with the result

$$L = \begin{bmatrix} \bar{p}_1^T C(A - \tilde{\bar{s}}_1 I)^{-1} \\ \vdots \\ \bar{p}_n^T C(A - \tilde{\bar{s}}_n I)^{-1} \end{bmatrix}^{-1} \begin{bmatrix} \bar{p}_1^T \\ \vdots \\ \bar{p}_n^T \end{bmatrix} \tag{5.142}$$

obtained in [55].

Remark 5.13. In the case of a completely reduced-order observer

$$\dot{\hat{\zeta}}(t) = TA\Theta\hat{\zeta}(t) + TA\Psi y(t) + TBu(t), \tag{5.143}$$

$$\hat{x}(t) = \Theta\hat{\zeta}(t) + \Psi y(t), \tag{5.144}$$

i.e., $\kappa = m$, $C_1 = 0$, $C_2 = C$, $L_1 = 0$ and $\Psi_2 = \Psi$ the parametric expression (5.125) takes the form

$$\Psi = \begin{bmatrix} \bar{p}_1^T C(A - \tilde{\bar{s}}_1 I)^{-1} \\ \vdots \\ \bar{p}_{n-m}^T C(A - \tilde{\bar{s}}_{n-m} I)^{-1} \\ C \end{bmatrix}^{-1} \begin{bmatrix} 0 \\ I_m \end{bmatrix}. \tag{5.145}$$

The matrix T in (5.143) can be obtained from solving $T\Psi = 0$ (see (3.21)) with $\operatorname{rank} T = n - \kappa$, such that the matrix Θ in (5.143) and (5.144) results from

$$\begin{bmatrix} \Psi & \Theta \end{bmatrix} = \begin{bmatrix} C \\ T \end{bmatrix}^{-1} \tag{5.146}$$

(see (3.22)).

6

Decoupling Control

A characteristic property of multivariable systems is that the input–output behaviour is coupled, *i.e.*, a change in one input affects several outputs. By designing a decoupled reference transfer behaviour each output variable is affected by only one reference signal. Therefore, each input–output pair can then be controlled by a SISO controller. A precise definition of this diagonal decoupling problem using static state feedback was first given in [48]. Later, a necessary and sufficient condition for the solvability of diagonal decoupling was established in [17]. If this condition is not satisfied or if the system is completely decouplable, but not in a stable scheme, at least partial decoupling can be achieved (for an overview see, *e.g.*, [66]). In this chapter a partial decoupling using static state feedback is considered, such that the reference transfer matrix contains one or more coupled rows. These coupled rows have non-zero elements outside the main diagonal, so that the corresponding output is affected by several inputs. Using the parametric approach this decoupling problem has been solved in the time domain in [44]. A frequency-domain formulation of this approach is presented in this chapter.

The decoupling problem can easily be stated and solved on the basis of the frequency-domain representation of state feedback control. Therefore, in this chapter the decoupling problem is only considered in the frequency domain. By using the relations connecting the parametric approach in the time and in the frequency domain (see Chapter 5) the equivalent time-domain results can also be derived.

It should be noted that a diagonally decoupled closed-loop system is always achievable if dynamics are added to the static state feedback. Results on this subject can, *e.g.*, be found in [47, 49].

In Section 6.1 the problem of diagonal decoupling is defined and solved. First, a very simple parameterization of a decoupling state feedback is presented and then a design procedure is developed using the parametric approach of Chapter 5. The decoupling with coupled rows of non-minimum phase and non-decouplable systems is considered in Section 6.2.

6.1 Diagonal Decoupling

6.1.1 Criterion for Diagonal Decoupling

Considered are systems with the state equations

$$\dot{x}(t) = Ax(t) + Bu(t), \tag{6.1}$$
$$y_c(t) = C_c x(t), \tag{6.2}$$

with the state $x \in \mathbb{R}^n$, the input $u \in \mathbb{R}^p$ and the controlled output $y_c \in \mathbb{R}^p$. It is assumed that the pair (A, B) is controllable and the pair (C_c, A) is observable. In the frequency domain the system is characterized by its transfer behaviour

$$y_c(s) = G_c(s)u(s). \tag{6.3}$$

The transfer matrix $G_c(s)$ in (6.3) is a $p \times p$ matrix represented by the right coprime MFD

$$G_c(s) = N_c(s)D^{-1}(s), \tag{6.4}$$

where $D(s)$ is column reduced. The problem of *diagonal decoupling by static state feedback* amounts to determining a state feedback controller

$$u(t) = -Kx(t) + Mr(t), \quad \det M \neq 0, \tag{6.5}$$

with the reference input $r \in \mathbb{R}^p$ and a constant matrix M, such that the reference transfer behaviour

$$y_c(s) = G_r(s)r(s) \tag{6.6}$$

is characterized by the diagonal transfer matrix

$$G_r(s) = N_c(s)\tilde{D}^{-1}(s)M = \operatorname{diag}(g_{ii}(s)) \tag{6.7}$$

(see (2.21)). In (6.7) the $p \times p$ polynomial matrix $\tilde{D}(s)$ parameterizes the state feedback controller in the frequency domain. The special form of the reference transfer matrix $G_r(s)$ in (6.7) implies that the reference transfer behaviour can be assigned independently for each input–output pair. To assure a vanishing steady-state error for stationary constant reference signals the reference transfer matrix also has to satisfy

$$G_r(0) = I. \tag{6.8}$$

The decoupled reference transfer behaviours

$$y_c^i(s) = g_{ii}(s)r_i(s), \quad i = 1, 2, \ldots, p \tag{6.9}$$

are characterized by the elements $g_{ii}(s)$ of $G_r(s)$. These transfer functions are assumed to be of the form

$$g_{ii}(s) = \frac{\tilde{d}_{ii}(0)}{\tilde{d}_{ii}(s)}, \quad i = 1, 2, \ldots, p, \tag{6.10}$$

where

$$\tilde{d}_{ii}(s) = (s - \tilde{s}_{i1}) \cdot \ldots \cdot (s - \tilde{s}_{i\delta_i}). \tag{6.11}$$

Obviously, the transfer functions (6.10) satisfy the Condition (6.8). However, the number of poles in each transfer function $g_{ii}(s)$ cannot be assigned arbitrarily since δ_i is a property of the open-loop system. The quantities δ_i are defined as follows.

Definition 6.1 (Relative degree of an output). *Consider the system (6.1) and (6.2) and let δ_i, $i = 1, 2, \ldots, p$, denote the smallest order of the time derivative $\frac{d^{\delta_i}}{dt^{\delta_i}} y_c^i$ of the ith output y_c^i with a direct feedthrough of some input u_j, $j = 1, 2, \ldots, p$. Then, δ_i is the* relative degree *associated with the output y_c^i of the system (6.1) and (6.2).*

Using Definition 6.1 the relative degree of a system can be defined.

Definition 6.2 (Relative degree of a system). *The* relative degree *δ, $p \leq \delta \leq n$, of a system of the order n with p outputs y_c^i, $i = 1, 2, \ldots, p$, is given by*

$$\delta = \sum_{i=1}^{p} \delta_i. \tag{6.12}$$

In the frequency domain the relative degrees can be determined on the basis of the MFD (6.4). The corresponding procedure is presented in the following theorem.

Theorem 6.1 (Relative degrees of the outputs computed from a right MFD). *For a system described by the MFD (6.4) the relative degree δ_i associated with the ith output y_c^i is given by*

$$\delta_i = \min_{\substack{j=1,2,\ldots,p \\ n_{ij}(s) \neq 0}} (\delta_{cj}[D(s)] - \deg[n_{ij}(s)]), \tag{6.13}$$

where $n_{ij}(s)$ is the polynomial in the ith row and jth column of $N_c(s)$.

Proof. Consider an integer $\lambda \geq 0$ and the transfer matrix

$$\bar{G}_c(s) = \operatorname{diag}(I_{i-1}, s^{\lambda}, I_{p-i}) G_c(s) = \operatorname{diag}(I_{i-1}, s^{\lambda}, I_{p-i}) N_c(s) D^{-1}(s), \tag{6.14}$$

where (6.4) has been used. This corresponds to taking the λth time derivative of y_c^i in the time domain. An input u_j appears in the λth time derivative of y_c^i if a non-zero element appears in the ith row of the direct feedthrough of $\bar{G}_c(s)$. This feedthrough can be computed in the frequency domain as

$$\bar{G}_c(\infty) = \Gamma_{\delta_c[D(s)]}[\operatorname{diag}(I_{i-1}, s^{\lambda}, I_{p-i}) N_c(s)] \Gamma_c^{-1}[D(s)] \tag{6.15}$$

(see (1.22)). Since $\Gamma_c^{-1}[D(s)]$ is a non-singular matrix ($D(s)$ is assumed to be column reduced), the ith row of $\bar{G}_c(\infty)$ will have a non-zero element iff the ith row of $\Gamma_{\delta_c[D(s)]}[\mathrm{diag}(I_{i-1}, s^{\lambda}, I_{p-i})N_c(s)]$ is not vanishing. Consequently, if λ is the smallest difference $\delta_{cj}[D(s)] - \deg[n_{ij}(s)]$, $j = 1, 2, \ldots, p$, where $n_{ij}(s)$ does not vanish, the output y_c^i has the relative degree $\delta_i = \lambda$. □

The relative degrees of the outputs of the closed-loop system can be determined by considering

$$\begin{aligned}\bar{G}_r(\infty) &= \lim_{s\to\infty} \mathrm{diag}(I_{i-1}, s^{\delta_i}, I_{p-i})G_r(s)\\ &= \lim_{s\to\infty} \mathrm{diag}(I_{i-1}, s^{\delta_i}, I_{p-i})N_c(s)\tilde{D}^{-1}(s)M\\ &= \Gamma_{\delta_c[\tilde{D}(s)]}[\mathrm{diag}(I_{i-1}, s^{\delta_i}, I_{p-i})N_c(s)]\Gamma_c^{-1}[\tilde{D}(s)]M. \qquad (6.16)\end{aligned}$$

In view of $\delta_{ci}[\tilde{D}(s)] = \delta_{ci}[D(s)]$, $i = 1, 2, \ldots, p$, and $\Gamma_c[\tilde{D}(s)] = \Gamma_c[D(s)]$ (see (2.28) and (2.29)) this can be written as

$$\bar{G}_r(\infty) = \Gamma_{\delta_c[D(s)]}[\mathrm{diag}(I_{i-1}, s^{\delta_i}, I_{p-i})N_c(s)]\Gamma_c^{-1}[D(s)]M. \qquad (6.17)$$

When comparing (6.15) and (6.17) the proof of Theorem 6.1 implies that the relative degrees of the outputs of the closed-loop system are given by (6.13) since M is non-singular. Therefore, the relative degrees are invariant under the state feedback control (6.5) and the number of poles in (6.10) is fixed by the relative degree δ_i of the output y_c^i.

Unfortunately, not every system can be decoupled by static state feedback (6.5). The next theorem characterizes the class of decouplable systems in the frequency domain.

Theorem 6.2 (Decoupling criterion). *A system described by the transfer behaviour (6.3) of the order n with z transmission zeros and relative degree δ is diagonally decouplable iff*

$$z = n - \delta. \qquad (6.18)$$

Proof. See [12]. □

Remark 6.1. Since the relation $z \leq n - \delta$ is always satisfied (see, *e.g.*, [45]) a non-decouplable system has not enough transmission zeros for diagonal decoupling.

Remark 6.2. Using the MFD (6.4) Condition (6.18) becomes

$$\deg[\det N_c(s)] = n - \delta, \qquad (6.19)$$

because the transmission zeros are a solution of $\det N_c(s) = 0$ (see Section 1.1).

Example 6.1. Decouplability of a system represented by its transfer behaviour
Consider a system of the order 3 with the transfer behaviour

$$y_c(s) = G_c(s)u(s), \tag{6.20}$$

where

$$G_c(s) = N_c(s)D^{-1}(s) = \begin{bmatrix} 4 & s-4 \\ -8 & 2 \end{bmatrix} \begin{bmatrix} 4s+4 & s^2+3s+2 \\ -s-3 & 0 \end{bmatrix}^{-1}. \tag{6.21}$$

The relative degrees δ_1 and δ_2 of the first and the second output can be computed using (6.13). This leads to

$$\begin{aligned} \delta_1 &= \min(\delta_{c1}[D(s)] - \deg[n_{11}(s)], \delta_{c2}[D(s)] - \deg[n_{12}(s)]) \\ &= \min(1-0, 2-1) = 1, \end{aligned} \tag{6.22}$$

and

$$\begin{aligned} \delta_2 &= \min(\delta_{c1}[D(s)] - \deg[n_{21}(s)], \delta_{c2}[D(s)] - \deg[n_{22}(s)]) \\ &= \min(1-0, 2-0) = 1. \end{aligned} \tag{6.23}$$

Since

$$\det N_c(s) = 8s - 24, \tag{6.24}$$

and therefore

$$\deg[\det N_c(s)] = 1 = n - \delta_1 - \delta_2 = 3 - 2, \tag{6.25}$$

the system with the transfer matrix (6.21) is diagonally decouplable (see Remark 6.2 and (6.12)).

6.1.2 A Simple Solution of the Diagonal Decoupling Problem

The next theorem shows that the diagonal decoupling problem has a simple solution in the frequency domain.

Theorem 6.3 (Design of state feedback controllers for diagonal decoupling in the frequency domain). *Consider the transfer behaviour*

$$y_c(s) = G_c(s)u(s) \tag{6.26}$$

of a decouplable system of the order n with p inputs u and p outputs y_c. Let the transfer matrix $G_c(s)$ in (6.26) be represented by a right coprime MFD

$$G_c(s) = N_c(s)D^{-1}(s), \tag{6.27}$$

with $D(s)$ column reduced. Assume that the outputs y_c^i have the relative degrees δ_i, $i = 1, 2, \ldots, p$, such that the system has the relative degree $\delta = \delta_1 + \ldots + \delta_p$. Additionally, it is assumed that no transmission zero of the system is located *at* $s = 0$, i.e., $\det N_c(0) \neq 0$. *The diagonal decoupling state feedback controller (6.5) is parameterized in the frequency domain by the polynomial matrix*

$$\tilde{D}(s) = M\Lambda(s)N_c(s). \tag{6.28}$$

In (6.28) the diagonal matrix

$$\Lambda(s) = \text{diag}\left(\frac{\tilde{d}_{ii}(s)}{\tilde{d}_{ii}(0)}\right) \tag{6.29}$$

contains the polynomials $\tilde{d}_{ii}(s)$ *as defined in (6.11), and the constant matrix* M *has the form*

$$M = \Gamma_c[D(s)]\Gamma_c^{-1}[\Lambda(s)N_c(s)]. \tag{6.30}$$

Proof. A simple rearrangement of (6.7) yields

$$\tilde{D}(s) = M\Lambda(s)N_c(s). \tag{6.31}$$

In order to prove that the polynomial matrix $\tilde{D}(s)$ in (6.31) parameterizes a state feedback controller in the frequency domain one has to verify that the conditions

$$\delta_{ci}[\tilde{D}(s)] = \delta_{ci}[D(s)], \quad i = 1, 2, \ldots, p, \tag{6.32}$$

$$\Gamma_c[\tilde{D}(s)] = \Gamma_c[D(s)] \tag{6.33}$$

hold (see (2.28) and (2.29)). First, it is shown that the polynomial matrix $\Lambda(s)N_c(s)$ is column reduced. By construction of $\Lambda(s)$ the transfer matrix $\Lambda(s)N_c(s)D^{-1}(s)$ is proper (see Definition 6.1 and (6.11)). Since $D(s)$ is column reduced, this leads to

$$\delta_{ci}[\Lambda(s)N_c(s)] \leq \delta_{ci}[D(s)], \quad i = 1, 2, \ldots, p \tag{6.34}$$

in the light of (1.21) and hence

$$\sum_{i=1}^{p} \delta_{ci}[\Lambda(s)N_c(s)] \leq \sum_{i=1}^{p} \delta_{ci}[D(s)] = n. \tag{6.35}$$

As the system is decouplable the Relation (6.19) is satisfied and with the aid of (6.11) and (6.12) the result

$$\deg\det\Lambda(s)N_c(s) = \deg\det\Lambda(s) + \deg\det N_c(s) = \delta + n - \delta = n \tag{6.36}$$

is obtained. This implies

$$\sum_{i=1}^{p} \delta_{ci}[\Lambda(s)N_c(s)] \geq n \tag{6.37}$$

in view of $\deg\det(\Lambda(s)N_c(s)) \leq \sum_{i=1}^{p} \delta_{ci}[\Lambda(s)N_c(s)]$ (see, *e.g.*, [36]). Consequently, (6.35) and (6.36) imply

$$\sum_{i=1}^{p} \delta_{ci}[\Lambda(s)N_c(s)] = n, \tag{6.38}$$

showing that the polynomial matrix $\Lambda(s)N_c(s)$ is column reduced (see (1.6)). Equation (6.38) shows that in (6.34)

$$\delta_{ci}[\Lambda(s)N_c(s)] = \delta_{ci}[D(s)] \tag{6.39}$$

holds, because

$$\sum_{i=1}^{p} \delta_{ci}[\Lambda(s)N_c(s)] = \sum_{i=1}^{p} \delta_{ci}[D(s)] = n. \tag{6.40}$$

Consider the representation

$$\Lambda(s)N_c(s) = \Gamma_c[\Lambda(s)N_c(s)]\,\mathrm{diag}(s^{\delta_{ci}[\Lambda(s)N_c(s)]}) + R(s) \tag{6.41}$$

of $\Lambda(s)N_c(s)$, where $\delta_{ci}[R(s)] < \delta_{ci}[\Lambda(s)N_c(s)]$, $i = 1, 2, \ldots, p$. Then, when choosing

$$M = \Gamma_c[D(s)]\Gamma_c^{-1}[\Lambda(s)N_c(s)] \tag{6.42}$$

the Conditions (6.32) and (6.33) are satisfied since pre-multiplying (6.41) with M yields

$$M\Lambda(s)N_c(s) = \Gamma_c[D(s)]\,\mathrm{diag}(s^{\delta_{ci}[D(s)]}) + MR(s), \tag{6.43}$$

in view of (6.39) and $\delta_{ci}[MR(s)] < \delta_{ci}[\Lambda(s)N_c(s)] = \delta_{ci}[D(s)]$, $i = 1, 2, \ldots, p$. The inverse $\Gamma_c^{-1}[\Lambda(s)N_c(s)]$ exists, because the polynomial matrix $\Lambda(s)N_c(s)$ is column reduced. In view of (6.5) the matrix M has to be non-singular, which is satisfied since $\det\Gamma_c[D(s)] \neq 0$ yields a non-singular matrix M in (6.42). □

Theorem 6.3 shows that the poles of the diagonally decoupled closed-loop system result from

$$\det\tilde{D}(s) = \det M \det\Lambda(s) \det N_c(s), \tag{6.44}$$

with

$$\deg\det N_c(s) = n - \delta \tag{6.45}$$

(see (6.19) and (6.28)). This implies that for $\delta < n$ the stability of the diagonally decoupled closed-loop system depends on the transmission zeros of the transfer matrix (6.4) (*i.e.*, the solutions of $\det N_c(s) = 0$). Therefore, the condition for the internal stability of the closed-loop system is that all transmission zeros have negative real parts, *i.e.*, the system (6.3) is a *minimum phase system*. An exception to this rule are *non-interconnecting zeros* (see [39]), which are a common factor of all elements in one row of $N_c(s)$. For convenience, all other transmission zeros are referred to as *interconnecting zeros*.

In what follows the stable decoupling of a non-minimum phase system with one non-interconnecting zero $\eta > 0$ is investigated. It is assumed that this transmission zero is a common factor in all polynomials $n_{kj}(s)$, $j = 1, 2, \ldots, p$, in the kth row of $N_c(s)$. It is further assumed that the system is decouplable and that all other transmission zeros lie in the open left-half complex plane. For this system the numerator matrix $N_c(s)$ in (6.4) can be written in the form

$$N_c(s) = N_0(s)\hat{N}_c(s), \tag{6.46}$$

where

$$N_0(s) = \mathrm{diag}(I_{k-1}, s - \eta, I_{p-k}), \tag{6.47}$$

and $\det \hat{N}_c(s)$ has only zeros in the open left-half complex plane. The decoupling controller is now determined for the minimum phase system with the transfer matrix

$$\hat{G}_c(s) = \hat{N}_c(s)D^{-1}(s). \tag{6.48}$$

This minimum phase system is decouplable if the original system (6.3) is decouplable since the kth relative degree of the minimum phase system is increased with respect to (6.3), but this is made up by a decrease in the number of transmission zeros, such that (6.18) is still satisfied. As a consequence, $\delta_k + 1$ closed-loop poles have to be assigned in the kth row. The reference transfer matrix of the minimum phase system is

$$\hat{G}_r(s) = \hat{N}_c(s)\tilde{D}^{-1}(s)M_1 = \mathrm{diag}(\hat{g}_{ii}(s)), \tag{6.49}$$

so that the polynomial matrix

$$\tilde{D}(s) = M_1 \left(\mathrm{diag}(\hat{g}_{ii}(s))\right)^{-1} \hat{N}_c(s), \tag{6.50}$$

and the constant matrix

$$M_1 = \Gamma_c[D(s)]\Gamma_c^{-1}[\left(\mathrm{diag}(\hat{g}_{ii}(s))\right)^{-1} \hat{N}_c(s)] \tag{6.51}$$

solve (6.49) (see (6.28) and (6.30)). Using the decoupling controller defined by (6.50) and (6.51) and introducing the matrix $M = M_1M_2$ the reference transfer matrix of the original system (6.3) takes the form

$$\begin{aligned} G_r(s) &= N_c(s)\tilde{D}^{-1}(s)M = N_0(s)\hat{N}_c(s)\tilde{D}^{-1}(s)M_1M_2 \\ &= N_0(s)\,\mathrm{diag}(\hat{g}_{ii}(s))M_2, \end{aligned} \tag{6.52}$$

where (6.46) and (6.49) have been used. The additional matrix M_2 is determined such that

$$G_r(0) = N_0(0)M_2 = I \tag{6.53}$$

in view of $\hat{G}_r(0) = \mathrm{diag}(\hat{g}_{ii}(0)) = I$. With (6.47) the diagonal matrix

$$M_2 = N_0^{-1}(0) = \mathrm{diag}(I_{k-1}, -\tfrac{1}{\eta}, I_{p-k}) \tag{6.54}$$

results, so that one obtains a diagonal transfer matrix in (6.52). This shows that diagonal decoupling with internal stability is also possible for non-minimum phase systems with one transmission zero η in the closed right-half complex plane provided that this zero is a non-interconnecting zero. By using the same reasoning the corresponding result can be obtained in the case of several real or conjugate complex non-interconnecting zeros in different rows.

Remark 6.3. The numerator matrix (6.46) and the reference transfer matrix in (6.52) show that the non-interconnecting zero η only affects the kth output y_c^k of the open-loop and the closed-loop systems. Consequently, this transmission zero need not be compensated for diagonal decoupling.

6.1.3 Diagonal Decoupling Using the Parametric Approach

An alternative solution to the diagonal decoupling problem is provided by the parametric approach (see Chapter 5). This design method has the advantage that the results of the diagonal decoupling can be extended to the partial decoupling of non-minimum phase and/or non-decouplable systems. Therefore, a parametric solution of the diagonal decoupling problem is presented in the following. In what follows conditions are formulated for the closed-loop poles and the corresponding pole directions of the decoupled closed-loop system. It is assumed that the matrix $A - BK$ of the closed-loop system is non-defective, *i.e.*, it is diagonalizable by a similarity transformation. Then, there exists a non-singular $n \times n$ matrix $\tilde{V} = [\tilde{v}_1 \ \ldots \ \tilde{v}_n]$ of the closed-loop eigenvectors $\tilde{v}_\nu$, $\nu = 1, 2, \ldots, n$, such that

$$A - BK = \tilde{V}\tilde{\Lambda}\tilde{V}^{-1} = \tilde{V}\tilde{\Lambda}\tilde{W}. \tag{6.55}$$

The $n \times n$ diagonal matrix $\tilde{\Lambda}$ contains the eigenvalues of the closed-loop system, *i.e.*,

$$\tilde{\Lambda} = \mathrm{diag}(\tilde{s}_\nu), \tag{6.56}$$

and $\tilde{W} = \tilde{V}^{-1}$. Using (6.55) the expansion

$$G_r(s) = C_c(sI - A + BK)^{-1}BM = C_c\tilde{V}(sI - \tilde{\Lambda})^{-1}\tilde{W}BM = \sum_{\nu=1}^{n} \frac{\tilde{R}_\nu}{s - \tilde{s}_\nu} \tag{6.57}$$

of the reference transfer matrix $G_r(s)$ can be obtained, where the $p \times p$ constant matrices $\tilde{R}_\nu$ have the form

$$\tilde{R}_\nu = C_c\tilde{v}_\nu\tilde{w}_\nu^T BM, \tag{6.58}$$

and $\tilde{w}_\nu^T$ denotes the $1 \times n$ row vectors of the matrix $\tilde{W}$. In terms of the closed-loop poles $\tilde{s}_\nu$ and the corresponding pole directions q_ν the matrices $\tilde{R}_\nu$ take the form

$$\tilde{R}_\nu = C_c N_x(\tilde{s}_\nu) q_\nu \tilde{w}_\nu^T BM = N_c(\tilde{s}_\nu) q_\nu \tilde{w}_\nu^T BM, \tag{6.59}$$

where $\tilde{v}_\nu = N_x(\tilde{s}_\nu)q_\nu$ and $C_c N_x(s) = N_c(s)$ (see (5.42)) and(2.14)) have been used. A non-zero element $\tilde{R}_{\nu,ij}$ in the matrix $\tilde{R}_\nu$ indicates that the closed-loop pole $\tilde{s}_\nu$ appears in the transfer function in the ith row and the jth column of $G_r(s)$ in (6.57). Therefore, a closed-loop pole $\tilde{s}_\nu$ appears solely in the ith row of $G_r(s)$ if the ith row of $\tilde{R}_\nu$ is not vanishing and all other rows only contain zero elements. With $\tilde{w}_\nu^T BM \neq 0^T$, the matrix

$$\tilde{R}_\nu = N_c(\tilde{s}_\nu) q_\nu \tilde{w}_\nu^T BM = e_i \tilde{w}_\nu^T BM = \begin{bmatrix} 0_{i-1 \times p} \\ \tilde{w}_\nu^T BM \\ 0_{p-i \times p} \end{bmatrix} \tag{6.60}$$

satisfies this condition. The property $\tilde{w}_\nu^T BM \neq 0^T$ is implied by the fact that the closed-loop system remains controllable under the feedback (6.5). Therefore, the Popov–Belevitch–Hautus eigenvector test for controllability (see, *e.g.*, [36]) shows that $\tilde{w}_\nu^T B \neq 0^T$. Because of $\det M \neq 0$ this leads to the result $\tilde{w}_\nu^T BM \neq 0^T$. Thus, the closed-loop poles $\tilde{s}_\nu$ and their pole directions q_ν must satisfy

$$N_c(\tilde{s}_\nu) q_\nu = e_i \tag{6.61}$$

in order to appear solely in the ith row. If the closed-loop poles $\tilde{s}_\nu$ to be assigned in the ith row of $G_r(s)$ are chosen different from the transmission zeros of the transfer behaviour (6.3), *i.e.*, $\det N_c(\tilde{s}_\nu) \neq 0$, then

$$q_\nu = N_c^{-1}(\tilde{s}_\nu) e_i. \tag{6.62}$$

In view of (6.11) and (6.12) only $\delta = \delta_1 + \ldots + \delta_p$ closed-loop poles can be assigned to the reference transfer behaviour of the closed-loop system. Consequently, $n - \delta$ closed-loop poles do not appear in the reference transfer matrix $G_r(s)$. If the system is decouplable (see Theorem 6.2), the system has $n - \delta$ transmission zeros, so that the remaining $n - \delta$ closed-loop poles can be cancelled by the $n - \delta$ transmission zeros of the transfer behaviour (6.3). If these zeros are located in the open left-half complex plane then a stable decoupling is possible. In order to determine the parameter vectors related to compensated closed-loop poles assume that a closed-loop pole $\tilde{s}_\nu$ does not appear in $G_r(s)$. Then, the corresponding matrix $\tilde{R}_\nu$ in (6.57) has to vanish, which means that the associated matrix (6.59) must satisfy

$$\tilde{R}_\nu = N_c(\tilde{s}_\nu) q_\nu \tilde{w}_\nu^T BM = 0. \tag{6.63}$$

With $\tilde{w}_\nu^T BM \neq 0^T$, this gives

$$N_c(\tilde{s}_\nu)q_\nu = 0, \tag{6.64}$$

where the non-zero vector q_ν is a solution of (6.64) if $\det N_c(\tilde{s}_\nu) = 0$. This means that the closed-loop pole $\tilde{s}_\nu$ has to coincide with a transmission zero of the transfer behaviour (6.3).

A necessary condition for stable diagonal decoupling is that the transfer behaviour (6.3) has no transmission zero at $s = 0$ (*i.e.*, $\det N_c(0) \neq 0$). If such a zero exists stationary constant reference signals can only be transferred independently to the controlled output y_c if this zero at $s = 0$ is compensated. This, however, leads to an unstable closed-loop system. Thus, for systems (6.1) and (6.2) which are decouplable in a stable scheme the matrix

$$M = \tilde{D}(0)N_c^{-1}(0) \tag{6.65}$$

exists (see (2.22)) and it can be used in (6.7) to assure vanishing steady-state errors (see (6.8)). This matrix also satisfies the assumption $\det M \neq 0$ in (6.5) since an asymptotically stable closed-loop system leads to $\det \tilde{D}(0) \neq 0$. Whereas (6.62) implies that a closed-loop pole is assigned to the ith row of the reference transfer matrix, the matrix M in (6.65) assures that the same closed-loop pole only appears in the ith column, *i.e.*, that the reference transfer matrix is diagonal. This can be verified by considering a non-diagonal element

$$g_{ij}(s) = \frac{\tilde{n}_{ij}(s)}{\tilde{d}_{ij}(s)}, \quad i \neq j \tag{6.66}$$

of the reference transfer matrix $G_r(s)$. The assignment of δ_i closed-loop poles to the ith row of $G_r(s)$ ensures that

$$\deg[\tilde{d}_{ij}(s)] = \delta_i, \quad j = 1, 2, \ldots, p, \tag{6.67}$$

and (6.8) leads to

$$\tilde{n}_{ij}(0) = 0, \quad i \neq j, \tag{6.68}$$

because (6.65) assures vanishing steady-state errors. Since the relative degrees are invariant under static state feedback the degrees of the numerator and the denominator polynomials of $g_{ij}(s)$ have to satisfy

$$\deg[\tilde{d}_{ij}(s)] - \deg[\tilde{n}_{ij}(s)] \geq \delta_i, \quad j = 1, 2, \ldots, p, \quad i \neq j. \tag{6.69}$$

By using (6.67) this gives

$$\deg[\tilde{n}_{ij}(s)] \leq \deg[\tilde{d}_{ij}(s)] - \delta_i = 0, \quad j = 1, 2, \ldots, p, \quad i \neq j, \tag{6.70}$$

after a simple rearrangement. This implies $\tilde{n}_{ij}(s) = \mathit{const}$, such that with (6.68) the result $\tilde{n}_{ij}(s) \equiv 0$ is obtained, which shows that the non-diagonal elements in $G_r(s)$ vanish.

The next theorem summarizes the results obtained so far.

Theorem 6.4 (Parametric solution of the diagonal decoupling problem). *Consider the transfer behaviour*

$$y_c(s) = G_c(s)u(s) \tag{6.71}$$

of a decouplable system of the order n with p inputs u and p outputs y_c. Let the transfer matrix $G_c(s)$ in (6.71) be represented by a right coprime MFD

$$G_c(s) = N_c(s)D^{-1}(s), \tag{6.72}$$

with $D(s)$ column reduced. Assume that the outputs y_c^i have the relative degrees δ_i, $i = 1, 2, \ldots, p$, so that the system has the relative degree $\delta = \delta_1 + \ldots + \delta_p$. Additionally, it is assumed that no transmission zero of the system is located at $s = 0$, i.e., $\det N_c(0) \neq 0$. All transmission zeros of the transfer behaviour (6.72) are distinct and lie in the open left-half complex plane. Let the closed-loop poles $\tilde{s}_\nu$, $\nu = 1, 2, \ldots, n$, and the corresponding pole directions q_ν, $\nu = 1, 2, \ldots, n$, be chosen according to the following rules:

- *Assign $n - \delta$ closed-loop poles $\tilde{s}_\nu$ such that they coincide with the transmission zeros of the transfer matrix $G_c(s)$, i.e., $\det N_c(\tilde{s}_\nu) = 0$, and the corresponding pole directions q_ν are the non-vanishing solutions of*

$$N_c(\tilde{s}_\nu)q_\nu = 0, \quad \nu = 1, 2, \ldots, n - \delta. \tag{6.73}$$

- *The remaining δ closed-loop poles $\tilde{s}_\nu$ are assigned to the reference transfer behaviour and satisfy $\det N_c(\tilde{s}_\nu) \neq 0$. The corresponding pole directions q_ν are given by*

$$q_\nu = N_c^{-1}(\tilde{s}_\nu)e_\mu, \quad \nu = n-\delta+1, n-\delta+2, \ldots, n, \quad \mu = 1, 2, \ldots, p, \tag{6.74}$$

  where (6.74) is used to assign δ_μ closed-loop poles to the μth row of the reference transfer matrix $G_r(s)$.

With the matrix $M = \tilde{D}(0)N_c^{-1}(0)$ the frequency-domain parameterization of the state feedback controller (6.5) achieving a diagonal reference transfer matrix $G_r(s)$ (see (6.7)) is given by

$$\tilde{D}(s) = D(s) - \left[D(\tilde{s}_1)q_1 \ldots D(\tilde{s}_n)q_n\right]\left[S(\tilde{s}_1)q_1 \ldots S(\tilde{s}_n)q_n\right]^{-1} S(s) \tag{6.75}$$

(see (5.14)).

In order to prove Theorem 6.4 it remains to be shown that under the stated assumptions the inverse matrix in (6.75) exists. This is verified in [45] by using time-domain arguments.

Remark 6.4. A diagonal decoupling can also be achieved if the system has a transmission zeros η with algebraic multiplicity k. Then, if k linearly independent solutions q_i of

$$N_c(\eta)q_i = 0, \quad i = 1, 2, \ldots, k \tag{6.76}$$

(see (6.73)) exist, Theorem 6.4 remains valid. Otherwise, the inverse matrix in (6.75) does not exist. Since $N_c(\eta)$ is a $p \times p$ matrix the number of linearly inpendent solutions of (6.76) is bounded by p. Consequently, the frequency-domain parameterization (6.75) of the decoupling controller can only be applied if $k \leq p$. However, if no k linearly independent solutions of (6.76) exist or if $k > p$ one can introduce *generalized pole directions* (see [13]) to obtain the decoupling controller in parametric form.

6.2 Decoupling with Coupled Rows

In this section the partial decoupling of the reference transfer behaviour with internal stability is considered. Partial decoupling by static state feedback is necessary in the following situations:

- The system (6.1) and (6.2) is decouplable, but it has interconnecting zeros in the closed right-half complex plane, so that a diagonal decoupling with internal stability is not possible.
- The system (6.1) and (6.2) is decouplable, but the interconnecting zeros to be cancelled for diagonal decoupling are located too far in the left-half complex plane. Due to the limited range of the input amplitude the decoupling control may not be implementable.
- The system (6.1) and (6.2) is not decouplable.

The aim of the partial decoupling is to keep the coupling small between the reference inputs and the controlled outputs. In the following, this is achieved by introducing a minimal number of *coupled rows* in the reference transfer matrix $G_r(s)$. For example, if the lth row of $G_r(s)$ is a coupled row the partially decoupled reference transfer matrix has the form

$$G_r(s) = \text{diag}(g_{ii}(s)) + e_l \left[g_{l1}(s) \ldots g_{l,l-1}(s)\ 0\ g_{l,l+1}(s) \ldots g_{lp}(s)\right]. \quad (6.77)$$

This type of coupling has the advantage that only the output y_c^l depends on several reference inputs. The corresponding decoupling controller is a static state feedback (6.5). For non-minimum phase and non-decouplable systems a frequency-domain parameterization of this controller is presented in the next sections by computing the closed-loop poles and the corresponding pole directions of the partially decoupled reference transfer behaviour.

6.2.1 Decoupling of Non-minimum Phase Systems

Consider a decouplable, non-minimum phase system (6.1) and (6.2) of the order n and assume that it has r_μ non-interconnecting zeros in the μth row and one interconnecting zero η in the closed right-half complex plane. In Section 6.1.2 it is shown that the $r_1 + r_2 + \ldots + r_p$ non-interconnecting zeros need not be compensated for diagonal decoupling. Consequently, right-half-plane non-interconnecting zeros do not lead to an unstable closed-loop system when applying the decoupling approach. However, the interconnecting zero needs to be compensated, so that diagonal decoupling with internal stability is no longer possible by the static state feedback (6.5). If the interconnecting zero η is not compensated, one output y_c^l is coupled to several reference inputs yielding the incompletely decoupled reference transfer matrix

$$G_r(s) = \text{diag}(g_{ii}(s)) + e_l \left[g_{l1}(s) \dots g_{l,l-1}(s)\; 0\; g_{l,l+1}(s) \dots g_{lp}(s)\right], \quad (6.78)$$

(see (6.77)) with $G_r(0) = I$. Different from Theorem 6.4 one has to assign $\delta_\mu + r_\mu$ closed-loop poles in the rows of $G_r(s)$ that do not coincide with the coupled row, *i.e.*, $\mu \neq l$. This is due to the fact that r_μ uncompensated non-interconnecting zeros appear in this row, so that the relative degree δ_μ with respect to this row is not changed by the assignment of $\delta_\mu + r_\mu$ closed-loop poles. Similiarly, $\delta_l + r_l + 1$ closed-loop poles are placed in the coupled row, since additionally to the r_l non-interconnecting zeros the interconnecting zero η appears in the corresponding transfer functions. Unfortunately, the number $l \in \{1, 2, \dots, p\}$ of the coupled row in (6.78) cannot be chosen freely in general. The next theorem provides a condition that shows which of the rows of $G_r(s)$ can be chosen as a coupled row. It also states when decoupling with one coupled row is possible even in case of several interconnecting right-half-plane transmission zeros.

Theorem 6.5 (Decoupling of non-minimum phase systems with one coupled row). *Consider the transfer behaviour*

$$y_c(s) = G_c(s)u(s) \quad (6.79)$$

of a decouplable system of the order n with p inputs u and p outputs y_c. Let the transfer matrix $G_c(s)$ in (6.79) be represented by a right coprime MFD

$$G_c(s) = N_c(s)D^{-1}(s), \quad (6.80)$$

with $D(s)$ column reduced. Assume that the outputs y_c^i have the relative degrees δ_i, $i = 1, 2, \dots, p$, so that the system has the relative degree $\delta = \delta_1 + \dots + \delta_p$. Additionally, it is assumed that no transmission zero of the system is located at $s = 0$, i.e., $\det N_c(0) \neq 0$. *Except for ρ distinct interconnecting right-half-plane transmission zeros η_i, $i = 1, 2, \dots, \rho$, $\text{Re}(\eta_i) \geq 0$, all transmission zeros of the transfer behaviour (6.79) in the closed right-half complex plane are distinct non-interconnecting zeros not coinciding with η_i, where r_μ of these transmission zeros appear in the μth row of the transfer matrix $G_c(s)$ in (6.80), such that $r_1 + \dots + r_p = r$.*

The lth row in the reference transfer matrix $G_r(s)$ can be chosen as the only coupled row iff for all $i = 1, 2, \dots, \rho$, there exists a $1 \times p$ solution vector a_i^T of the equation

$$a_i^T N_c(\eta_i) = 0^T, \quad i = 1, 2, \ldots, \rho, \tag{6.81}$$

whose lth element is non-zero. Let the closed-loop poles $\tilde{s}_\nu$, $\nu = 1, 2, \ldots, n$, and the corresponding pole directions q_ν, $\nu = 1, 2, \ldots, n$, be chosen according to the following rules:

- *Assign $n - \delta - r - \rho$ closed-loop poles $\tilde{s}_\nu$ such that they coincide with the transmission zeros of the transfer matrix $G_c(s)$,* i.e., $\det N_c(\tilde{s}_\nu) = 0$, *having negative real parts, and the corresponding pole directions q_ν satisfy*

$$N_c(\tilde{s}_\nu) q_\nu = 0, \quad \nu = 1, 2, \ldots, n - \delta - r - \rho. \tag{6.82}$$

- *The remaining $\delta + r + \rho$ closed-loop poles $\tilde{s}_\nu$ are assigned to the reference transfer behaviour and satisfy $\det N_c(\tilde{s}_\nu) \neq 0$. The corresponding pole directions q_ν are given by*

$$q_\nu = N_c^{-1}(\tilde{s}_\nu) e_\mu, \quad \nu = n - \delta - r - \rho + 1, \ldots, n, \quad \mu = 1, 2, \ldots, p, \tag{6.83}$$

  where (6.83) is used to assign $\delta_\mu + r_\mu$, $\mu \neq l$, closed-loop poles to the μth row and $\delta_l + r_l + \rho$ closed-loop poles to the lth row of the reference transfer matrix $G_r(s)$.

Then, the frequency-domain parameterization of the state feedback controller (6.5) achieving a partially decoupled reference transfer matrix $G_r(s)$ (see (6.78)) with one coupled row is given by

$$\tilde{D}(s) = D(s) - \left[D(\tilde{s}_1)q_1 \ldots D(\tilde{s}_n)q_n\right] \left[S(\tilde{s}_1)q_1 \ldots S(\tilde{s}_n)q_n\right]^{-1} S(s), \tag{6.84}$$

(see (5.14)) provided that the matrix M is chosen as $M = \tilde{D}(0) N_c^{-1}(0)$.

Proof. The choice of M, of the closed-loop poles and of their pole directions directly follows from the derivation of Theorem 6.4.

In order to prove that the Condition (6.81) is necessary for the lth row to be the only coupled row, pre-multiply $G_r(s)$ in (6.78) by the vectors $a_i^T = [a_{i1} \ a_{i2} \ \ldots \ a_{ip}]$ defined in (6.81). In view of $G_r(s) = N_c(s)\tilde{D}^{-1}(s)M$ (see (2.21)) and (6.81) the relation

$$a_i^T G_r(\eta_i) = a_i^T N_c(\eta_i) \tilde{D}^{-1}(\eta_i) M = 0^T \tag{6.85}$$

holds, which leads to

$$\begin{aligned} a_i^T G_r(\eta_i) &= \left[a_{i1} g_{11}(\eta_i) + a_{il} g_{l1}(\eta_i) \ldots a_{il} g_{ll}(\eta_i) \ldots a_{ip} g_{pp}(\eta_i) + a_{il} g_{lp}(\eta_i)\right] \\ &= 0^T. \end{aligned} \tag{6.86}$$

Now, suppose that the assertion of Theorem 6.5 is not true, *i.e.*, that $a_{il} = 0$. Then,

$$a_i^T G_r(\eta_i) = \left[a_{i1} g_{11}(\eta_i) \ldots 0 \ldots a_{ip} g_{pp}(\eta_i)\right] = 0^T \tag{6.87}$$

follows from (6.86). Since by assumption $a_i^T \neq 0^T$ and $g_{ii}(\eta_i) \neq 0$, $i \neq l$, (the numerators of $g_{ii}(s)$ are constant or only contain non-interconnecting zeros different from η_i) (6.87) cannot vanish, which contradicts (6.86). Consequently, $a_{il} \neq 0$ must be satisfied for the lth row to be a feasible candidate for a coupled row.

In order to show that the matrix $M = \tilde{D}(0)N_c^{-1}(0)$ and the denominator polynomial matrix (6.84) exist such that the reference transfer matrix $G_r(s)$ has the form (6.78) the results in [45] can be used. □

Remark 6.5. The vectors a_i^T in (6.86) indicate whether a non-diagonal element in the row of $G_r(s)$, where a zero is not compensated, is vanishing or not. In particular, a zero element $a_{ij} = 0$, $j \neq l$, in all solutions a_i^T, $a_{il} \neq 0$, $i = 1, 2, \ldots, \rho$, of (6.81) implies that the reference transfer matrix $G_r(s)$ in (6.78) has a zero non-diagonal element $g_{lj}(s) \equiv 0$ in the coupled row. In order to prove this consider the element

$$a_{ij}g_{jj}(\eta_i) + a_{il}g_{lj}(\eta_i) = 0, \quad a_{il} \neq 0, \quad j \neq l, \quad i = 1, 2, \ldots, \rho \tag{6.88}$$

in the row vector (6.86), where $\eta_i \neq 0$ by assumption (see Theorem 6.5). If $a_{ij} = 0$, then

$$g_{lj}(\eta_i) = 0, \quad i = 1, 2, \ldots, \rho \tag{6.89}$$

in view of $a_{il} \neq 0$ and (6.88). On the one hand $G_r(0) = I$ implies

$$g_{lj}(0) = 0, \quad j \neq l. \tag{6.90}$$

On the other hand, r_l non-interconnecting zeros appear in the lth row of $G_r(s)$, and they are different from zero by assumption. Therefore, (6.89) and (6.90) show that the transfer functions $g_{lj}(s)$ must either vanish identically or they must have at least $\rho + r_l + 1$ transmission zeros. In the latter case the assignment of $\delta_l + r_l + \rho$ closed-loop poles in the lth row (see Theorem 6.5) would lead to a relative degree that is less than or equal to $\delta_l - 1$ with respect to y_c^l. However, the relative degree cannot be changed by the static state feedback (6.5), so that $g_{lj}(s) \equiv 0$. Now assume that $a_{ij} \neq 0$. Then, because of (6.88) the relations

$$a_{ij}g_{jj}(\eta_i) = -a_{il}g_{lj}(\eta_i) \neq 0, \quad a_{il} \neq 0, \quad j \neq l, \quad i = 1, 2, \ldots, \rho \tag{6.91}$$

are satisfied, as $g_{jj}(\eta_i) \neq 0$ (uncompensated interconnecting zeros η_i only appear in the lth row of $G_r(s)$). Consequently, since $a_{il} \neq 0$ also $g_{lj}(\eta_i) \neq 0$, which shows that the non-diagonal element $g_{lj}(s)$ does not vanish. These results verify that r_μ uncompensated non-interconnecting zeros $\eta_{niz,i}$, $i = 1, 2, \ldots, r_\mu$, in the μth row do not lead to a coupled row in $G_r(s)$. In order to show this, observe that (6.81) has the solution $a_{niz,i}^T = e_\mu^T$ since the non-interconnecting zeros $\eta_{niz,i}$, $i = 1, 2, \ldots, r_\mu$, are common factors of all numerators in the μth row of $G_c(s)$. In view of the previous discussion this indicates that all non-diagonal elements in the μth row of $G_r(s)$ are zero elements since $a_{niz,ij} = 0$, $j \neq \mu$, $i = 1, 2, \ldots, r_\mu$. If the right-half-plane zero is an interconnecting zero η_i, then for some $j \neq l$ the corresponding solution a_i^T of (6.81) has the property $a_{ij} \neq 0$, $j \neq l$. Consequently, the non-diagonal element $g_{lj}(s)$ is not vanishing and therefore, the lth row is a coupled row.

Remark 6.6. Iff all right-half-plane zeros are non-interconnecting zeros a stable diagonal decoupling of decouplable systems is possible. This condition for diagonal decoupling with internal stability is provided in reference [45] (see also Remark 6.5).

Remark 6.7. Only if a common non-zero element does not exist in the different vectors a_i^T in (6.81), does the decoupling approach require more than one coupled row. However, in most practical cases at least one common non-zero element exists in the different vectors a_i^T (see [45]).

Example 6.2. Diagonal decoupling of a non-minimum phase system with internal stability

Consider a system of the order $n = 3$ with the transfer matrix

$$G_c(s) = N_c(s)D^{-1}(s) = \begin{bmatrix} -4s+12 & 0 \\ -4s-12 & -34 \end{bmatrix} \begin{bmatrix} -2s^2-8s-6 & 17s+17 \\ -0.5s^2-2s-1.5 & -4.25s-12.75 \end{bmatrix}^{-1}. \tag{6.92}$$

In order to check whether this system is decouplable the relative degrees $\delta_1 = \delta_2 = 1$ are computed using (6.13). The determinant

$$\det N_c(s) = 136(s-3) \tag{6.93}$$

shows that the system is decouplable since the Condition (6.19) is satisfied. However, the system is non-minimum phase as it has a transmission zero $\eta = 3$ (see (6.93)). Inspection of the numerator matrix $N_c(s)$ in (6.92) reveals that (6.81) has the solution $a_{niz}^T = e_1^T$, *i.e.*, the transmission zero at $\eta = 3$ is a non-interconnecting zero, which need not be compensated for diagonal decoupling. Therefore, $\delta_1 + r_1 = 2$ closed-loop poles $\tilde{s}_\nu$, $\nu = 1, 2$, can be assigned to the first row of the reference transfer matrix $G_r(s)$. Choosing $\tilde{s}_1 = -1$ and $\tilde{s}_2 = -3$ the corresponding pole directions can be computed from

$$q_\nu = N_c^{-1}(\tilde{s}_\nu)e_1, \quad \nu = 1, 2 \tag{6.94}$$

(see (6.74)). The results are $q_1^T = [0.0625 \quad -0.0147]$ and $q_2^T = [0.0417 \quad 0]$. The pole direction

$$q_3 = N_c^{-1}(\tilde{s}_3)e_2 = \begin{bmatrix} 0 & -0.0295 \end{bmatrix}^T \tag{6.95}$$

assures that the third closed-loop pole $\tilde{s}_3 = -2$ is assigned to the second row. These results completely parameterize the polynomial matrix $\tilde{D}(s)$, which is obtained from

$$\begin{aligned}\tilde{D}(s) &= D(s) - \left[D(\tilde{s}_1)q_1\ D(\tilde{s}_2)q_2\ D(\tilde{s}_3)q_3\right]\left[S(\tilde{s}_1)q_1\ S(\tilde{s}_2)q_2\ S(\tilde{s}_3)q_3\right]^{-1}S(s)\\ &= \begin{bmatrix} -2s^2-6s & 17s+34\\ -0.5s^2-2.5s-3 & -4.2s-8.5\end{bmatrix},\end{aligned} \tag{6.96}$$

where

$$S(s) = \begin{bmatrix} s & 0\\ 1 & 0\\ 0 & 1\end{bmatrix}. \tag{6.97}$$

Vanishing steady-state errors are assured by the matrix

$$M = \tilde{D}(0)N_c^{-1}(0) = \begin{bmatrix} -1 & -1\\ 0 & 2.5\end{bmatrix} \tag{6.98}$$

(see (6.65)). Since the non-interconnecting zero $\eta = 3$ only affects the first output of the closed-loop system it appears in the first row of the reference transfer matrix $G_r(s)$ (see (6.52)) that takes the form

$$G_r(s) = \begin{bmatrix} \frac{-(s-3)}{(s+1)(s+3)} & 0\\ 0 & \frac{2}{s+2}\end{bmatrix}. \tag{6.99}$$

6.2.2 Decoupling of Non-decouplable Systems

If the condition of Theorem 6.2 is not satisfied, *i.e.*, if less than $n-\delta$ transmission zeros exist (see Remark 6.1), the system is not diagonally decouplable by the static state feedback (6.5). However, a decoupling with coupled rows can be obtained yielding the reference transfer matrix $G_r(s)$ in (6.78). The next theorem presents the decoupling approach with coupled rows for such non-decouplable systems. In order to keep the derivations simple only distinct non-interconnecting zeros in the right-half complex plane are considered. This theorem also provides a condition for the choice of the coupled row when decoupling with only one coupled row is feasible.

Theorem 6.6 (Decoupling of non-decouplable systems with one coupled row). *Consider the transfer behaviour*

$$y_c(s) = G_c(s)u(s) \tag{6.100}$$

of a non-decouplable system of the order n with p inputs u and p outputs y_c. Let the transfer matrix $G_c(s)$ in (6.100) be represented by a right coprime MFD

$$G_c(s) = N_c(s)D^{-1}(s), \tag{6.101}$$

with $D(s)$ column reduced. Assume that the outputs y_c^i have the relative degrees δ_i, $i = 1, 2, \ldots, p$, so that the system has the relative degree $\delta = \delta_1 + \ldots + \delta_p$. Additionally, it is assumed that no transmission zero of the system is located at $s = 0$, i.e., $\det N_c(0) \neq 0$. For diagonal decoupling of the system there are

*σ transmission zeros missing (*i.e.,* it has $n-\delta-\sigma$ transmission zeros; see Theorem 6.2) and assume that all transmission zeros of the transfer behaviour (6.100) in the closed right-half complex plane are distinct non-interconnecting zeros, where r_μ of these transmission zeros appear in the μ*th *row of the transfer matrix $G_c(s)$ in (6.101), such that $r_1+\ldots+r_p=r$.*

*The l*th *row in the reference transfer matrix $G_r(s)$ can be chosen as the only coupled row iff there exists a $1\times p$ solution vector $\bar{a}^T$ of*

$$\bar{a}^T \Gamma_{\delta_c[D(s)]}[\mathrm{diag}(s^{\delta_i})N_c(s)] = 0^T, \tag{6.102}$$

*whose l*th *element is non-zero and*

$$\mathrm{rank}\,\Gamma_{\delta_c[D(s)]}[\mathrm{diag}(s^{\delta_i})N_c(s)] = p-1 \tag{6.103}$$

holds. Let the closed-loop poles $\tilde{s}_\nu$, $\nu=1,2,\ldots,n$, and the corresponding pole directions q_ν, $\nu=1,2,\ldots,n$, be chosen according to the following rules:

- *Assign $n-\delta-r-\sigma$ closed-loop poles $\tilde{s}_\nu$ such that they coincide with the transmission zeros of the transfer matrix $G_c(s)$,* i.e., $\det N_c(\tilde{s}_\nu)=0$, *having negative real parts, and the corresponding pole directions q_ν satisfy*

$$N_c(\tilde{s}_\nu)q_\nu = 0, \quad \nu=1,2,\ldots,n-\delta-r-\sigma. \tag{6.104}$$

- *The remaining $\delta+r+\sigma$ closed-loop poles $\tilde{s}_\nu$ are assigned to the reference transfer behaviour and satisfy* $\det N_c(\tilde{s}_\nu)\neq 0$. *The corresponding pole directions q_ν are given by*

$$q_\nu = N_c^{-1}(\tilde{s}_\nu)e_\mu, \quad \nu=n-\delta-r-\sigma+1,\ldots,n, \quad \mu=1,2,\ldots,p, \tag{6.105}$$

  *where (6.105) is used to assign $\delta_\mu+r_\mu$, $\mu\neq l$, closed-loop poles to the μ*th *row and $\delta_l+r_l+\sigma$ closed-loop poles to the l*th *row of the reference transfer matrix $G_r(s)$.*

Then, the frequency-domain parameterization of the state feedback controller (6.5) achieving a partially decoupled reference transfer matrix $G_r(s)$ (see (6.78)) with one coupled row is given by

$$\tilde{D}(s) = D(s) - \left[D(\tilde{s}_1)q_1 \ldots D(\tilde{s}_n)q_n\right]\left[S(\tilde{s}_1)q_1 \ldots S(\tilde{s}_n)q_n\right]^{-1} S(s) \tag{6.106}$$

(see (5.14)), provided that the matrix M is chosen as $M=\tilde{D}(0)N_c^{-1}(0)$.

Proof. The choice of M, of the closed-loop poles and of their pole directions directly follows from the derivation of Theorem 6.4. Using the results of [45] it is easy to verify that with $M=\tilde{D}(0)N_c^{-1}(0)$ all non-diagonal elements of $G_r(s)$ vanish except for the lth row. In order to prove that the Condition (6.102) is necessary for the lth row to be a coupled row, consider

$$\lim_{s\to\infty}\left(\mathrm{diag}(s^{\delta_i})G_r(s)\right) = \lim_{s\to\infty}\left(\mathrm{diag}(s^{\delta_i})N_c(s)\tilde{D}^{-1}(s)M\right). \tag{6.107}$$

By using (1.22), (2.28) and (2.29) this can be written as

$$\begin{aligned}&\lim_{s\to\infty}\left(\mathrm{diag}(s^{\delta_i})G_r(s)\right)\\&=\lim_{s\to\infty}\left(\mathrm{diag}(s^{\delta_i})N_c(s)\tilde{D}^{-1}(s)\right)M\\&=\Gamma_{\delta_c[D(s)]}[\mathrm{diag}(s^{\delta_i})N_c(s)]\Gamma_c^{-1}[D(s)]M.\end{aligned}\tag{6.108}$$

Equation (6.108) shows that (6.102) can equivalently be represented by

$$\bar{a}^T\lim_{s\to\infty}\left(\mathrm{diag}(s^{\delta_i})G_r(s)\right)=0^T\tag{6.109}$$

when postmultipyling (6.102) with the non-singular matrix $\Gamma_c^{-1}[D(s)]M$. With $\bar{a}^T=\left[\bar{a}_1\ldots\bar{a}_p\right]$ and (6.78), (6.109) can be written as

$$\begin{aligned}&\bar{a}^T\lim_{s\to\infty}\left(\mathrm{diag}(s^{\delta_i})G_r(s)\right)=\\&\lim_{s\to\infty}\left[\bar{a}_1s^{\delta_1}g_{11}(s)+\bar{a}_ls^{\delta_l}g_{l1}(s)\ldots\bar{a}_ls^{\delta_l}g_{ll}(s)\ldots\bar{a}_ps^{\delta_p}g_{pp}(s)+\bar{a}_ls^{\delta_l}g_{lp}(s)\right]\\&=0^T.\end{aligned}\tag{6.110}$$

In order to prove that the Condition (6.102) is necessary for the lth row to be a coupled row, suppose that the assertion of Theorem 6.6 is not true, *i.e.*, $\bar{a}_l=0$ holds. Then, (6.110) takes the form

$$\begin{aligned}&\lim_{s\to\infty}\left[\bar{a}_1s^{\delta_1}g_{11}(s)\ldots0\ldots\bar{a}_ps^{\delta_p}g_{pp}(s)\right]\\&=\left[\bar{a}_1\lim_{s\to\infty}s^{\delta_1}g_{11}(s)\ldots0\ldots\bar{a}_p\lim_{s\to\infty}s^{\delta_p}g_{pp}(s)\right]=0^T.\end{aligned}\tag{6.111}$$

The constant $p\times p$ decouplability matrix D^* (see [17]) of the system can be represented by

$$D^*=\lim_{s\to\infty}\left(\mathrm{diag}(s^{\delta_i})G_r(s)\right)=\Gamma_{\delta_c[D(s)]}[\mathrm{diag}(s^{\delta_i})N_c(s)]\Gamma_c^{-1}[D(s)]M\tag{6.112}$$

(see (6.108)). Since the non-singularity of the matrix D^* is a necessary and sufficient condition for diagonal decoupling (see [17]), there exists a solution $\bar{a}^T\neq0^T$ of (6.109) for non-decouplable systems. The relative degrees of the diagonal elements $g_{ii}(s)$, $i\neq l$, are δ_i, which implies

$$\lim_{s\to\infty}\left(s^{\delta_i}g_{ii}(s)\right)\neq0,\quad i=1,2,\ldots,l-1,l+1,\ldots,p,\tag{6.113}$$

so that, given $\bar{a}^T\neq0^T$, the right-hand side of (6.111) cannot vanish. This contradicts (6.110), so that $\bar{a}_l\neq0$ must be satisfied if the lth row is a coupled row. Sufficiency of the proof, namely that the polynomial matrix $\tilde{D}(s)$ in (6.106) exists if the conditions of the Theorem 6.6 are satisfied, can be shown by using the results in [45]. □

Remark 6.8. If $\mathrm{rank}\,\Gamma_{\delta_c[D(s)]}[\mathrm{diag}(s^{\delta_i})N_c(s)]<p-1$ the decoupling approach for non-decouplable systems requires more than one coupled row. However, in most practical cases the rank condition (6.103) is satisfied, so that only one coupled row is needed (see [45]).

6.2.3 Decoupling of Non-minimum Phase and Non-decouplable Systems

If the system (6.1) and (6.2) is not decouplable and has interconnecting zeros in the right-half complex plane one can achieve a decoupling with coupled rows by using the results of the preceding sections. This is due to the fact that the missing transmission zeros for satisfying the decoupling criterion (6.18) can be regarded as transmission zeros being located at infinity (see [45]). In order to motivate this consider a decouplable system with

$$\det N_c(s) = (1 - \frac{s}{\eta_1})(1 - \frac{s}{\eta_2}) \cdot \ldots \cdot (1 - \frac{s}{\eta_{n-\delta}}), \tag{6.114}$$

i.e., it has the transmission zeros $\eta_1, \eta_2, \ldots, \eta_{n-\delta}$. Then, obviously

$$\lim_{\eta_{n-\delta} \to \infty} \det N_c(s) = (1 - \frac{s}{\eta_1})(1 - \frac{s}{\eta_2}) \cdot \ldots \cdot (1 - \frac{s}{\eta_{n-\delta-1}}). \tag{6.115}$$

In this case, the system is no longer decouplable since it has only $n - \delta - 1$ transmission zeros, *i.e.*, it violates the decoupling criterion (6.18). This indicates that non-decouplable systems can be regarded as the limiting case of decouplable systems where transmission zeros are shifted to infinity. Obviously such transmission zeros cannot be compensated by closed-loop poles. By means of the results in [45] it can be verified that the framework derived in Section 6.2.1 can also be applied to non-decouplable systems, where now the transmission zeros that are not compensated by the closed-loop poles are the finite transmission zeros in the open right-half complex plane plus the zeros at infinity. This results in a reference transfer matrix $G_r(s)$ as defined in (6.78). Theorem 6.5 can be used to compute the closed-loop poles and their corresponding pole directions. Also, the choice of the coupled row can be generalized in a common framework by considering

$$a_i^T \lim_{s \to \eta_i} (\operatorname{diag}(s^{\delta_i}) G_c(s)) = 0^T. \tag{6.116}$$

If η_i is a finite transmission zero this yields

$$a_i^T \operatorname{diag}(\eta_i^{\delta_i}) G_c(\eta_i) = 0^T. \tag{6.117}$$

With $G_c(s) = N_c(s)D^{-1}(s)$ (see (6.4)) this is equivalent to (6.81) provided that $\det D(\eta_i) \neq 0$, *i.e.*, no transmission zero coincides with an open-loop pole. If η_i is an infinite transmission zero, *i.e.*, $\eta_i \to \infty$ then (6.116) becomes

$$a_i^T \lim_{s \to \infty} (\operatorname{diag}(s^{\delta_i}) G_c(s)) = a_i^T \Gamma_{\delta_c[D(s)]} [\operatorname{diag}(s^{\delta_i}) N_c(s)] \Gamma_c^{-1}[D(s)] = 0^T \tag{6.118}$$

(see (1.22)), which is equivalent to (6.102) since $\det \Gamma_c^{-1}[D(s)] \neq 0$. In addition to (6.118) it is assumed that the infinite zero $\eta_i \to \infty$ is a simple zero which means that (6.103) has to hold. This result, which has been derived in [45] in the time domain, is summarized in the next theorem.

Theorem 6.7 (Generalized condition for the coupled row). *Consider the transfer behaviour*

$$y_c(s) = G_c(s)u(s) \tag{6.119}$$

of a non-minimum phase and non-decouplable system of the order n with p inputs u and p outputs y_c. Let the transfer matrix $G_c(s)$ in (6.119) be represented by a right coprime MFD

$$G_c(s) = N_c(s)D^{-1}(s), \tag{6.120}$$

*with $D(s)$ column reduced. The transfer matrix (6.120) has ρ_0 finite right-half-plane interconnecting zeros η_i and ρ_∞ transmission zeros missing for decoupling (*i.e., *ρ_∞ zeros are infinite transmission zeros since $n-\delta = \rho_0+\rho_\infty$). All ρ_0 finite right-half-plane interconnecting zeros are different from the poles of (6.120). The* lth *row in the reference transfer matrix $G_r(s)$ can be chosen as the only coupled row iff for all $i = 1, 2, \ldots, \rho_0 + \rho_\infty$ there exists a $1 \times p$ solution vector a_i^T of the equation*

$$a_i^T \lim_{s\to\eta_i} \left(\operatorname{diag}(s^{\delta_i})N_c(s)D^{-1}(s)\right) = 0^T, \quad i = 1, 2, \ldots, \rho_0 + \rho_\infty, \tag{6.121}$$

whose lth *element is non-zero. In case of an infinite transmission zero (6.121) must be evaluated for $\eta_i \to \infty$, which yields*

$$\bar{a}^T \Gamma_{\delta_c[D(s)]}[\operatorname{diag}(s^{\delta_i})N_c(s)] = 0^T, \tag{6.122}$$

where the rank condition

$$\operatorname{rank} \Gamma_{\delta_c[D(s)]}[\operatorname{diag}(s^{\delta_i})N_c(s)] = p - 1 \tag{6.123}$$

must additionally be satisfied.

Example 6.3. Decoupling of a non-minimum phase and non-decouplable system with one coupled row

Consider a system of the order $n = 6$ with the transfer matrix

$$\begin{aligned} G_c(s) &= N_c(s)D^{-1}(s) \\ &= \begin{bmatrix} 1 & 1 & 0 \\ 0 & 1 & 0 \\ 0 & 1 & -0.2s+0.2 \end{bmatrix} \\ &\quad \cdot \begin{bmatrix} s^3+6s^2+11s+6 & 0 & 0 \\ -s-1 & s+1 & -s-2 \\ 0 & 0 & s^2+6s+8 \end{bmatrix}^{-1}. \end{aligned} \tag{6.124}$$

In view of

$$\det N_c(s) = -0.2(s-1) \tag{6.125}$$

the system has a right-half-plane transmission zero at $\eta_1 = 1$ that is an interconnecting zero (*i.e.*, it is not a common element in all elements of a row of $N_c(s)$), so that $\rho_0 = 1$ (see Theorem 6.7). Using (6.13) one obtains the relative degrees $\delta_1 = \delta_2 = \delta_3 = 1$ from the numerator and denominator matrices in (6.124), so that the system has $\rho_\infty = n - (\delta_1 + \delta_2 + \delta_3) - \rho_0 = 2$ transmission zeros at infinity and hence it is not decouplable by constant state feedback control. However, a partially decoupled reference transfer behaviour in the form (6.78) can be obtained with static state feedback. First, the number of the coupled row has to be computed. Applying the criterion (6.121) and (6.122) for the right-half-plane transmission zero and for the transmission zeros at infinity yields

$$a_1^T = \begin{bmatrix} 0 & -1 & 1 \end{bmatrix}, \quad a_2^T = a_3^T = \begin{bmatrix} 1 & -1 & 0 \end{bmatrix}. \tag{6.126}$$

Thus, only the second row (*i.e.*, $l = 2$) of the reference transfer matrix $G_r(s)$ can be chosen as a coupled row. Since $\delta_1 = 1$ one closed-loop pole $\tilde{s}_1 = -1$ can be assigned to the first row. Given the relative degree $\delta_2 = 1$ of the second output a number of $\delta_2 + \rho_0 + \rho_\infty = 4$ closed-loop poles $\tilde{s}_2 = -2$, $\tilde{s}_3 = -3$, $\tilde{s}_4 = -4$, $\tilde{s}_5 = -5$ can be assigned to the second row. Finally, with $\delta_3 = 1$ one closed-loop pole $\tilde{s}_6 = -6$ is assigned to the third row. The corresponding pole directions result by solving the equations in (6.83). Then, the polynomial matrix

$$\tilde{D}(s) = \begin{bmatrix} s^3 + 15s^2 + 86s + 72 & -\frac{1632}{14} & -\frac{144}{14}s - \frac{864}{14} \\ s+1 & s+1 & 0 \\ 5s+5 & -25 & s^2 + 5s - 6 \end{bmatrix} \tag{6.127}$$

can be computed from (6.84) and the matrix M is given by

$$M = \tilde{D}(0)N_c^{-1}(0) = \begin{bmatrix} 72 & 120 & -\frac{432}{14} \\ 1 & 0 & 0 \\ 5 & 0 & -30 \end{bmatrix}. \tag{6.128}$$

The resulting reference transfer matrix has one coupled row and it takes the form

$$G_r(s) = \begin{bmatrix} \frac{1}{s+1} & 0 & 0 \\ \frac{s(s^2+13s-14)}{(s+2)(s+3)(s+4)(s+5)} & \frac{-120(s-1)}{(s+2)(s+3)(s+4)(s+5)} & \frac{2160s}{7(s+2)(s+3)(s+4)(s+5)} \\ 0 & 0 & \frac{6}{s+6} \end{bmatrix}. \tag{6.129}$$

Whereas the reference transients of y_1 and y_3 are not affected by the other two reference inputs there is a pronounced influence of all three reference inputs on the transients of y_2. This becomes obvious by inspection of Figure 6.1 which shows the reaction of the three outputs to reference inputs, where only one $r_i(t)$, $i = 1, 2, 3$ is applied as a unit step, whereas the remaining two $r_j(t)$, $j = 1, 2, 3$, $j \neq i$ are kept at zero. In [45] it is shown by time-domain arguments that additional degrees of freedom exist in the design of the partial decoupling controller which can be used to reduce the existing couplings. On the basis of the decoupling approach presented here these results can also be obtained in the frequency domain.

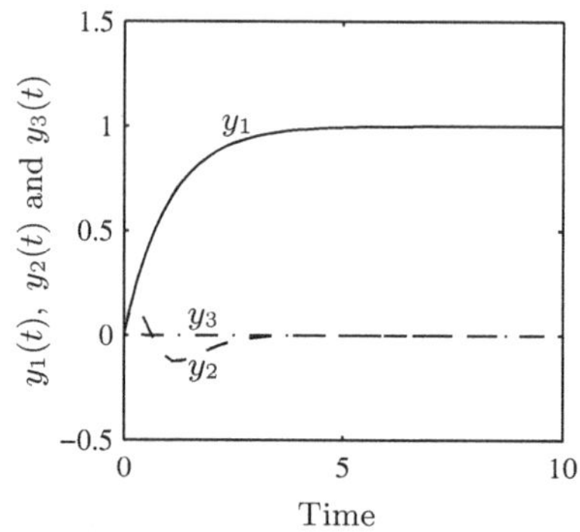

(a) Response to a reference input with $r_1(t) = 1(t)$ and $r_2 = r_3 = 0$

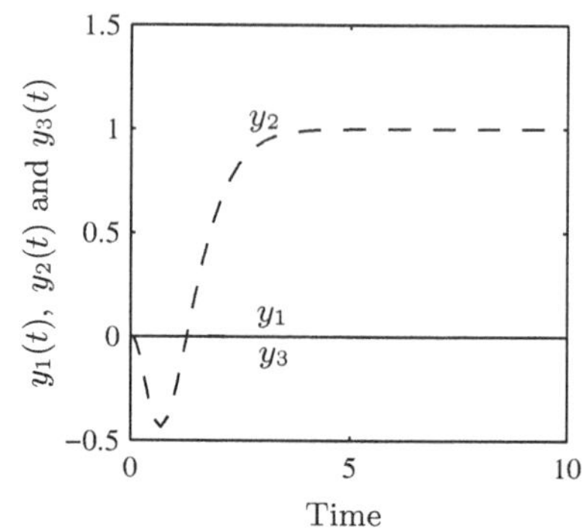

(b) Response to a reference input with $r_2(t) = 1(t)$ and $r_1 = r_3 = 0$

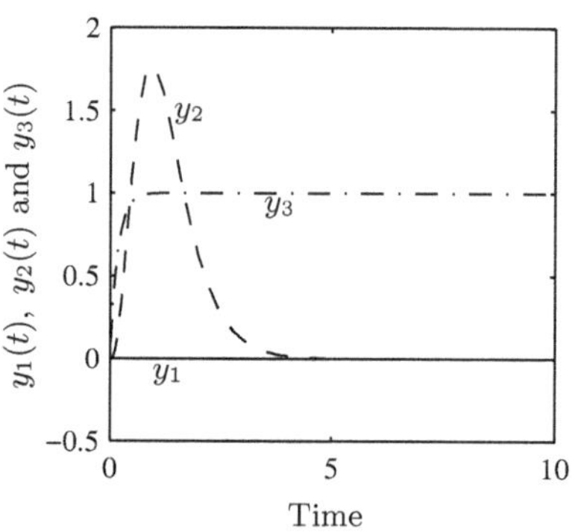

(c) Response to a reference input with $r_3(t) = 1(t)$ and $r_1 = r_2 = 0$

Figure 6.1. Transients of the partially decoupled reference transfer behaviour

7

Disturbance Rejection Using the Internal Model Principle

Persistently acting disturbances are not compensated asymptotically by a constant state feedback controller. Therefore, the design of disturbance rejecting controllers requires additional measures. If the disturbances can be modelled by a suitable signal process (*i.e.*, the disturbances can be represented as the solutions of linear differential equations with constant coefficients) two approaches exist for their rejection.

The first one was suggested by Johnson in [35]. By a feedforward of the states of the disturbance signal process its eigenvalues are rendered unobservable in the controlled outputs. This achieves an asymptotic rejection of the disturbances. As the states of the signal process are not measurable in general, an observer is required for their reconstruction. However, due to the feedforward control this approach is neither robust to changing input locations of the disturbances nor to uncertain parameters of the system.

The second approach to disturbance rejection was suggested by Davison (see [10, 11]). It constitutes a generalization of the classical controller with integral action, which only rejects constant disturbances, to general forms of disturbance signals. A model of the disturbances is added to the plant and this disturbance model is driven by the difference between the tracking signal and the corresponding output of the system. If this augmented plant is stabilized by a feedback controller the tracking error does not contain the modelled signal forms in the steady state. If it did, resonance would cause increasing states of the disturbance model, thus contradicting the stability of the closed-loop system. Therefore, independent of where they enter the system, the disturbances are rejected asymptotically. Asymptotic disturbance rejection is also assured for modelling errors that do not endanger closed-loop stability. Consequently, Davison's approach assures robust disturbance rejection. It requires, however, that the controlled variables are part of the measurements, which is not necessary when using Johnson's approach. Since the disturbance model is a part of the closed-loop system Davison's approach is called the *internal model principle*. Davison's approach not only guarantees the asymptotic rejection of all modelled disturbance signals, but also the asymptotic tracking of such reference signals. This can be a drawback, because asymptotic tracking and disturbance rejection are rarely required for the same type of signals and it may entail undesired side effects.

In Section 7.1 an extension of Davison's approach is presented in the time domain. To circumvent the possibly existing negative side effects of a joint disturbance and reference model, the signal model is not driven by the tracking error but by the controlled outputs and a suitable feedforward of the reference inputs. This additional degree of freedom is used to formulate the driven signal model as an observer. This has the effect that all modelled disturbance signals are asymptotically rejected in the controlled outputs, but the signal model does not influence the tracking behaviour. After the initial observer errors have vanished, the tracking behaviour is the same as if constant state feedback had been applied. This also allows a simple prevention of controller windup due to input saturation (see Section 4.5). In Section 7.2 the frequency-domain parameterization of the stabilizing state feedback of the augmented system is presented. Section 7.3 contains the frequency-domain parameterization of the observer for the non-augmented system, which provides sufficient information as the states of the signal model are directly measurable. The frequency-domain design of observer-based compensators with signal models for disturbance rejection can be found in Section 7.4.

7.1 Time-domain Approach to Disturbance Rejection

Considered are systems with the state equations

$$\dot{x}(t) = Ax(t) + Bu(t) + B_d d(t), \tag{7.1}$$

$$y(t) = Cx(t) + D_d d(t), \tag{7.2}$$

where $x \in \mathbb{R}^n$ is the state, $u \in \mathbb{R}^p$ is the input, $y \in \mathbb{R}^m$ with $m \geq p$ is the measurement, and $d \in \mathbb{R}^\rho$ is an unmeasurable disturbance. It is assumed that the pair (A, B) is controllable and the pair (C, A) is observable. In view of designing reduced-order observers of the order $n_O = n - \kappa$, $0 \leq \kappa \leq m$, the output vector y of these systems is arranged according to

$$y(t) = \begin{bmatrix} y_1(t) \\ y_2(t) \end{bmatrix} = \begin{bmatrix} C_1 \\ C_2 \end{bmatrix} x(t) + \begin{bmatrix} D_{d1} \\ D_{d2} \end{bmatrix} d(t). \tag{7.3}$$

Here, $y_2 \in \mathbb{R}^\kappa$ with $0 \leq \kappa \leq m$ contains the measurements directly used in the construction of the estimate $\hat{x}$ and $y_1 \in \mathbb{R}^{m-\kappa}$ contains the remaining $m - \kappa$ measurements (see Section 3.1). As in Chapter 2 the controlled output $y_c \in \mathbb{R}^p$ is assumed to be measurable. This is a necessity in Davison's approach because the model of the assumed signal process is driven by y_c. Therefore, a $p \times m$ selection matrix Ξ (see (2.3)) exists, so that

$$y_c(t) = \Xi y(t) = \Xi Cx(t) + \Xi D_d d(t) = C_c x(t) + D_{cd} d(t). \tag{7.4}$$

Remark 7.1. If the controlled output y_c is not measurable, the non-robust approach of Johnson in [35] must be applied for the asymptotic rejection of a persistently acting disturbance d.

It is assumed that the disturbance d in (7.1) and (7.2) can be modelled in a signal process

$$\dot{v}^*(t) = S^* v^*(t), \tag{7.5}$$

$$d(t) = H v^*(t), \tag{7.6}$$

with $v^* \in \mathbb{R}^q$, $v^*(0) = v_0^*$ unknown, and (H, S^*) observable. Using this method constant disturbances $d(t) = v_0^*$ can be modelled by

$$\dot{v}^*(t) = 0, \tag{7.7}$$

$$d(t) = v^*(t). \tag{7.8}$$

When using Davison's approach this model is part of the controller and it leads to the well-known integral action. Sinusoidal signals $d(t) = d_0 \sin(\omega_0 t + \varphi_0)$ can, *e.g.*, be modelled by the process

$$\dot{v}^*(t) = \begin{bmatrix} 0 & 1 \\ -\omega_0^2 & 0 \end{bmatrix} v^*(t), \tag{7.9}$$

$$d(t) = \begin{bmatrix} 1 & 0 \end{bmatrix} v^*(t). \tag{7.10}$$

These examples show that the eigenvalues of S^* are usually located at the imaginary axis of the complex plane and that they characterize the signal forms generated by this process. Therefore, the type of signal is defined by the characteristic polynomial

$$\det(sI - S^*) = s^q + \psi_{q-1} s^{q-1} + \cdots + \psi_1 s + \psi_0 \tag{7.11}$$

of the disturbance process (7.5). The actual amplitudes and phases of the disturbance signals depend on the unknown initial condition v_0^*, which must be reconstructed. Table 7.1 contains an overview of the most commonly occurring types of signals and their characteristic polynomials.

Table 7.1. Typical types of signals and corresponding characteristic polynomials

Signal form	$\det(sI - S^*)$
step	s
ramp	s^2
parabola	s^3
sine	$s^2 + \omega_0^2$
sine with non-zero mean	$s(s^2 + \omega_0^2)$
exponential increase	$s - \alpha, \alpha > 0$

Remark 7.2. It is assumed that the modelled disturbance is observable in the output y_c because an unobservable disturbance would be asymptotically compensated in y_c by constant state feedback control.

Similar to the classical controller with integral action, the original approach of Davison adds signal models

$$\dot{\bar{v}}_i(t) = S^*\bar{v}_i(t) + b_\varepsilon(y_c^i(t) - r_i(t)), \quad i = 1, 2, \ldots, p, \tag{7.12}$$

with $\bar{v}_i \in \mathbb{R}^q$ to the compensator and these models are driven by the ith element $y_c^i - r_i$ of the tracking error $y_c - r$.

Remark 7.3. If one is not interested in a suppression of all disturbances in all controlled outputs y_c one can use different disturbance models in different channels. For simplicity it is assumed that all modelled disturbances are compensated in all controlled outputs, so that the same disturbance model is present in each channel.

In (7.12) the matrix S^* and the vector b_ε can be chosen as

$$S^* = \begin{bmatrix} 0 & 1 & \cdots & 0 \\ \vdots & \vdots & \ddots & \vdots \\ 0 & 0 & \cdots & 1 \\ -\psi_0 & -\psi_1 & \cdots & -\psi_{q-1} \end{bmatrix} \quad \text{and} \quad b_\varepsilon = \begin{bmatrix} 0 \\ \vdots \\ 0 \\ 1 \end{bmatrix}, \tag{7.13}$$

where S^* satisfies (7.11) and the controller form of (S^*, b_ε) assures that the signal models are controllable by the tracking errors. These subsystems can be aggregated to obtain the composite model

$$\dot{v}(t) = Sv(t) + B_\varepsilon(y_c(t) - r(t)), \tag{7.14}$$

where the vector $v \in \mathbb{R}^{pq}$ and the matrices S and B_ε have the forms

$$v(t) = \begin{bmatrix} \bar{v}_1(t) \\ \bar{v}_2(t) \\ \vdots \\ \bar{v}_p(t) \end{bmatrix}, \quad S = \begin{bmatrix} S^* & & \\ & S^* & \\ & & \ddots & \\ & & & S^* \end{bmatrix}, \quad B_\varepsilon = \begin{bmatrix} b_\varepsilon & & \\ & b_\varepsilon & \\ & & \ddots & \\ & & & b_\varepsilon \end{bmatrix}. \tag{7.15}$$

Adding this composite model to the system and applying a stabilizing control to the resulting augmented system assures an asymptotic rejection of all modelled disturbance signals in y_c. This becomes plausible by the following argument. If the closed-loop system is asymptotically stable the tracking error $y_c - r$ in (7.14) cannot contain the modelled disturbance signals in the steady state, because an excitation of the system (7.14) by the modelled signals would lead to an unbounded state v. Consequently, the steady-state error $y_c(\infty) - r(\infty)$ vanishes for all modelled signals. This, however, not only assures an asymptotic rejection of all modelled disturbances but also an asymptotic tracking of y_c for the same class of reference signals. Generally, the applied reference signals do not have the same signal forms as the external disturbances (one exception may be constant signals) and it turns out that a joint signal model for disturbance and reference signals may have an unfavourable influence on the tracking transients (see also the discussions in [31]).

This drawback of Davison's approach can be removed when treating the internal signal model as an observer, which does not influence the tracking behaviour. To this end, consider the modified signal model

$$\dot{v}(t) = Sv(t) + B_\varepsilon y_c(t) - B_\sigma Mr(t), \tag{7.16}$$

where the constant $pq \times p$ matrix B_σ will be used to assign the observer property to this signal model. Obviously, the disturbance behaviour of the closed-loop system is the same whether one uses the driven signal model (7.14) or the modified form (7.16), because for $r \equiv 0$ both models coincide.

In a first step, a stabilizing state feedback is considered for the system (7.1) and (7.2) augmented by the signal model (7.16). Adding the driven signal model (7.16) to the system (7.1) and (7.2) and neglecting the disturbance input d, one obtains the state equations

$$\dot{x}_{aug}(t) = A_{aug}x_{aug}(t) + B_{aug}u(t) + B_{r,aug}r(t), \tag{7.17}$$

$$y_{aug}(t) = C_{aug}x_{aug}(t) \tag{7.18}$$

of the augmented system with

$$x_{aug}(t) = \begin{bmatrix} x(t) \\ v(t) \end{bmatrix}, \ A_{aug} = \begin{bmatrix} A & 0 \\ B_\varepsilon C_c & S \end{bmatrix}, \ B_{aug} = \begin{bmatrix} B \\ 0 \end{bmatrix}, \ B_{r,aug} = \begin{bmatrix} 0 \\ -B_\sigma M \end{bmatrix},$$

$$y_{aug}(t) = \begin{bmatrix} y(t) \\ v(t) \end{bmatrix}, \ C_{aug} = \begin{bmatrix} C & 0 \\ 0 & I \end{bmatrix}. \tag{7.19}$$

The controlled output y_c in (7.4) can be represented in terms of the augmented state x_{aug} as

$$y_c(t) = [C_c \quad 0]\, x_{aug}(t). \tag{7.20}$$

To obtain a stabilizing control for the augmented system (7.17) consider the state feedback

$$u(t) = -\tilde{u}_{aug}(t) + Mr(t) = -K_{aug}x_{aug}(t) + Mr(t), \tag{7.21}$$

where

$$K_{aug} = [K_x \quad K_v]. \tag{7.22}$$

To assign an observer property to the driven signal model solve (7.21) for

$$Mr(t) = u(t) + K_x x(t) + K_v v(t), \tag{7.23}$$

and insert the result in (7.16), which yields

$$\dot{v}(t) = (S - B_\sigma K_v)v(t) + B_\varepsilon y_c(t) - B_\sigma(K_x x(t) + u(t)). \tag{7.24}$$

In what follows the matrix B_σ is determined such that (7.24) constitutes an observer for the quantity Σx. In Σx the matrix Σ has the dimension $pq \times n$ and x is the state of the controlled system

$$\dot{x}(t) = (A - BK_x)x(t) - BK_v v(t) + BMr(t), \tag{7.25}$$

which results from inserting (7.21) in (7.1). Then, the dynamics of $\Sigma x - v$ have the form

$$\begin{aligned} \Sigma\dot{x}(t) - \dot{v}(t) &= \Sigma(A - BK_x)x(t) - \Sigma BK_v v(t) + \Sigma BMr(t) \qquad (7.26)\\ &\quad -(S - B_\sigma K_v)v(t) - B_\varepsilon C_c x(t) + B_\sigma(K_x x(t) + u(t)) \\ &= (S - B_\sigma K_v)(\Sigma x(t) - v(t)) \qquad (7.27)\\ &\quad +(\Sigma(A - BK_x) - (S - B_\sigma K_v)\Sigma - B_\varepsilon C_c)x(t) \\ &\quad +(\Sigma B + B_\sigma)(K_x x(t) + u(t)), \end{aligned}$$

where the Equations (7.24), (7.25) and (7.23) have been used. One obtains the homogeneous error dynamics

$$\Sigma\dot{x}(t) - \dot{v}(t) = (S - B_\sigma K_v)(\Sigma x(t) - v(t)), \tag{7.28}$$

if the two equations

$$\Sigma(A - BK_x) - (S - B_\sigma K_v)\Sigma = B_\varepsilon C_c, \tag{7.29}$$

and

$$B_\sigma = -\Sigma B \tag{7.30}$$

are satisfied. Provided that (7.29) and (7.30) are satisfied, it can be shown that the driven signal model has the usual properties of a state observer, namely

- the reference transfer behaviour of the closed-loop system is not influenced by the dynamics of the observer,
- the separation principle is satisfied.

The first property assures that the reference behaviour of the closed-loop system is not influenced by the driven signal model and the second property shows that the dynamics of the controlled plant and of the controlled signal model can be chosen independently. In order to verify these properties consider the state equations of the closed-loop system in the form

$$\begin{bmatrix} \dot{x}(t) \\ \Sigma\dot{x}(t) - \dot{v}(t) \end{bmatrix} = \begin{bmatrix} A - B(K_x + K_v\Sigma) & BK_v \\ 0 & S - B_\sigma K_v \end{bmatrix} \begin{bmatrix} x(t) \\ \Sigma x(t) - v(t) \end{bmatrix} + \begin{bmatrix} BM \\ 0 \end{bmatrix} r(t), \tag{7.31}$$

where (7.25) and (7.28) have been used. Obviously

$$y_c(t) = \begin{bmatrix} C_c & 0 \end{bmatrix} \begin{bmatrix} x(t) \\ \Sigma x(t) - v(t) \end{bmatrix}. \tag{7.32}$$

Due to the triangular structure of (7.31) the eigenvalues of the closed-loop system are the eigenvalues of the matrices $A - B(K_x + K_v \Sigma)$ and $S - B_\sigma K_v$, *i.e.*, they consist of the eigenvalues of the controlled signal model and of the eigenvalues assigned by the state feedback

$$u(t) = -Kx(t) + Mr(t) \tag{7.33}$$

applied to the system (7.1), where

$$K = K_x + K_v \Sigma. \tag{7.34}$$

This shows that the separation principle is satisfied. Provided that the feedback gain K is stabilizing and the system (C_c, A, B) has no invariant zero at $s = 0$ (see Chapter 2)

$$M = \left[C_c(-A + BK)^{-1}B\right]^{-1} \tag{7.35}$$

in (7.33) assures a vanishing steady-state error $y_c(\infty) - r(\infty)$ for stationary constant reference signals (*i.e.*, $r(\infty) = const$). In view of (7.34) the reference transfer behaviour of (7.31) and (7.32) has the form

$$y_c(s) = C_c(sI - A + BK)^{-1}BMr(s), \tag{7.36}$$

and it is not affected by the disturbance model, *i.e.*, the observer (7.24) does not influence the tracking behaviour.

Remark 7.4. It should be noted, however, that the reference behaviour (7.36) is not robust since $B_\sigma = -\Sigma B$ depends on the parameters of the system.

Remark 7.5. Inserting $K_x = K - K_v \Sigma$ in (7.21) yields $u = -Kx - K_v(v - \Sigma x) + Mr$. After the initial errors $\Sigma x(0) - v(0)$ in (7.28) have vanished an observation error $\Sigma x - v$ only appears in the presence of disturbances. This verifies that the driven signal model only affects the disturbance behaviour.

The next theorem summarizes the results obtained so far. Given the desired dynamics of the driven signal model and the controlled plant it presents the design of the state feedback (7.21) and the driven signal model (7.16).

Theorem 7.1 (Disturbance rejection using the internal model principle). *Consider the system (7.1) and (7.4)*

$$\dot{x}(t) = Ax(t) + Bu(t) + B_d d(t), \tag{7.37}$$

$$y_c(t) = C_c x(t) + D_{cd} d(t), \tag{7.38}$$

and assume that the disturbance d can be modelled by the signal process (7.5) and (7.6), namely

$$\dot{v}^*(t) = S^* v^*(t), \tag{7.39}$$

$$d(t) = Hv^*(t). \tag{7.40}$$

Given further a driven model

$$\dot{v}(t) = Sv(t) + B_\varepsilon y_c(t) - B_\sigma Mr(t) \tag{7.41}$$

of the disturbance signal process. Let the following assumptions be satisfied.

(a1) *The pairs* (A, B) *and* (S^*, b_ε) *are controllable.*

(a2) *The condition* $\operatorname{rank}\begin{bmatrix} \lambda_i I - A & B \\ C_c & 0 \end{bmatrix} = n + p$ *holds for all eigenvalues* λ_i, $i = 1, 2, \ldots, q$, *of* S^*, i.e., *no eigenvalue of the signal model (7.39) is a transmission zero of the transfer behaviour of the system (7.37) and (7.38) between* u *and* y_c.

Compute a feedback gain K *such that* $A - BK$ *is Hurwitz and find the unique solution* Σ *of*

$$\Sigma(A - BK) - S\Sigma = B_\varepsilon C_c, \tag{7.42}$$

to obtain

$$B_\sigma = -\Sigma B. \tag{7.43}$$

Now determine a gain K_v, *which always exists, such that* $S - B_\sigma K_v$ *is Hurwitz and finally compute*

$$K_x = K - K_v \Sigma. \tag{7.44}$$

Then, the feedback

$$u(t) = -\begin{bmatrix} K_x & K_v \end{bmatrix} \begin{bmatrix} x(t) \\ v(t) \end{bmatrix} + Mr(t) \tag{7.45}$$

of the states x *of the system (7.37) and the states* v *of the signal model (7.41) with* M *as defined in (7.35) stabilizes the system (7.37) and (7.38) and (7.41) and assures an asymptotic rejection of the disturbance modelled by (7.39) and (7.40). The reference transfer matrix of the closed-loop system is given by*

$$G_r(s) = C_c(sI - A + BK)^{-1} BM, \tag{7.46}$$

i.e., *the reference transfer behaviour is not affected by the signal model for disturbance rejection.*

Proof. Since (A, B) is controllable there exists a matrix K, such that $A - BK$ is Hurwitz. The Sylvester equation (7.42) results from a simple rearrangement of (7.29) when inserting (7.43) and (7.44). A unique solution Σ exists if the eigenvalues of $A - BK$ and S are disjoint (see [9]). This is the case because the eigenvalues of S are located in the closed right-half complex plane (see, *e.g.*, Table 7.1) and $A - BK$ is Hurwitz. A stabilizing gain K_v in $S - B_\sigma K_v$ can be computed if the pair (S, B_σ) is controllable. The Assumptions (a1) and (a2) assure that the pair (A_{aug}, B_{aug}) of (7.17) is controllable (see [10]). To show that this also leads to a controllable pair (S, B_σ) recall that a controllable pair (A_{aug}, B_{aug}) implies that the associated MFD $(sI - A_{aug})^{-1}B_{aug}$ is left coprime (see [36]) and that, consequently, there exist solutions $\bar{X}_i(s)$, $i = 1, 2, 3, 4$, and $\bar{Y}_j(s)$, $j = 1, 2$, of the Bezout identity

$$\begin{bmatrix} sI - A & 0 \\ -B_\varepsilon C_c & sI - S \end{bmatrix} \begin{bmatrix} \bar{X}_1(s) & \bar{X}_2(s) \\ \bar{X}_3(s) & \bar{X}_4(s) \end{bmatrix} + \begin{bmatrix} B \\ 0 \end{bmatrix} \begin{bmatrix} \bar{Y}_1(s) & \bar{Y}_2(s) \end{bmatrix} = \begin{bmatrix} I_n & 0 \\ 0 & I_{pq} \end{bmatrix} \tag{7.47}$$

(see (7.19) and Theorem 1.2). The pair (S, B_σ) is controllable if the MFD $(sI - S)^{-1}B_\sigma$ is left coprime and this is the case if there exist solutions $\bar{X}_S(s)$ and $\bar{Y}_S(s)$ of the Bezout identity

$$(sI - S)\bar{X}_S(s) + B_\sigma \bar{Y}_S(s) = I_{pq}. \tag{7.48}$$

From the (1,2)-element of (7.47) one obtains the equation

$$B\bar{Y}_2(s) = -(sI - A)\bar{X}_2(s), \tag{7.49}$$

and from the (2,2)-element of (7.47) follows

$$-B_\varepsilon C_c \bar{X}_2(s) + (sI - S)\bar{X}_4(s) = I_{pq}. \tag{7.50}$$

By inserting (7.42) in (7.50) one obtains

$$S\Sigma\bar{X}_2(s) - \Sigma(A - BK)\bar{X}_2(s) + (sI - S)\bar{X}_4(s) = I_{pq}. \tag{7.51}$$

With (7.43) and by adding $s\Sigma\bar{X}_2(s) - s\Sigma\bar{X}_2(s)$ and $-\Sigma B\bar{Y}_2(s) + \Sigma B\bar{Y}_2(s)$ on the left-hand side, (7.51) can be written as

$$\begin{aligned} (sI - S)(\bar{X}_4(s) - \Sigma\bar{X}_2(s)) + B_\sigma(\bar{Y}_2(s) - K\bar{X}_2(s)) \\ + \Sigma((sI - A)\bar{X}_2(s) + B\bar{Y}_2(s)) = I_{pq}. \end{aligned} \tag{7.52}$$

Replacing the term $B\bar{Y}_2(s)$ in (7.52) by the right-hand side of (7.49) one obtains the Bezout identity

$$(sI - S)(\bar{X}_4(s) - \Sigma\bar{X}_2(s)) + B_\sigma(\bar{Y}_2(s) - K\bar{X}_2(s)) = I_{pq}. \tag{7.53}$$

This shows that there exist solutions

$$\bar{X}_S(s) = \bar{X}_4(s) - \Sigma\bar{X}_2(s), \tag{7.54}$$

and

$$\bar{Y}_S(s) = \bar{Y}_2(s) - K\bar{X}_2(s), \tag{7.55}$$

of the Bezout identity (7.48). Therefore, the pair (S, B_σ) is controllable if the pair (A_{aug}, B_{aug}) is controllable. The characteristic polynomial of the closed-loop system that consists of (7.37), (7.41) and (7.45) is given by $\det(sI - A + BK)\det(sI - S + B_\sigma K_v)$ (see (7.31) and (7.34)), so that (7.45) is a stabilizing feedback. Then, the internal model principle assures asymptotic rejection of all disturbances modelled by (7.39) and (7.40) (see [10] for a formal proof). The reference transfer behaviour (7.46) has been determined in (7.36).

□

When implementing the state feedback (7.45) the states x of the system have to be estimated. An estimate $\hat{x}$ can be obtained with the aid of an observer for the system (7.1) and (7.2) having the order $n_O = n - \kappa$, $0 \leq \kappa \leq m$. Its design has been described in Section 3.1. The state equations of this observer have the form

$$\dot{\hat{\zeta}}(t) = T(A - L_1 C_1)\Theta\hat{\zeta}(t) + [TL_1 \quad T(A - L_1 C_1)\Psi_2]\begin{bmatrix} y_1(t) \\ y_2(t) \end{bmatrix} + TBu(t), \quad (7.56)$$

$$\hat{x}(t) = \Theta\hat{\zeta}(t) + \Psi_2 y_2(t) \quad (7.57)$$

(see (3.15) and (3.16)). If x is replaced by the estimate (7.57) the control signal (7.45) takes the form

$$u(t) = -K_x\Theta\hat{\zeta}(t) - K_x\Psi_2 y_2(t) - K_v v(t) + Mr(t). \quad (7.58)$$

In the case of observer-based controllers (see Chapter 4) the order n_C of the compensator is the same as the order of the state observer, namely $n_C = n_O$. Here, the signal model is also part of the compensator, so that the order of the observer-based compensator with a signal model for disturbance rejection is

$$n_C = n_O + pq. \quad (7.59)$$

Remark 7.6. Johnson's approach to the rejection of disturbances only requires a state-plus-disturbance observer of the order $n_O = n + q$ that coincides with the order n_C of the corresponding observer-based compensator. Thus, in general, Davison's approach leads to compensators of higher orders, but it assures a robust rejection of all modelled disturbance signals.

It was shown in Section 4.5 that input saturation can cause undesired effects like large and badly decaying tracking errors if the compensator contains eigenvalues in the vicinity of the imaginary axis or in the closed right-half complex plane. It was also shown how such a controller windup can be prevented in an observer-based compensator. In Davison's approach the compensator contains the dynamics of the disturbance signal model, and these signal models are usually unstable (see Table 7.1). Therefore, input saturation can cause a severe controller windup. Due to the presented modification of Davison's original approach a straightforward prevention of controller windup is feasible. In order to apply the method presented in Section 4.5 the observer formulation (7.24) of the driven signal model is joined with the state observer (7.56) and (7.57) to obtain the *overall observer*, which uses the saturated input u_s instead of u. This yields the disturbance rejecting state feedback controller

$$\dot{\hat{\zeta}}(t) = F\hat{\zeta}(t) + [TL_1 \quad T(A - L_1 C_1)\Psi_2]\begin{bmatrix} y_1(t) \\ y_2(t) \end{bmatrix} + TBu_s(t), \quad (7.60)$$

$$\dot{v}(t) = (S - B_\sigma K_v)v(t) + B_\varepsilon y_c(t) - B_\sigma(K_x\Theta\hat{\zeta}(t) + K_x\Psi_2 y_2(t) + u_s(t)), \quad (7.61)$$

$$\hat{\bar{u}}_{aug}(t) = K_x\Theta\hat{\zeta}(t) + K_x\Psi_2 y_2(t) + K_v v(t) \quad (7.62)$$

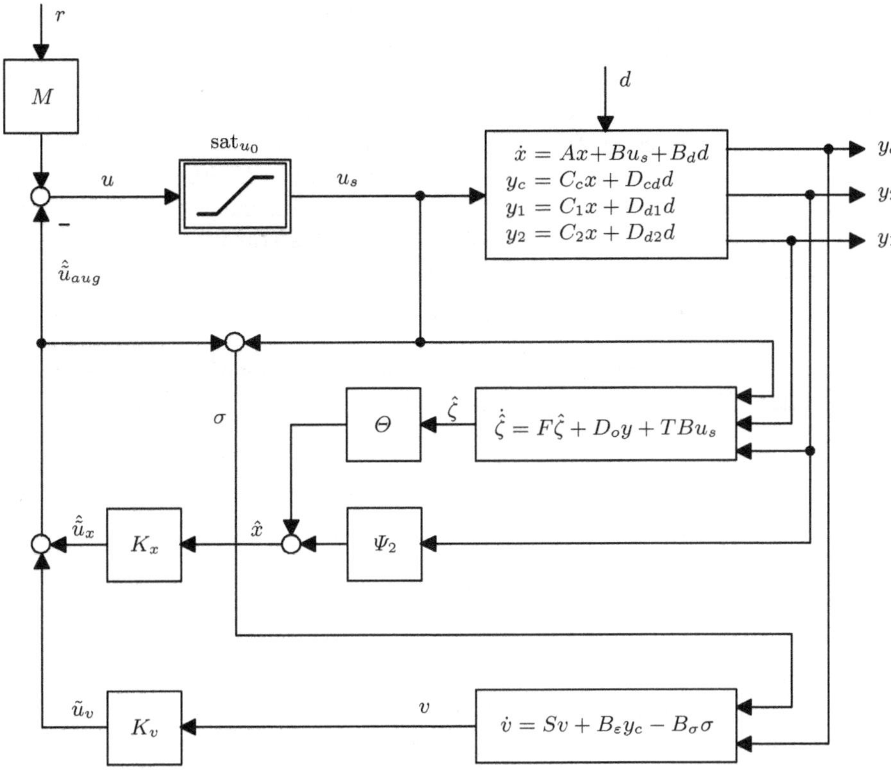

Figure 7.1. Closed-loop system with signal model for disturbance rejection and measures for the prevention of controller windup

on the basis of the overall observer (7.60) and (7.61). Here, the abbreviation

$$F = T(A - L_1C_1)\Theta \tag{7.63}$$

has been used. The controller (7.60)–(7.62) is in the observer structure. The observer structure contains a model of the input saturation whose restricted output u_s is used to drive the observer (its frequency-domain version was introduced in Section 4.5). Thus, a systematic prevention of the controller windup is achieved. Figure 7.1 shows the corresponding compensator, where the additional abbreviation

$$D_o = [TL_1 \quad T(A - L_1C_1)\Psi_2] \tag{7.64}$$

has been used.

7.2 State Feedback Control of the Augmented System in the Frequency Domain

In this section a frequency-domain parameterization of the state feedback (7.45) of the augmented system (7.17)–(7.19) is presented.

In the time domain the first step of the compensator design is to determine the feedback gain K such that $A - BK$ is a Hurwitz matrix (see Theorem 7.1). The frequency-domain parameterization of this state feedback is based on the right coprime MFD

$$(sI - A)^{-1}B = N_x(s)D^{-1}(s), \tag{7.65}$$

with $D(s)$ column reduced. In Chapter 2 it has been shown that a $p \times p$ polynomial matrix $\tilde{D}(s)$ parameterizes the state feedback control in the frequency domain and that it is related with the gain K by

$$\tilde{D}(s) = D(s) + KN_x(s) \tag{7.66}$$

(see (2.18)). This polynomial matrix has to be chosen such that

$$\delta_{ci}[\tilde{D}(s)] = \delta_{ci}[D(s)], \quad i = 1, 2, \ldots, p, \tag{7.67}$$

and

$$\Gamma_c[\tilde{D}(s)] = \Gamma_c[D(s)] \tag{7.68}$$

are satisfied (see Theorem 2.1) and $\det \tilde{D}(s)$ is a Hurwitz polynomial. The next step in the time-domain design of the compensator is to determine the feedback gain K_v such that $S - B_\sigma K_v$ is Hurwitz. Similar to the frequency-domain parameterization of the state feedback gain K one uses the right coprime MFD

$$(sI - S)^{-1}B_\sigma = N_{Sr}(s)D_S^{-1}(s), \tag{7.69}$$

where $N_{Sr}(s)$ is a $pq \times p$ polynomial matrix and $D_S(s)$ is a $p \times p$ polynomial matrix. Without loss of generality $D_S(s)$ in (7.69) can be chosen as

$$D_S(s) = \det(sI - S^*)I_p \tag{7.70}$$

(see also (7.15)), *i.e.*, $D_S(s)$ is a diagonal matrix with identical entries on the main diagonal. Such polynomial matrices commute with any polynomial matrix $R(s)$, *i.e.*,

$$\begin{aligned} R(s)D_S(s) &= R(s)\det(sI - S^*) = \det(sI - S^*)R(s) \\ &= D_S(s)R(s). \end{aligned} \tag{7.71}$$

An application of Theorem 2.1 shows that the dynamics of the controlled signal model, which are assigned by the feedback gain K_v in the time domain, are parameterized by a $p \times p$ polynomial matrix $\tilde{D}_S(s)$ in the frequency domain and that this matrix satisfies the equation

$$\tilde{D}_S(s) = D_S(s) + K_v N_{Sr}(s). \tag{7.72}$$

In view of this and of (7.70) the polynomial matrix $\tilde{D}_S(s)$ has the properties

$$\delta_{ci}[\tilde{D}_S(s)] = \delta_{ci}[D_S(s)] = q, \quad i = 1, 2, \ldots, p, \tag{7.73}$$

and

$$\Gamma_c[\tilde{D}_S(s)] = \Gamma_c[D_S(s)] = I_p. \tag{7.74}$$

Equation (7.31) shows that K (see (7.34)) and K_v specify the dynamics of the controlled augmented plant. Consequently, the polynomial matrices $\tilde{D}(s)$ and $\tilde{D}_S(s)$ already parameterize the dynamics of the closed-loop system in the frequency domain. However, the feedback controller (7.45) cannot directly be computed from these polynomial matrices. In what follows a frequency-domain parameterization of the state feedback (7.45) is derived and it is shown how it is related with $\tilde{D}(s)$ and $\tilde{D}_S(s)$. To this end, consider the transfer behaviour

$$\begin{bmatrix} x(s) \\ v(s) \end{bmatrix} = \begin{bmatrix} (sI - A)^{-1}B \\ (sI - S)^{-1}B_\varepsilon C_c(sI - A)^{-1}B \end{bmatrix} u(s) \tag{7.75}$$

of (7.17)–(7.19). Due to the controller form (7.13) and (7.15) the transfer matrix of the driven signal model in $v(s) = G_v(s)y_c(s)$ (see (7.16)) can be represented in the right and left coprime MFDs

$$G_v(s) = (sI - S)^{-1}B_\varepsilon = N_S(s)D_S^{-1}(s), \tag{7.76}$$

where the $pq \times p$ polynomial matrix $N_S(s)$ has the form

$$N_S(s) = \begin{bmatrix} 1 & & \\ s & & 0 \\ \vdots & & \\ s^{q-1} & & \\ & \ddots & \\ & & 1 \\ & & s \\ 0 & & \vdots \\ & & s^{q-1} \end{bmatrix}, \tag{7.77}$$

with $D_S(s)$ as defined in (7.70). When using $N_c(s) = C_c N_x(s)$ (see also (2.14)) and the factorizations (7.65) and (7.76) the transfer behaviour (7.75) can also be expressed by

$$\begin{bmatrix} x(s) \\ v(s) \end{bmatrix} = N_{x,aug}(s)D_{aug}^{-1}(s)u(s), \tag{7.78}$$

where

$$N_{x,aug}(s) = \begin{bmatrix} N_x(s)D_S(s) \\ N_S(s)N_c(s) \end{bmatrix}, \tag{7.79}$$

and

$$D_{aug}(s) = D_S(s)D(s). \tag{7.80}$$

Since the augmented system is controllable and $\deg \det D_{aug}(s)$ equals the order of the augmented system, the pair $(N_{x,aug}(s), D_{aug}(s))$ is right coprime. In (7.79) and (7.80) the property was used that $D_S(s)$ is a diagonal polynomial matrix with identical entries on the main diagonal (see (7.70)) and that this polynomial matrix commutes with any polynomial matrix (see (7.71)).

Following Theorem 2.1 the $p \times p$ polynomial matrix $\tilde{D}_{aug}(s)$ that parameterizes the state feedback (7.45) in the frequency domain satisfies

$$\tilde{D}_{aug}(s) = D_{aug}(s) + K_{aug}N_{x,aug}(s), \tag{7.81}$$

(see also (2.18)) where K_{aug} is defined in (7.22). The column-degree structure of the polynomial matrix $\tilde{D}_{aug}(s)$ is fixed by the column-degree structure of $D_{aug}(s)$ that has the form

$$\begin{aligned} D_{aug}(s) &= D_S(s)D(s) \\ &= \det(sI - S^*)\left(\Gamma_c[D(s)]\mathrm{diag}\left(s^{\delta_{ci}[D(s)]}\right) + R(s)\right), \end{aligned} \tag{7.82}$$

with a polynomial matrix $R(s)$ satisfying $\delta_{ci}[R(s)] < \delta_{ci}[D(s)]$, $i = 1, 2, \ldots, p$. This can be rewritten as

$$D_{aug}(s) = \Gamma_c[D(s)]\mathrm{diag}\left(s^{\delta_{ci}[D(s)]+q}\right) + \bar{R}(s), \tag{7.83}$$

with a polynomial matrix $\bar{R}(s)$, such that $\delta_{ci}[\bar{R}(s)] < \delta_{ci}[D(s)] + q$. Therefore, the polynomial matrix $D_{aug}(s)$ has the properties

$$\delta_{ci}[D_{aug}(s)] = \delta_{ci}[D(s)] + q, \quad i = 1, 2, \ldots, p, \tag{7.84}$$

and

$$\Gamma_c[D_{aug}(s)] = \Gamma_c[D(s)], \tag{7.85}$$

so that, following Theorem 2.1, the parameterizing polynomial matrix $\tilde{D}_{aug}(s)$ has the column-degree structure

$$\delta_{ci}[\tilde{D}_{aug}(s)] = \delta_{ci}[D_{aug}(s)] = \delta_{ci}[D(s)] + q, \quad i = 1, 2, \ldots, p, \tag{7.86}$$

and

$$\Gamma_c[\tilde{D}_{aug}(s)] = \Gamma_c[D_{aug}(s)] = \Gamma_c[D(s)]. \tag{7.87}$$

In order to clarify the relation between the polynomial matrices $\tilde{D}(s)$, $\tilde{D}_S(s)$ and $\tilde{D}_{aug}(s)$ consider the separation property

$$\det(sI - A_{aug} + B_{aug}K_{aug}) = \det(sI - S + B_\sigma K_v)\det(sI - A + BK) \quad (7.88)$$

implied by (7.17), (7.21), (7.31) and (7.34). With

$$\det \tilde{D}(s) = \det(sI - A + BK), \quad (7.89)$$

$$\det \tilde{D}_S(s) = \det(sI - S + B_\sigma K_v), \quad (7.90)$$

$$\det \tilde{D}_{aug}(s) = \det(sI - A_{aug} + B_{aug}K_{aug}), \quad (7.91)$$

(compare with (2.20)) this leads to

$$\det \tilde{D}_{aug}(s) = \det \tilde{D}_S(s) \det \tilde{D}(s). \quad (7.92)$$

Therefore, it seems reasonable to set

$$\tilde{D}_{aug}(s) = \tilde{D}_S(s)\tilde{D}(s). \quad (7.93)$$

This polynomial matrix has the properties (7.86) and (7.87) since

$$\tilde{D}_S(s)\tilde{D}(s) = \Big(\Gamma_c[\tilde{D}_S(s)]\mathrm{diag}(s^q) + R_{\tilde{D}_S}(s)\Big)\Big(\Gamma_c[\tilde{D}(s)]\mathrm{diag}(s^{\delta_{ci}[D(s)]}) + R_{\tilde{D}}(s)\Big). \quad (7.94)$$

With $\delta_{ci}[R_{\tilde{D}_S}(s)] < q$ and $\delta_{ci}[R_{\tilde{D}}(s)] < \delta_{ci}[D(s)]$, $i = 1, 2, \ldots, p$, this can be written as

$$\tilde{D}_S(s)\tilde{D}(s) = \Gamma_c[\tilde{D}_S(s)]\Gamma_c[\tilde{D}(s)]\mathrm{diag}(s^{\delta_{ci}[D(s)]+q}) + \bar{R}(s), \quad (7.95)$$

with $\delta_{ci}[\bar{R}(s)] < \delta_{ci}[D(s)] + q$, $i = 1, 2, \ldots, p$. This shows that (7.93) has the properties (7.86) and (7.87) because $\Gamma_c[\tilde{D}_S(s)] = I$ and $\Gamma_c[\tilde{D}(s)] = \Gamma_c[D(s)]$ (see (7.74) and (7.68)).

For the derivation of the observer-based controller in the next section the frequency-domain parameterization of the feedback $\hat{\tilde{u}}_x = K_x\hat{x}$ is a prerequisite. The feedback $\tilde{u}_x = K_x x$ for the system (7.1) is characterized by the polynomial matrix $\tilde{D}_x(s)$ satisfying the equation

$$\tilde{D}_x(s)D^{-1}(s) = I + K_x(sI - A)^{-1}B \quad (7.96)$$

(see (2.23)). With (7.65) this is equivalent to

$$\tilde{D}_x(s) = D(s) + K_x N_x(s), \quad (7.97)$$

and it can be used to obtain K_x for a given $\tilde{D}_x(s)$. It also shows that the $p \times p$ polynomial matrix $\tilde{D}_x(s)$ has the properties

$$\delta_{ci}[\tilde{D}_x(s)] = \delta_{ci}[D(s)], \quad i = 1, 2, \ldots, p, \quad (7.98)$$

and

$$\Gamma_c[\tilde{D}_x(s)] = \Gamma_c[D(s)] \quad (7.99)$$

(see (7.67) and (7.68)).

In the time domain (see Theorem 7.1) the eigenvalues of $A - BK$ and $S - B_\sigma K_v$ are freely assignable if the Assumptions (a1) and (a2) are satisfied. This also assures the controllability of the augmented system. In the frequency domain, a desired $\tilde{D}_S(s)$ having the properties (7.73) and (7.74) is chosen. Given this $\tilde{D}_S(s)$ a gain K_v needs to be computed. The frequency-domain equivalent of the assumptions in Theorem 7.1 are

- The pairs $(N_x(s), D(s))$ and $(N_S(s), D_S(s))$ are right coprime.
- The polynomials $\det N_c(s)$ and $\det D_S(s) = \det(sI - S)$ do not have a common zero.

These assumptions can also be directly derived in the frequency domain by computing

$$\begin{aligned}\tilde{D}_S(s)\tilde{D}(s) &= K_x N_x(s) D_S(s) + K_v N_S(s) N_c(s) + D_S(s) D(s) \\ &= D_S(s)(K_x N_x(s) + D(s)) + K_v N_S(s) N_c(s) \\ &= \tilde{D}_x(s) D_S(s) + K_v N_S(s) N_c(s),\end{aligned} \tag{7.100}$$

which follows from inserting (7.93), (7.80), (7.22), (7.97) and (7.79) in (7.81) and by using (7.71). Obviously, for arbitrary left-hand sides there only exists a solution $(\tilde{D}_x(s), K_v)$ of the Diophantine equation (7.100) if the above conditions are satisfied.

After specifying the dynamics of the reference behaviour by $\tilde{D}(s)$ and the observer dynamics by $\tilde{D}_S(s)$ the feedback gain K_v in (7.45) can be obtained by determining a solution $(\tilde{D}_x(s), K_v)$ of the Diophantine equation (7.100). Then, the state feedback gain K_x can be obtained by solving (7.97). This state feedback leads to the block diagram shown in Figure 7.2.

The unknown polynomial matrix $N_{Sr}(s)$ in Figure 7.2 can be obtained as a solution of a polynomial matrix equation, which is the frequency-domain equivalent of the Sylvester equation (7.42). To derive this polynomial matrix equation add $s\Sigma - s\Sigma$ to the Sylvester equation (7.42), which then takes the form

$$-\Sigma(sI - A + BK) + (sI - S)\Sigma = B_\varepsilon C_c. \tag{7.101}$$

Now pre-multiply this result by $(sI - S)^{-1}$ and postmultiply it by $(sI - A)^{-1}B$ to obtain

$$\begin{aligned}-(sI - S)^{-1}\Sigma B - (sI - S)^{-1}\Sigma BK(sI - A)^{-1}B + \Sigma(sI - A)^{-1}B \\ = (sI - S)^{-1}B_\varepsilon C_c(sI - A)^{-1}B.\end{aligned} \tag{7.102}$$

When observing (7.43), (7.69), (2.23), (7.65), (7.76) and (2.14) this leads to

$$N_{Sr}(s)D_S^{-1}(s)\tilde{D}(s)D^{-1}(s) + \Sigma N_x(s)D^{-1}(s) = N_S(s)D_S^{-1}N_c(s)D^{-1}(s). \tag{7.103}$$

Postmultiplying this result by $D(s)D_S(s)$ and observing (7.71) the frequency-domain equivalent of the Sylvester equation (7.42) becomes

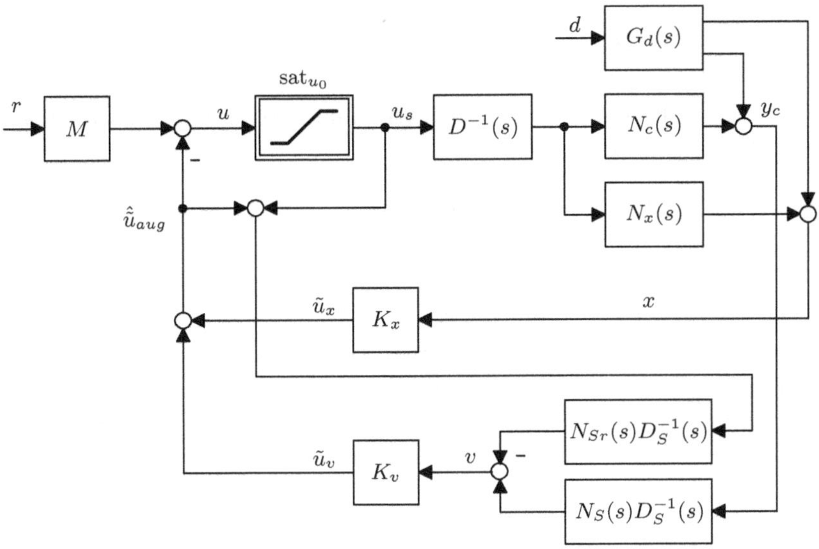

Figure 7.2. Closed-loop system with signal model for disturbance rejection and measures for the prevention of controller windup when using state feedback control

$$N_{Sr}(s)\tilde{D}(s) + \Sigma N_x(s)D_S(s) = N_S(s)N_c(s). \tag{7.104}$$

Given the known polynomial matrices $\tilde{D}(s)$, $D_S(s)$, $N_S(s)$ and $N_c(s)$ it can be solved for the unknown polynomial matrices $N_{Sr}(s)$ and $\Sigma N_x(s)$.

In the next sections a frequency-domain version of the overall observer is considered, which consists of the state observer and the controlled signal model. This overall observer can be parameterized without the knowledge of $N_{Sr}(s)$.

7.3 State Observer for the Non-augmented System in the Frequency Domain

Since the driven model of the disturbance process is part of the compensator, its pq states v can be directly measured. Therefore, it is sufficient to use an observer (7.56) and (7.57) of the order $n_O = n - \kappa$ with $0 \leq \kappa \leq m$ for the reconstruction of the states x of the non-augmented system. This observer can be designed in the frequency domain, as described in Chapter 3.

Given a representation of the transfer matrix $G(s)$ in $y(s) = G(s)u(s)$ by

$$G(s) = \bar{D}^{-1}(s)\bar{N}(s), \tag{7.105}$$

such that

$$\bar{D}_\kappa(s) = \Pi\left\{\bar{D}(s)\begin{bmatrix} I_{m-\kappa} & 0 \\ 0 & s^{-1}I_\kappa \end{bmatrix}\right\} \tag{7.106}$$

is row reduced, the parameterizing polynomial matrix $\bar{\bar{D}}(s)$ of this observer has the properties

$$\delta_{rj}[\bar{\bar{D}}(s)] = \delta_{rj}[\bar{D}_\kappa(s)], \quad j = 1, 2, \ldots, m, \tag{7.107}$$

and

$$\Gamma_r[\bar{\bar{D}}(s)] = \Gamma_r[\bar{D}_\kappa(s)] \tag{7.108}$$

(see Theorem 3.1). The $m(n-\kappa)$ degrees of freedom in $\bar{\bar{D}}(s)$ must be chosen to obtain a Hurwitz polynomial $\det \bar{\bar{D}}(s)$.

The next section shows how this observer can be joined with the driven signal model to obtain the overall observer in the frequency domain.

7.4 Design of the Observer-based Compensator with an Internal Signal Model in the Frequency Domain

In this section, the frequency-domain design of the observer-based compensator (7.60)–(7.62) is presented. This compensator consists of the observer for the states of the non-augmented system, the driven signal model and the feedback of the states of the augmented system. In what follows it will be shown how the state observer and the driven signal model can be joined in the frequency domain to obtain the overall observer. The resulting controller has the structure of the observer-based compensator introduced in Chapter 4. As a consequence of this the methods of Section 4.5 for the prevention of windup are directly applicable.

An inspection of the compensator in Figure 7.1 shows that it consists of two parts, namely a feedback $\tilde{u}_v = K_v v$ of the measurable states of the driven disturbance model and a feedback $\hat{\tilde{u}}_x = K_x\hat{x}$ of the observed state of the plant. The transfer behaviour of the compensator with respect to the latter part is characterized by

$$\hat{\tilde{u}}_x(s) = \{K_x\Theta(sI-F)^{-1}D_o + [0 \quad K_x\Psi_2]\}\begin{bmatrix} y_1(s) \\ y_2(s) \end{bmatrix} + K_x\Theta(sI-F)^{-1}TBu(s) \tag{7.109}$$

(see also (7.63) and (7.64)), where here and in what follows the input saturation is omitted, *i.e.*, one has $u_s = u$. By introducing the left coprime MFDs

$$K_x\Theta(sI-F)^{-1}D_o + [0 \quad K_x\Psi_2] = \hat{\Delta}^{-1}(s)\hat{N}_C(s), \tag{7.110}$$

and

$$K_x\Theta(sI-F)^{-1}TB = \hat{\Delta}^{-1}(s)\hat{N}_u(s), \tag{7.111}$$

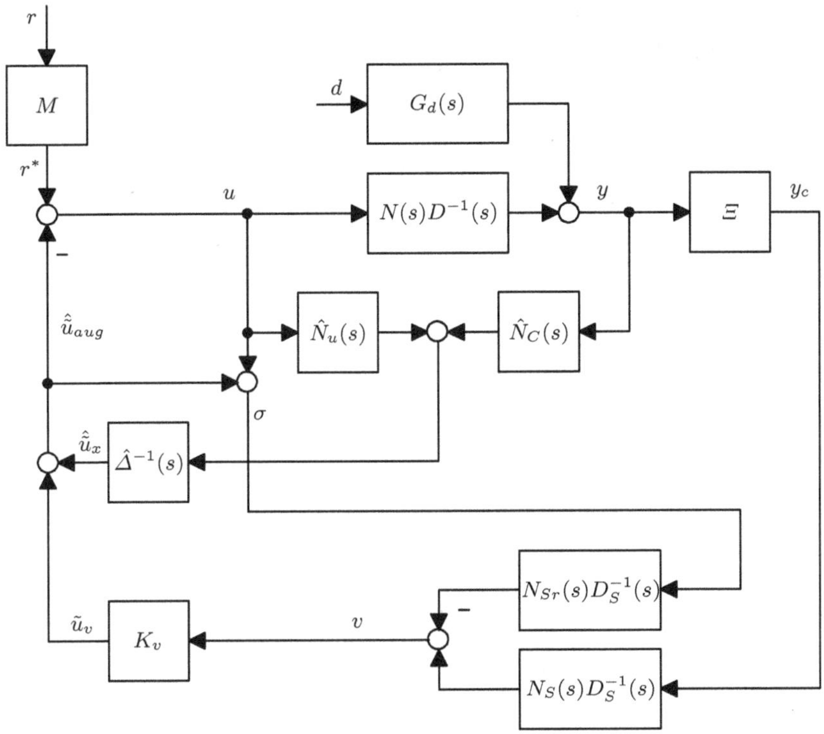

Figure 7.3. Closed-loop system of Figure 7.1 in the frequency domain

the transfer behaviour of the observer-based compensator with signal model for disturbance rejection is characterized by

$$\hat{\hat{u}}_{aug}(s) = \hat{\Delta}^{-1}(s) \left[\hat{N}_u(s) \;\; \hat{N}_C(s) \;\; \hat{\Delta}(s)K_v \right] \begin{bmatrix} u(s) \\ y(s) \\ v(s) \end{bmatrix}. \tag{7.112}$$

Now using the MFDs (7.69) and (7.76) for the driven signal model (7.16) the closed-loop system of Figure 7.1 can be represented in the frequency domain as shown in Figure 7.3. There are two problems in the frequency-domain design of the controller in Figure 7.3. First, the observer-based compensator characterized by the polynomial matrices $\hat{N}_u(s)$, $\hat{N}_C(s)$ and $\hat{\Delta}(s)$ can only be designed if $\tilde{D}_x(s)$ is known, which parameterizes the feedback of x (see (7.97)). This polynomial matrix, however, cannot be obtained directly from $\tilde{D}_{aug}(s)$ (see (7.93)). Second, the feedback gain K_v is not known in the frequency domain.

To compute the polynomial matrix $\tilde{D}_x(s)$ and the feedback gain K_v for given parameterizing matrices $\tilde{D}_S(s)$ and $\tilde{D}(s)$ consider (7.100), namely

$$K_v N_S(s) N_c(s) + \tilde{D}_x(s) D_S(s) = \tilde{D}_S(s) \tilde{D}(s). \tag{7.113}$$

This shows that $\tilde{D}_x(s)$ and K_v can be obtained by solving

$$P(s) N_S(s) N_c(s) + Q(s) D_S(s) = \tilde{D}_S(s) \tilde{D}(s), \tag{7.114}$$

where the solution satisfies

$$\delta_{ci}[Q(s)] = \delta_{ci}[D(s)], \quad i = 1, 2, \ldots, p, \tag{7.115}$$

$$\Gamma_c[Q(s)] = \Gamma_c[D(s)], \tag{7.116}$$

(see (7.98) and (7.99)) and $P(s) = const.$

When the parameterizing polynomial matrices $\tilde{D}_x(s)$ (state feedback for the non-augmented plant) and $\tilde{\bar{D}}(s)$ (state observer for the non-augmented plant) are known the observer-based controller with the transfer behaviour

$$\hat{\tilde{u}}_x(s) = \hat{\Delta}^{-1}(s)(\hat{N}_u(s)u(s) + \hat{N}_C(s)y(s)) \tag{7.117}$$

can be computed using Theorem 4.1. This design procedure is based on the right and left coprime MFDs

$$C(sI - A)^{-1}B = CN_x(s)D^{-1}(s) = N(s)D^{-1}(s) = \bar{D}^{-1}(s)\bar{N}(s) \tag{7.118}$$

of the original non-augmented system. In what follows it is assumed that $D(s)$ is column reduced and that $\bar{D}_\kappa(s)$ (see (7.106)) is row reduced. One first determines the pair $Y(s)$ and $X(s)$ from the Bezout identity

$$Y(s)N(s) + X(s)D(s) = I_p \tag{7.119}$$

(see (4.40)). Then, with

$$\bar{V}(s) = \Pi\left\{\tilde{D}_x(s)Y(s)\bar{D}^{-1}(s)\tilde{\bar{D}}(s)\right\} \tag{7.120}$$

and with the prime right-to-left conversion

$$\bar{V}(s)\tilde{\bar{D}}^{-1}(s) = \hat{\Delta}^{-1}(s)V(s) \tag{7.121}$$

(see (4.44)) the numerator and the denominator matrices of the output feedback structure (see Figure 4.4) of this observer-based controller have the forms

$$\hat{N}_C(s) = \hat{\Delta}(s)\tilde{D}_x(s)Y(s) - V(s)\bar{D}(s), \tag{7.122}$$

and

$$\hat{D}_C(s) = \hat{\Delta}(s)\tilde{D}_x(s)X(s) + V(s)\bar{N}(s) \tag{7.123}$$

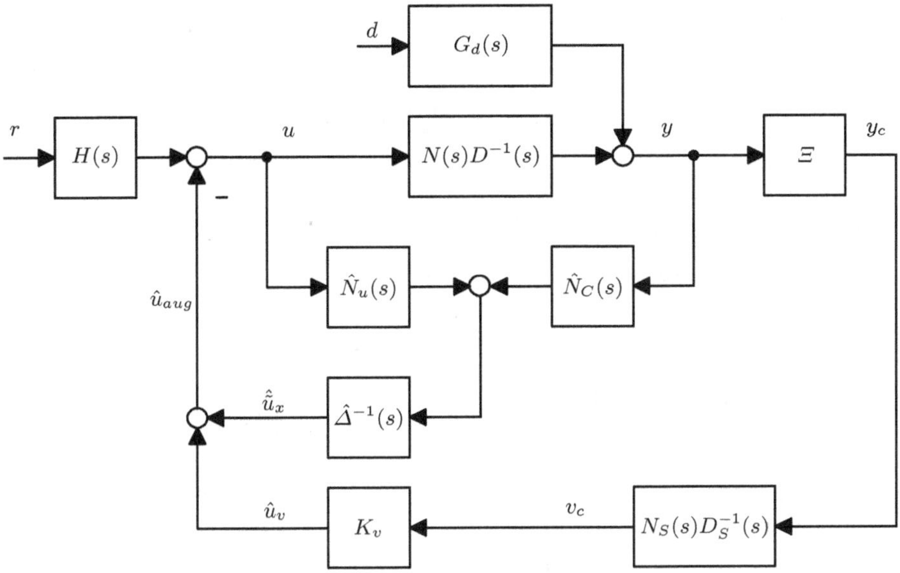

Figure 7.4. Equivalent representation of the block diagram in Figure 7.3

(see (4.45) and (4.46)), *i.e.*, $\hat{\tilde{u}}_x(s) = \hat{D}_C^{-1}(s)\hat{N}_C(s)y(s)$ holds. The numerator matrix $\hat{N}_u(s)$ in the observer structure (7.117) of this controller can be obtained from

$$\hat{N}_u(s) = \hat{D}_C(s) - \hat{\Delta}(s) \tag{7.124}$$

(see (4.27)).

To join the observer-based controller (7.117) with the driven signal model (7.16) consider Figure 7.4 that shows an equivalent representation of the block diagram in Figure 7.3. By comparing Figures 7.3 and 7.4 and in view of

$$\sigma(s) - u(s) + \hat{\tilde{u}}_{uuy}(s) - r^*(s), \tag{7.125}$$

it is straightforward to verify that the transfer matrix $H(s)$ in Figure 7.4 has the form

$$H(s) = (I + K_v N_{Sr}(s) D_S^{-1}(s))M. \tag{7.126}$$

To obtain the output feedback structure (see Figure 4.4) of the compensator in Figure 7.4 first compute

$$\hat{u}_{aug}(s) = \hat{\Delta}^{-1}(s)\hat{N}_u(s)u(s) + \hat{\Delta}^{-1}(s)\hat{N}_C(s)y(s) + K_v N_S(s) D_S^{-1}(s)\Xi y(s). \tag{7.127}$$

Now, using

$$u(s) = -\hat{u}_{aug}(s) + H(s)r(s), \tag{7.128}$$

one obtains

$$\begin{aligned} u(s) = &- \hat{\Delta}^{-1}(s)\hat{N}_u(s)u(s) \\ &- (\hat{\Delta}^{-1}(s)\hat{N}_C(s) + K_v N_S(s)D_S^{-1}(s)\Xi)y(s) + H(s)r(s), \end{aligned} \tag{7.129}$$

or

$$\begin{aligned} &(I + \hat{\Delta}^{-1}(s)\hat{N}_u(s))u(s) \\ &= -(\hat{\Delta}^{-1}(s)\hat{N}_C(s) + K_v N_S(s)D_S^{-1}(s)\Xi)y(s) + H(s)r(s). \end{aligned} \tag{7.130}$$

Pre-multiplying this by $D_S(s)\hat{\Delta}(s)$ leads to

$$\begin{aligned} D_S(s)\hat{D}_C(s)u(s) = -(D_S(s)\hat{N}_C(s) + \hat{\Delta}(s)K_v N_S(s)\Xi)y(s) \\ +D_S(s)\hat{\Delta}(s)H(s)r(s), \end{aligned} \tag{7.131}$$

where the properties (7.71) and (7.124) have been used. Solving this for $u(s)$ and introducing the abbreviations

$$D_C(s) = D_S(s)\hat{D}_C(s), \tag{7.132}$$

$$N_C(s) = D_S(s)\hat{N}_C(s) + \hat{\Delta}(s)K_v N_S(s)\Xi, \tag{7.133}$$

and

$$\begin{aligned} N_{Cr}(s) &= \hat{\Delta}(s)D_S(s)H(s) \\ &= \hat{\Delta}(s)D_S(s)(I + K_v N_{Sr}(s)D_S^{-1}(s))M \\ &= \hat{\Delta}(s)(D_S(s) + K_v N_{Sr}(s))M \\ &= \hat{\Delta}(s)\tilde{D}_S(s)M \end{aligned} \tag{7.134}$$

(where (7.72) and (7.126) have been used) leads to the output feedback structure

$$u(s) = -D_C^{-1}(s)N_C(s)y(s) + D_C^{-1}(s)N_{Cr}(s)r(s) \tag{7.135}$$

of the observer-based compensator for disturbance rejection as shown in Figure 7.5. To obtain the observer structure (see Figure 4.2) of this compensator, the polynomial matrix $\Delta(s)$ characterizing the dynamics of the overall observer needs to be defined. To this end, consider the characteristic equation

$$N_C(s)N(s) + D_C(s)D(s) = \tilde{D}'(s) \tag{7.136}$$

of the feedback loop in Figure 7.5. Inserting (7.132) and (7.133) in (7.136) leads to

$$\tilde{D}'(s) = (D_S(s)\hat{N}_C(s) + \hat{\Delta}(s)K_v N_S(s)\Xi)N(s) + D_S(s)\hat{D}_C(s)D(s), \tag{7.137}$$

so that with (7.122) and (7.123) one obtains

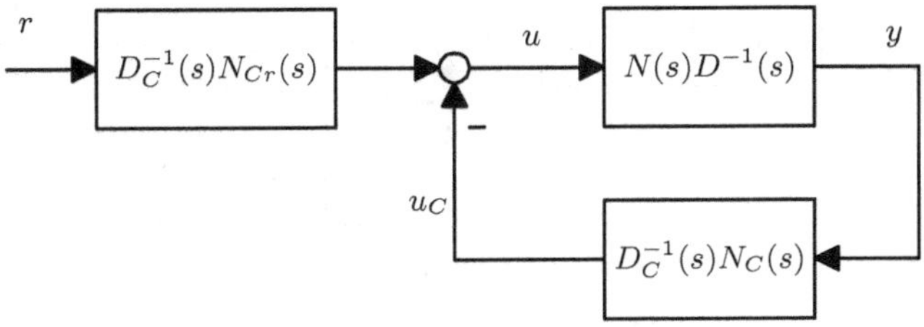

Figure 7.5. Output feedback structure of the observer-based compensator

$$\begin{aligned}\tilde{D}'(s) &= (D_S(s)\hat{\Delta}(s)\tilde{D}_x(s)Y(s) - D_S(s)V(s)\bar{D}(s) + \hat{\Delta}(s)K_v N_S(s)\Xi)N(s)\\ &\quad + D_S(s)(\hat{\Delta}(s)\tilde{D}_x(s)X(s) + V(s)\bar{N}(s))D(s)\\ &= \hat{\Delta}(s)(D_S(s)\tilde{D}_x(s)(Y(s)N(s) + X(s)D(s)) + K_v N_S(s)N_c(s))\\ &\quad + D_S(s)V(s)(\bar{N}(s)D(s) - \bar{D}(s)N(s)),\end{aligned} \tag{7.138}$$

where $N_c(s) = \Xi N(s)$ and (7.71) have been used. Thus, with the prime right-to-left conversion $N(s)D^{-1}(s) = \bar{D}^{-1}(s)\bar{N}(s)$, (7.119) and (7.100) the polynomial matrix in (7.138) simplifies to

$$\tilde{D}'(s) = \hat{\Delta}(s)\tilde{D}_S(s)\tilde{D}(s). \tag{7.139}$$

Consequently, the polynomial matrix $\Delta(s)$ of the overall observer has the form

$$\Delta(s) = \hat{\Delta}(s)\tilde{D}_S(s). \tag{7.140}$$

This leads to the Diophantine equation

$$N_C(s)N(s) + D_C(s)D(s) = \Delta(s)\tilde{D}(s) \tag{7.141}$$

by substituting (7.139) and (7.140) in (7.136). The polynomial matrix $N_u(s)$ that appears in the observer structure of the observer-based compensator in the frequency domain (see (4.27)) results as

$$N_u(s) = D_C(s) - \Delta(s). \tag{7.142}$$

By elementary operations the output feedback structure of Figure 7.5 can be transformed to the block diagram in Figure 7.6 (also compare Figures 4.2 and 4.4), which shows the observer structure of the compensator for the rejection of persistently acting disturbances. The transfer behaviour between r^* and r, namely

$$r^*(s) = \Delta^{-1}(s)D_C(s)D_C^{-1}(s)N_{Cr}(s)r(s) = \Delta^{-1}(s)N_{Cr}(s)r(s) \tag{7.143}$$

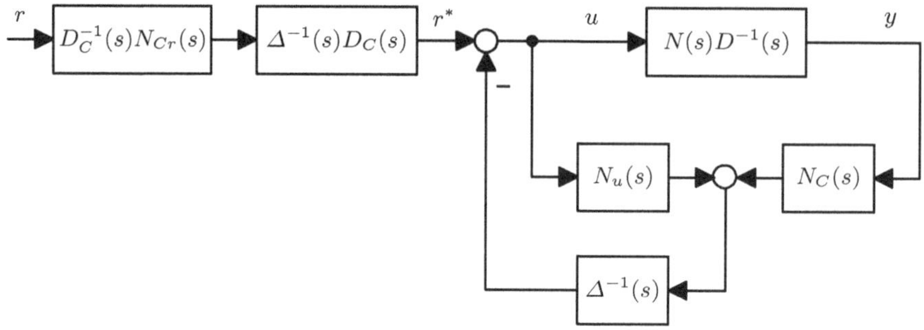

Figure 7.6. Observer structure of the observer-based compensator

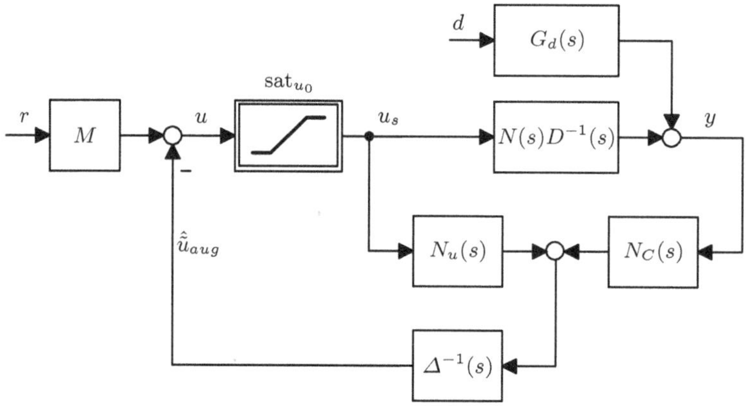

Figure 7.7. Frequency-domain representation of the observer-based compensator in Figure 7.1

can be further simplified by considering

$$N_{Cr}(s) = \Delta(s)M \tag{7.144}$$

(which follows from (7.134) by inserting (7.140)). Using this in (7.143) one obtains

$$r^*(s) = Mr(s). \tag{7.145}$$

Since the resulting compensator with overall observer is in the observer structure the prevention of controller windup of Section 4.5 can be directly applied. This leads to the frequency-domain representation of the observer-based compensator of Figure 7.1, which is shown in Figure 7.7. These results can be summarized in the following theorem.

Theorem 7.2 (Design of observer-based compensators with disturbance rejection in the frequency domain). *Given is the system (7.37) and (7.38) described by its transfer behaviour* $y(s) = G(s)u(s)$ *and by* $y_c(s) = G_c(s)u(s)$, *where the* $m \times p$ *transfer matrix* $G(s)$ *is represented in a right coprime and a left coprime MFD*

$$G(s) = N(s)D^{-1}(s) = \bar{D}^{-1}(s)\bar{N}(s), \tag{7.146}$$

and the $p \times p$ *transfer matrix* $G_c(s)$ *in a right MFD*

$$G_c(s) = N_c(s)D^{-1}(s) = \Xi N(s)D^{-1}(s). \tag{7.147}$$

In (7.146) and (7.147) the polynomial matrix $D(s)$ *is column reduced and* $\bar{D}(s)$ *is such that*

$$\bar{D}_\kappa(s) = \Pi\left\{\bar{D}(s)\begin{bmatrix} I_{m-\kappa} & 0 \\ 0 & s^{-1}I_\kappa \end{bmatrix}\right\} \tag{7.148}$$

is row reduced, where $n_O = n - \kappa$ *with* $0 \leq \kappa \leq m$ *is the order of the state observer for the system. The transfer behaviour between* $y_c \in \mathbb{R}^p$ *and* $v \in \mathbb{R}^{pq}$ *of the driven signal model (see (7.41)) is represented by the right coprime MFD*

$$v(s) = N_S(s)D_S^{-1}(s)y_c(s) \tag{7.149}$$

(see (7.76)), where $D_S(s)$ *has the form* $D_S(s) = \det(sI - S^*)I_p$ *(see (7.13)). It is assumed that* $\det N_c(s) = \det(\Xi N(s))$ *(see (7.4)) and* $\det D_S(s)$ *do not have a common zero.*

Also given are the $p \times p$ *parameterizing polynomial matrices* $\tilde{D}(s)$ *(state feedback having the properties (7.67) and (7.68)),* $\tilde{D}_S(s)$ *(controlled signal model having the properties (7.73) and (7.74)), and* $\tilde{\bar{D}}(s)$ *(reduced-order state observer having the properties (7.107) and (7.108)).*

Determine solutions $Q(s) = \tilde{D}_x(s)$ *with*

$$\delta_{ci}[Q(s)] = \delta_{ci}[D(s)], \quad i = 1, 2, \ldots, p, \tag{7.150}$$

$$\Gamma_c[Q(s)] = \Gamma_c[D(s)], \tag{7.151}$$

and $P(s) = P = const$ *of the Diophantine equation*

$$P(s)N_S(s)N_c(s) + Q(s)D_S(s) = \tilde{D}_S(s)\tilde{D}(s), \tag{7.152}$$

and find solutions $Y(s)$ *and* $X(s)$ *of the Bezout identity*

$$Y(s)N(s) + X(s)D(s) = I_p. \tag{7.153}$$

Then, with

$$\bar{V}(s) = \Pi\left\{Q(s)Y(s)\bar{D}^{-1}(s)\tilde{\bar{D}}(s)\right\}, \tag{7.154}$$

and with the prime right-to-left conversion

$$\bar{V}(s)\tilde{\bar{D}}^{-1}(s) = \hat{\Delta}^{-1}(s)V(s) \tag{7.155}$$

the numerator and the denominator matrices of the output feedback structure of this observer-based controller have the forms

$$N_C(s) = D_S(s)[\hat{\Delta}(s)Q(s)Y(s) - V(s)\bar{D}(s)] + \hat{\Delta}(s)PN_S(s)\Xi, \tag{7.156}$$

and

$$D_C(s) = D_S(s)[\hat{\Delta}(s)Q(s)X(s) + V(s)\bar{N}(s)]. \tag{7.157}$$

The error dynamics of the overall observer are characterized by the polynomial matrix

$$\Delta(s) = \hat{\Delta}(s)\tilde{D}_S(s). \tag{7.158}$$

The numerator matrix $N_u(s)$ in the observer structure of this controller as shown in the Figures 7.6 and 7.7 can be obtained from

$$N_u(s) = D_C(s) - \Delta(s). \tag{7.159}$$

This compensator asymptotically rejects all modelled disturbances in the controlled output y_c and it assures a reference behaviour

$$y_c(s) = N_c(s)\tilde{D}^{-1}(s)Mr(s). \tag{7.160}$$

Remark 7.7. In the derivation of Theorem 7.2 it has been shown that the observer-based compensator in Figure 7.3 is characterized by the polynomial matrices $\hat{\Delta}(s)$, $\hat{N}_u(s)$ and $\hat{N}_C(s)$ and that the feedback gain K_v can be obtained by solving the Diophantine equation (7.114). Then, the controller in Figure 7.3 could be obtained by additionally solving the Diophantine equation (7.104) to obtain $N_{Sr}(s)$. The results of Theorem 7.2, however, show that the solution of the Diophantine equation (7.104) can be circumvented by formulating the driven signal model and the state observer as a joint observer for x and Σx.

Example 7.1. Design of an observer-based compensator with signal model for step-like and sinusoidal disturbances for an unstable system

Considered is a system of the order 4 with two inputs (*i.e.*, $p = 2$), three measured outputs (*i.e.*, $m = 3$) and two disturbance inputs (*i.e.*, $\rho = 2$). A reduced-order observer of the order $n_O = 2$ (*i.e.*, $\kappa = 2$) will be used.

The transfer matrix $G(s)$ of the system is represented either by its right coprime MFD

$$G(s) = N(s)D^{-1}(s) = \begin{bmatrix} 0 & s+1 \\ s+2 & -s-2 \\ -2 & s-1 \end{bmatrix} \begin{bmatrix} s^2+3s+2 & 0 \\ 0 & s^2+s-2 \end{bmatrix}^{-1}, \tag{7.161}$$

or by its left coprime MFD $G(s) = \bar{D}^{-1}(s)\bar{N}(s)$ with

$$\bar{D}(s) = \begin{bmatrix} -s-2 & -s^2-s & 0 \\ s+2 & 2 & s+2 \\ 0 & -s+1 & s+2 \end{bmatrix} \text{ and } \bar{N}(s) = \begin{bmatrix} -s & s+1 \\ 0 & 2 \\ -1 & 2 \end{bmatrix}. \tag{7.162}$$

This form of $\bar{D}(s)$ assures that for $\kappa = 2$ the resulting $\bar{D}_\kappa(s)$ (see (7.148)) is row reduced. The controlled output is defined as

$$y_c(t) = \begin{bmatrix} y^2(t) \\ y^1(t) \end{bmatrix}, \tag{7.163}$$

where $y^i(t)$ denotes the ith component of $y(t)$. Therefore, the selection matrix Ξ in $y_c = \Xi y$ has the form

$$\Xi = \begin{bmatrix} 0 & 1 & 0 \\ 1 & 0 & 0 \end{bmatrix}, \tag{7.164}$$

and therefore

$$N_c(s) = \Xi N(s) = \begin{bmatrix} s+2 & -s-2 \\ 0 & s+1 \end{bmatrix}. \tag{7.165}$$

The disturbances act at the input of the system, *i.e.*, $G_d(s) = G(s)$ and it is known that they have the form

$$d_1(t) = d_{01}1(t-\tau_{01}),$$

and

$$d_2(t) = d_{02}1(t-\tau_{02})\sin(3t+\varphi_{S2}),$$

where $1(t)$ denotes the unit step function and d_{01}, d_{02}, τ_{01}, τ_{02} and φ_{S2} are unknown constants. Therefore, the disturbances can be modelled in a signal process having the characteristic polynomial $\det(sI - S^*) = s(s+\mathrm{j}3)(s-\mathrm{j}3) = s^3 + 9s$ (see (7.11)). The frequency-domain representation (7.76) of the signal process consequently contains the polynomial matrices

$$N_S(s) = \begin{bmatrix} 1 & 0 \\ s & 0 \\ s^2 & 0 \\ 0 & 1 \\ 0 & s \\ 0 & s^2 \end{bmatrix} \quad \text{and} \quad D_S(s) = \begin{bmatrix} s^3+9s & 0 \\ 0 & s^3+9s \end{bmatrix}. \tag{7.166}$$

The relative degrees of the outputs of the system are $\delta_1 = 1$ and $\delta_2 = 1$ (see Theorem 6.1) and therefore, the relative degree of the system is $\delta = 2$ (see Definition 6.2). The transfer matrix $G_c(s) = N_c(s)D^{-1}(s)$ of the system has $z = 2$ zeros at $s = -1$ and $s = -2$, so that the decoupling criterion (see Theorem 6.2) is satisfied and the system is decouplable in a stable scheme.

Assuming that the freely assignable eigenvalues are placed at $\tilde{s}_1 = -5$ and $\tilde{s}_2 = -10$ one can choose, *e.g.*,

$$\Lambda(s) = \begin{bmatrix} \frac{s+5}{5} & 0 \\ 0 & \frac{s+10}{10} \end{bmatrix} \tag{7.167}$$

(see (6.29)). Applying Theorem 6.3 this leads to the decoupling controller parameterization

$$M = \Gamma_c[D(s)]\Gamma_c^{-1}[\Lambda(s)N_c(s)] = \begin{bmatrix} 5 & 10 \\ 0 & 10 \end{bmatrix}, \tag{7.168}$$

(see (6.30)) and

$$\tilde{D}(s) = M\Lambda(s)N_c(s) = \begin{bmatrix} (s+2)(s+5) & 4s \\ 0 & (s+1)(s+10) \end{bmatrix} \tag{7.169}$$

(see (6.28)). The dynamics of the controlled signal model are assigned with eigenvalues at $s = -4$ and $s = -5$ and are parameterized by

$$\tilde{D}_S(s) = \begin{bmatrix} (s+4)^3 & 0 \\ 0 & (s+5)^3 \end{bmatrix}. \tag{7.170}$$

Given the orders $n_O = 2$ (*i.e.*, $\kappa = 2$) of the state observer and $pq = 6$ of the signal model, the order of the resulting compensator is $n_C = 8$. The eigenvalues of the state observer are assigned at $s = -4$. For $\kappa = 2$ the row-reduced polynomial matrix $\bar{D}_\kappa(s)$ (see (7.148)) has the form

$$\bar{D}_2(s) = \begin{bmatrix} -s-2 & -s-1 & 0 \\ s+2 & 0 & 1 \\ 0 & -1 & 1 \end{bmatrix}. \tag{7.171}$$

The condition $\det \tilde{\bar{D}}(s) = (s+4)^2$ and the Equations (7.107) and (7.108) are, *e.g.*, satisfied by

$$\tilde{\bar{D}}(s) = \begin{bmatrix} s+4 & 0 & 0 \\ 0 & s+4 & 0 \\ 0 & 0 & 1 \end{bmatrix} \Gamma_r[\bar{D}_2(s)] = \begin{bmatrix} -s-4 & -s-4 & 0 \\ s+4 & 0 & 0 \\ 0 & -1 & 1 \end{bmatrix} \tag{7.172}$$

(see Remark 3.4). Now, all parameters for the design of the compensator are defined. Following Theorem 7.2, the solutions

$$P = K_v = \begin{bmatrix} 320 & 151 & 99 & 640 & 346 & 159 \\ 0 & 0 & 0 & 1250 & 650 & 216 \end{bmatrix}, \tag{7.173}$$

and

$$Q(s) = \tilde{D}_x(s) = \begin{bmatrix} s^2 + 19s + 34 & 4s - 12 \\ 0 & s^2+26s+25 \end{bmatrix} \tag{7.174}$$

of the Diophantine equation (7.152) have to be computed first. Then, using the solutions

$$Y(s) = \begin{bmatrix} -0.5s + 0.5 & 0 & -0.5 \\ 0.5s & 0 & 0 \end{bmatrix}, \tag{7.175}$$

and

$$X(s) = \begin{bmatrix} 0 & 0.5 \\ 0 & -0.5 \end{bmatrix} \tag{7.176}$$

of the Bezout identity (7.153) one can compute

$$\bar{V}(s) = \begin{bmatrix} -0.5s^3-8s^2-25.5s+14 & 2s + 25 & 0.5s^2 + 6s - 1 \\ 0.5s^3+14s^2+36.5s-23 & -2s - 50 & -0.5s^2 - 12.5s - 1 \end{bmatrix}, \tag{7.177}$$

(see (7.154)) and carry out the right-to-left conversion (7.155), which leads to

$$V(s) = \tag{7.178}$$
$$\begin{bmatrix} -0.5s^2 - 8s - 24 & -0.5s^3 - 8.5s^2 - 33.5s - 10 & 0.5s^3 + 8s^2 + 23s - 4 \\ 0.5s^2 + 14.5s + 51 & 0.5s^3 + 14.5s^2 + 51s + 28 & -0.5s^3-14.5s^2-51s-4 \end{bmatrix},$$

and

$$\hat{\Delta}(s) = \begin{bmatrix} s + 4 & 0 \\ 0 & s+4 \end{bmatrix}. \tag{7.179}$$

Inserting the above results in (7.156) and (7.157) finally gives

$$N_C(s) = \begin{bmatrix} 199s^3 + 982s^2 + 2384s + 2560 \\ 262s^3 + 1514s^2 + 4264s + 5000 \end{bmatrix.$$
$$\begin{matrix} 16s^4 + 123s^3 + 691s^2 + 1140s + 1280 & -20(s + 2)(s^3 + 9s) \\ -4(s + 13)(s^3 + 9s) & -24(s + 2)(s^3 + 9s) \end{matrix}\Bigg], \tag{7.180}$$

and

$$D_C(s) = \begin{bmatrix} (s + 4)(s^3 + 9s) & 40(s^3 + 9s) \\ 4(s^3 + 9s) & (s+49)(s^3+9s) \end{bmatrix}. \tag{7.181}$$

The polynomial matrix (7.140) that characterizes the overall observer (*i.e.*, the dynamics of the controlled signal model and the state observer) is

$$\Delta(s) = \hat{\Delta}(s)\tilde{D}_S(s) = \begin{bmatrix} (s + 4)^4 & 0 \\ 0 & (s+4)(s+5)^3 \end{bmatrix}, \tag{7.182}$$

so that one obtains

$$N_u(s) = D_C(s) - \Delta(s)$$
$$= \begin{bmatrix} -12s^3 - 87s^2 - 220s - 256 & 40s^3 + 360s \\ 4s^3 + 36s & 30s^3 - 126s^2 + 16s - 500 \end{bmatrix}. \quad (7.183)$$

Figure 7.8a shows the reaction of the closed-loop system with the above compensator when both disturbances ($d_{01} = 1, \tau_{01} = 0, d_{02} = -1, \tau_{02} = 0$ and $\varphi_{S2} = 0$) act at the same time and Figure 7.8b shows the corresponding input signals. The disturbances are asymptotically compensated in both components of the controlled output y_c.

The solid lines in Figure 7.9a show the responses of the closed-loop system to reference inputs $r_1(t) = 1(t)$ and $r_2(t) = -1(t)$. They exhibit a decoupled behaviour without overshoots. If one had applied Davison's original approach the signal model would have the form (7.14) in a time-domain approach. This compensator does not only assure an asymptotic tracking of constant but also of sinusoidal reference inputs. The dashed lines in Figure 7.9b show the reference signals $r_1(t) = \sin(3t)$ and $r_2(t) = 0.5\sin(3t + \pi/2)$ and the solid lines the resulting components y_c^1 and y_c^2 of the controlled output y_c that demonstrates the asymptotic tracking of sinusoidal reference inputs with angular frequency $\omega_0 = 3$. This, however, has an unfavourable influence on the reference step transients. The dashed lines in Figure 7.9a depict the reaction of the closed-loop system to the same reference steps as above. Though also an asymptotic tracking of step-like reference signals is assured, Davison's original approach leads to an unfavourable influence of the sinusoidal part of the signal model on the reference step responses. This is not the case, if one applies the modified design presented in this chapter.

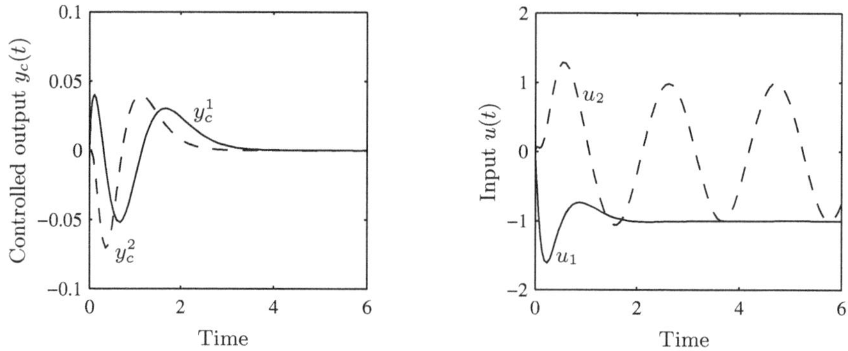

(a) Controlled output of the system (b) Corresponding input

Figure 7.8. Disturbance behaviour of the closed-loop system for $d_{01} = 1$, $\tau_{01} = 0$, $d_{02} = -1$, $\tau_{02} = 0$ and $\varphi_{S2} = 0$

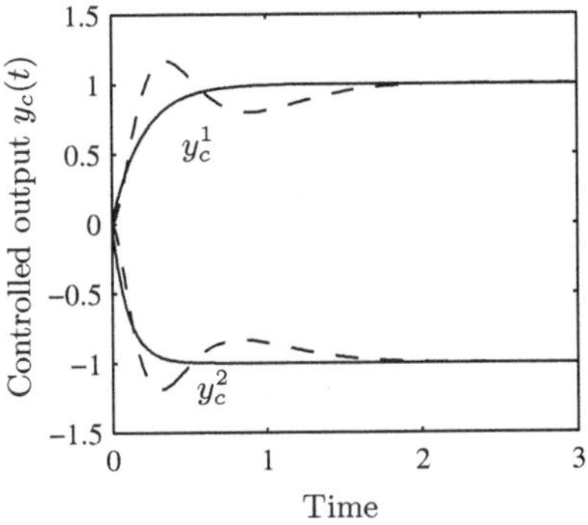

(a) Reference step response of the closed-loop system (*solid lines*: new approach and *dashed lines*: Davison's classical approach)

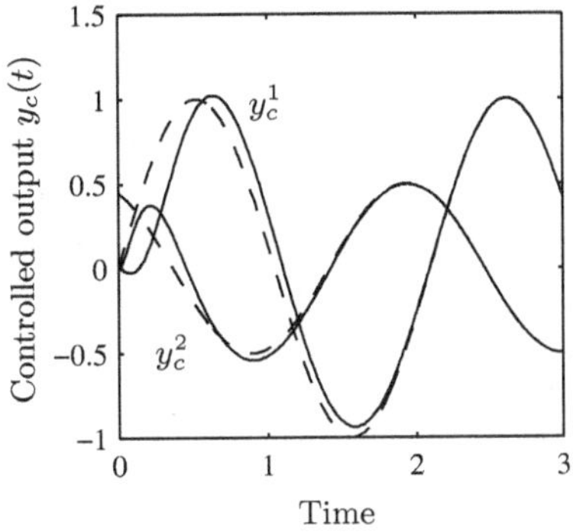

(b) Response to sinusoidal reference inputs when using Davison's classical approach (*solid lines*: controlled output and *dashed lines*: reference input)

Figure 7.9. Reference behaviour of the closed-loop system when compared to Davison's classical approach

Example 7.2. Observer-based compensator with integral action giving rise to controller and plant windup

Considered is the design of an observer-based compensator with integral action for a system of the order $n = 5$, with two inputs (*i.e.*, $p = 2$) and two outputs (*i.e.*, $m = 2$). The two controlled outputs are the two measurements, *i.e.*, $y_c = y$, so that the selection matrix Ξ in $y_c = \Xi y$ is the identity matrix. The right coprime MFD of the transfer matrix $G(s)$ in $y(s) = N(s)D^{-1}(s)u(s)$ of the system is characterized by

$$N(s) = \begin{bmatrix} s+20 & 0 \\ 1 & 1 \end{bmatrix}, \tag{7.184}$$

and

$$D(s) = \begin{bmatrix} s^3 + 3s^2 + 6s + 4 & s + 1 \\ s^2 + 2s + 1 & s^2 + 2s + 1 \end{bmatrix}, \tag{7.185}$$

and with $\Xi = I$ one obtains $N_c(s) = N(s)$. The polynomial matrices of the left coprime MFD $G(s) = \bar{D}^{-1}(s)\bar{N}(s)$ are

$$\bar{N}(s) = \begin{bmatrix} s+20 & -1 \\ 0 & 1 \end{bmatrix}, \tag{7.186}$$

and

$$\bar{D}(s) = \begin{bmatrix} s^3 + 3s^2 + 5s + 3 & 19s + 19 \\ 0 & s^2 + 2s + 1 \end{bmatrix}. \tag{7.187}$$

The relative degrees of the outputs of the system are $\delta_1 = 2$ and $\delta_2 = 2$ (see Theorem 6.1) and therefore, the relative degree of the system is $\delta = 4$ (see Definition 6.2). Since $\det N_c(s) = s + 20$ the transfer matrix $G_c(s) = N_c(s)D^{-1}(s)$ of the system has $z = 1$ transmission zero $s = -20$, so that $z = n - \delta$, *i.e.*, the decoupling criterion (see Theorem 6.2) is satisfied and the system is decouplable in a stable scheme. Assuming that the four freely assignable eigenvalues are placed at $s = -15$ the matrix $\Lambda(s)$ in (6.30) takes the form

$$\Lambda(s) = \begin{bmatrix} \frac{(s+15)^2}{225} & 0 \\ 0 & \frac{(s+15)^2}{225} \end{bmatrix}. \tag{7.188}$$

Applying Theorem 6.3 leads to the frequency-domain parameterization

$$M = \Gamma_c[D(s)]\Gamma_c^{-1}[\Lambda(s)N_c(s)] = \begin{bmatrix} 225 & 0 \\ 0 & 225 \end{bmatrix}, \tag{7.189}$$

and

$$\begin{aligned} \tilde{D}(s) &= M\Lambda(s)N_c(s) \\ &= \begin{bmatrix} s^3 + 50s^2 + 825s + 4500 & 0 \\ s^2 + 30s + 225 & s^2+30s+225 \end{bmatrix} \end{aligned} \tag{7.190}$$

of the decoupling controller.

Since only two measured outputs exist a state observer is required. Using a completely reduced-order observer with $\kappa = 2$ or $n_O = 3$, the row-reduced polynomial matrix $\bar{D}_2(s)$ has the form

$$\bar{D}_2(s) = \begin{bmatrix} s^2+3s+5 & 19 \\ 0 & s+2 \end{bmatrix}. \tag{7.191}$$

A completely reduced-order observer having eigenvalues at $s = -15$ is, *e.g.*, specified by the polynomial matrix

$$\tilde{\bar{D}}(s) = \begin{bmatrix} s^2+30s+225 & 0 \\ 0 & s+15 \end{bmatrix}. \tag{7.192}$$

The transfer behaviour $v(s) = N_S(s)D_S^{-1}(s)y_c(s)$ of the signal model (integral action) is characterized by the two polynomial matrices

$$N_S(s) = \begin{bmatrix} 1 & 0 \\ 0 & 1 \end{bmatrix} \text{ and } D_S(s) = \begin{bmatrix} s & 0 \\ 0 & s \end{bmatrix}. \tag{7.193}$$

Placing the eigenvalues of the controlled signal model also at $s = -15$ the parameterizing polynomial matrix $\tilde{D}_S(s)$ can be chosen as

$$\tilde{D}_S(s) = \begin{bmatrix} s+15 & 0 \\ 0 & s+15 \end{bmatrix}. \tag{7.194}$$

Thus, all parameters for the design of the compensator are known. Following Theorem 7.2 the solutions

$$P = \begin{bmatrix} 3375 & 0 \\ 0 & 3375 \end{bmatrix}, \tag{7.195}$$

and

$$Q(s) = \tilde{D}_x(s) = \begin{bmatrix} s^3 + 65s^2 + 1575s + 13500 & 0 \\ s^2 + 45s + 675 & s^2 + 45s + 675 \end{bmatrix} \tag{7.196}$$

of the Diophantine equation (7.152) can be computed. Then, using the solutions

$$Y(s) = \frac{1}{6897} \begin{bmatrix} s^2 - 17s + 345 & s+1 \\ -s^2 + 17s - 345 & -s + 6896 \end{bmatrix}, \tag{7.197}$$

and

$$X(s) = \frac{1}{6897} \begin{bmatrix} -1 & 0 \\ 1 & 0 \end{bmatrix} \tag{7.198}$$

of the Bezout identity (7.153) one can compute

$$\bar{V}(s) = \frac{1}{6897} \begin{bmatrix} s^4 + 75s^3 + 2250s^2 + 37272s + 648459 & s^3 + 60s^2 + 1350s + 10125 \\ 0 & 6897s + 400026 \end{bmatrix}, \tag{7.199}$$

(see (7.154)) and carry out the right-to-left conversion (7.155), which then leads to

$$V(s) = \frac{1}{6897} \begin{bmatrix} s^4 + 75s^3 + 2250s^2 + 37272s + 648459 & s^4 + 75s^3 + 2250s^2 + 30375s + 151875 \\ 0 & 6897s + 400026 \end{bmatrix}, \tag{7.200}$$

and

$$\hat{\Delta}(s) = \begin{bmatrix} s^2 + 30s + 225 & 0 \\ 0 & s+15 \end{bmatrix}. \tag{7.201}$$

Inserting the above results in (7.156) and (7.157) finally yields the polynomial matrices

$$N_C(s) = \begin{bmatrix} 2029s^3+33387s^2+252909s+759375 & -19s^3-1387s^2-1368s \\ 0 & 1233s^2+13442s+50625 \end{bmatrix}, \tag{7.202}$$

and

$$D_C(s) = \begin{bmatrix} s^3+92s^2+1440s & -s^2-72s \\ 0 & s^2+58s \end{bmatrix} \tag{7.203}$$

of the left MFD of the compensator. With the above results for $\hat{\Delta}(s)$ and $\tilde{D}_S(s)$ the characterizing polynomial matrix $\Delta(s)$ of the overall observer obtains the form

$$\Delta(s) = \hat{\Delta}(s)\tilde{D}_S(s) = \begin{bmatrix} s^3+45s^2+675s+3375 & 0 \\ 0 & s^2+30s+225 \end{bmatrix}. \tag{7.204}$$

This compensator contains integral action, so that step-like disturbances are asymptotically compensated in $y_c = y$.

The reference behaviour is decoupled and the reference transfer matrix of the closed-loop system is

$$G_r(s) = N_c(s)\tilde{D}^{-1}(s)M = \begin{bmatrix} 225 & 0 \\ 0 & 225 \end{bmatrix} \frac{1}{(s+15)^2}, \tag{7.205}$$

i.e., the responses to reference steps $r_i(t) = r_{Si}1(t)$, $i = 1,2$, do not exhibit an overshoot behaviour for arbitrary amplitudes $r_{Si} = const.$

When an input saturation (4.108) with $u_{01} = 3$ and $u_{02} = 3$ is introduced, this is only the case for very small reference input amplitudes not causing saturating output signals of the compensator. If one uses Davison's original approach and applies the observer technique to prevent input saturation causing estimation errors in the state observer, reference step inputs with $r_{S1} = 1$ and $r_{S1} = -1$ give rise to the transients shown in Figure 7.10. They are impaired by a severe windup that causes an unstable behaviour of the transients, so that limit cycles occur. This windup is mainly due to the integrators in the compensator, which give rise to a controller windup when using the driven disturbance model according to (7.14).

When using the modified signal model (7.16), assigning to it the property of an observer, and adding a model of the input saturation (which was carried out in the time domain in Section 7.1) controller windup can be prevented systematically.

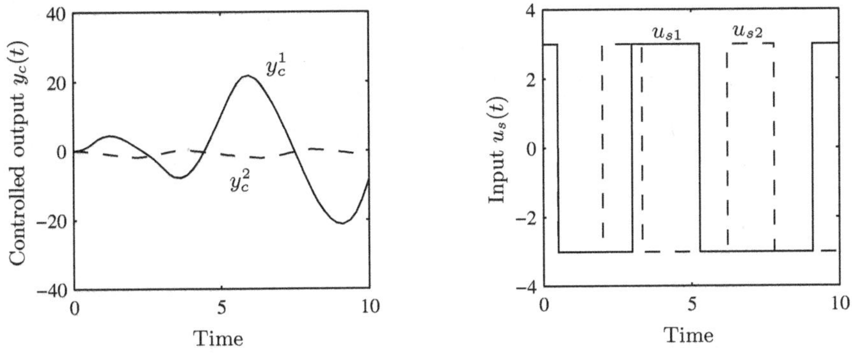

(a) Controlled output of the system (b) Corresponding input

Figure 7.10. Reference behaviour of the closed-loop system without the prevention of controller windup

The frequency-domain equivalent is to compute

$$N_u(s) = D_C(s) - \Delta(s) = \begin{bmatrix} 47s^2 + 765s - 3375 & -s^2 - 72s \\ 0 & 28s - 225 \end{bmatrix}, \tag{7.206}$$

(see (7.159)) and realizing the compensator in the observer structure as shown in Figure 7.7. Then, the reference transients shown in Figure 7.11a result. The transient response of the first output y_c^1 is developing into a limit cycle. Since a realization of the compensator in the observer structure prevents the controller windup the undesired oscillating behaviour of the reference transient response is due to plant windup (see Section 4.5). The danger of plant windup can be investigated on the basis of the transfer matrix

$$G_L(s) = \tilde{D}(s)D^{-1}(s) - I \tag{7.207}$$

(see (4.111)). Given the system used in this example and the above parameterizing polynomial matrix $\tilde{D}(s)$, the transfer matrix $G_L(s)$ does not satisfy the Kalman–Yacubovich lemma. Therefore, the limit cycle that can be observed in the reference transients shown in Figure 7.11a can be attributed to a severe plant windup.

An existing plant windup can be prevented with the aid of an ADE as discussed in Section 4.5. The block diagram in Figure 4.8 shows the closed-loop system consisting of the plant, the observer-based compensator and the ADE. The saturation non-linearity at the input of the system is not shown, because due to the realization of the compensator in the observer structure the input u_s to the constrained system never exceeds the saturation limits.

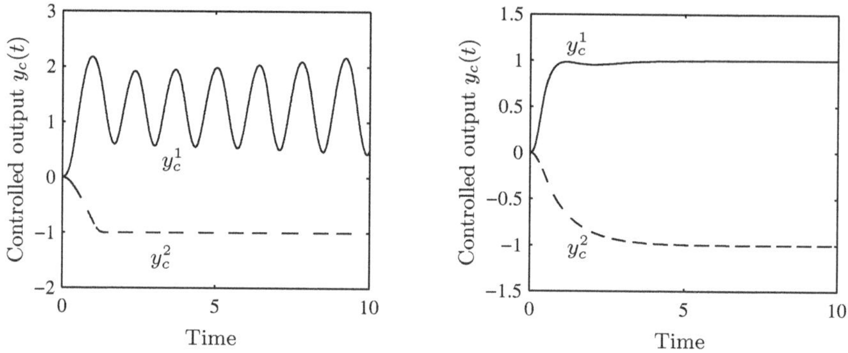

(a) Controlled output after the prevention of controller windup

(b) Controlled output after the controller windup and the plant windup are prevented

Figure 7.11. Reference behaviour of the closed-loop system with controller in the observer structure when input saturation is active

A transfer behaviour of the linear part of the closed-loop system that satisfies the Kalman–Yacubovich lemma can be obtained by assigning $\tilde{\Omega}(s)$ and $\Omega(s)$ in the ADE, such that $\tilde{\Omega}(s) = \tilde{D}(s)$ and the polynomial matrix $\Omega(s)$ solves the bilateral symmetric polynomial matrix equation (4.119) for some matrices $N_Q(s)$ and W that satisfy (4.117) and (4.118).

When choosing

$$N_Q(s) = \begin{bmatrix} s^2 + 23s + 120 & 1.05s + 2.1 \\ 0 & s + 5 \end{bmatrix}, \tag{7.208}$$

and

$$W = \sqrt{2}I \tag{7.209}$$

the polynomial matrix equation (4.119) is solved by

$$\Omega(s) = \begin{bmatrix} s^3+595s^2+1442s+2137 & -365s - 354 \\ -367s^2 - 520s - 651 & s^2 + 374s + 381 \end{bmatrix}. \tag{7.210}$$

Using this $\Omega(s)$ in the scheme of Figure 4.8 the reference transients shown in Figure 7.11b result. Input saturation no longer causes an oscillating behaviour in the transients of the closed-loop system, because now both controller windup and plant windup are prevented.

8

Optimal Control and Estimation

By constant state feedback the dynamics of a linear, controllable, time-invariant system can be assigned in an arbitrary manner. But is there an assignment that is best, or optimal? Whether a solution is optimal or not depends on the criterion to be minimized. A well-known and widely used method is the design of so-called *linear quadratic regulators* (LQR) (see, *e.g.*, [4, 42]). The optimization criterion used is the integral of quadratic functions of the output (or the state) and the inputs. This allows a tradeoff between achieving fast dynamics of the closed-loop system and the corresponding control effort needed. The optimal controller is a constant state feedback, where the feedback gain is obtained from solving an algebraic Riccati equation (ARE) after specifying the matrices weighting the input and the output (or the state) of the linear system.

An equivalent frequency-domain formulation of the linear quadratic control problem leads to a polynomial matrix equation, which is obtained from the ARE by simple manipulations using the connecting relations between the time- and the frequency-domain design of state feedback control. A solution of this equation is possible by spectral factorization and it yields the polynomial matrix $\tilde{D}(s)$ characterizing the LQR.

Also, the dynamics of the state observer can be assigned in an arbitrary manner provided that the system is observable. This observer generates an estimate $\hat{x}$ for the state x of the system, where $\hat{x}$ converges asymptotically to x if no disturbances are present. However, if the inputs and the measurements are corrupted by noise, persistent observation errors occur. In the presence of such random disturbances one can look for a state estimate assuring the smallest mean-square estimation error. Given Gaussian white noise with zero mean and known covariances, an estimate with minimal stationary variance of the estimation error can be obtained by a *stationary Kalman filter*. This optimal observer can also be obtained from a Riccati equation in the time domain (see, *e.g.*, [3, 42]).

The equivalent frequency-domain version of the stationary Kalman filter is parameterized by the polynomial matrix $\bar{\tilde{D}}(s)$, which can also be obtained by spectral factorization of a polynomial matrix equation. This polynomial matrix equation is determined from the related ARE by use of simple manipulations and the connecting relations between the time- and the frequency-domain design of observers. If κ measurements are not corrupted by noise, the order of the resulting Kalman filter is reduced by the number of perfect measurements, *i.e.*, it is an observer of the order $n - \kappa$. The equations for the design of this reduced-order Kalman filter become especially simple in the frequency domain.

The solution of the optimal control problem by means of a dynamic output feedback in the presence of noise is the *linear quadratic Gaussian* (LQG) problem (see, *e.g.*, [8]). It is solved by designing a LQR that is implemented using a Kalman filter. Consequently, the results of this chapter and the design of observer-based compensators in Chapter 4 provide a frequency-domain solution of the LQG problem.

In Section 8.1 the time-domain solution of the LQR is briefly revisited and in Section 8.2 the frequency-domain solution is derived from the time-domain results.

Section 8.3 contains a formulation of the reduced-order Kalman filter on the basis of the observer representation already introduced in Chapter 3. It contains the full-order filter as a special case. In Section 8.4 the frequency-domain formulation of this optimal filter is derived from the time-domain results.

8.1 The Linear Quadratic Regulator in the Time Domain

Considered are linear, time-invariant systems with the state equations

$$\dot{x}(t) = Ax(t) + Bu(t), \tag{8.1}$$

$$y(t) = Cx(t), \tag{8.2}$$

where $x \in \mathbb{R}^n$ is the state, $u \in \mathbb{R}^p$ is the input, and $y \in \mathbb{R}^m$ is the output. It is assumed that the pair (A, B) is controllable and that the pair (C, A) is observable.

Given a constant positive-semidefinite matrix $Q = Q^T$ and a constant positive-definite matrix $R = R^T$ the LQR problem consists of determining a stabilizing control u that minimizes the performance index

$$V = \int_0^\infty \left(y^T(t)Qy(t) + u^T(t)Ru(t) \right) dt, \tag{8.3}$$

when starting at the initial condition $x(0) = x_0$. The first term y^TQy in (8.3) assures that the minimization process leads to a decay of the outputs. The second term $u^T Ru$ is required to obtain a solution with finite control effort. Obviously, increasing elements of Q with $R = const$ lead to faster dynamics, whereas smaller elements of Q yield slower transient responses of the closed-loop system. If a weighting of the controlled output $y_c = \Xi y$ is desired, this can be achieved by choosing the matrix Q in (8.3) as $Q = \Xi^T Q_c \Xi$ with Q_c being a constant positive-semidefinite matrix.

It is shown in, *e.g.*, [4, 42] that the solution of this problem is a constant state feedback

$$u(t) = -Kx(t). \tag{8.4}$$

If $Q = Q_0^T Q_0$ is such that $(Q_0 C, A)$ has no unobservable eigenvalues on the imaginary axis the stabilizing feedback gain in (8.4) is given by

$$K = R^{-1} B^T P, \tag{8.5}$$

where $P = P^T$ is the unique positive-semidefinite solution of the ARE

$$PA + A^T P - PBR^{-1} B^T P + C^T QC = 0 \tag{8.6}$$

(see [4, 22]).

8.2 The Linear Quadratic Regulator in the Frequency Domain

The LQR problem can also be solved on the basis of the transfer behaviour

$$y(s) = C(sI - A)^{-1} Bu(s) = G(s)u(s) \tag{8.7}$$

of the system (8.1) and (8.2). In (8.7) the transfer matrix $G(s)$ is represented by the right coprime MFD

$$G(s) = N(s)D^{-1}(s), \tag{8.8}$$

with $D(s)$ column reduced.

In the frequency domain the state feedback (8.4) is characterized by the denominator polynomial matrix $\tilde{D}(s)$ that has the properties (2.28) and (2.29). Therefore, to solve the optimal control problem in the frequency domain a polynomial matrix equation needs to be determined that yields the optimal solution $\tilde{D}(s)$ of the LQR problem. This polynomial equation can be directly derived from the ARE (8.6) by using the connecting relation between the time- and the frequency-domain representations of state feedback control. This underlines the usefulness of the equivalent representations of the same control problem in the time and in the frequency domains, because it allows derivation of an optimal frequency-domain solution on the basis of the optimization results in the time domain. The polynomial matrix equation that characterizes the optimal $\tilde{D}(s)$ leads to a $p \times p$ polynomial matrix $H(s)$ in the complex variable $s = \sigma + \mathrm{j}\omega$, which is symmetric and positive, *i.e.*,

$$H^T(-s) = H(s), \tag{8.9}$$

and

$$H(\mathrm{j}\omega) > 0 \text{ for all real } \omega. \tag{8.10}$$

Due to the Condition (8.9) the zeros of $\det H(s)$ are located symmetrically with respect to the imaginary axis and (8.10) implies that no zero of $\det H(s)$ lies on the imaginary axis.

The task of *spectral factorization* consists of finding a $p \times p$ polynomial matrix $\tilde{\tilde{D}}(s)$ that satisfies

$$H(s) = \tilde{\tilde{D}}^T(-s)\tilde{\tilde{D}}(s), \tag{8.11}$$

with

$$\det \tilde{\tilde{D}}(s) \neq 0 \text{ for all } \mathrm{Re}(s) \geq 0, \tag{8.12}$$

i.e., the determinant $\det \tilde{\tilde{D}}(s)$ is a Hurwitz polynomial and $\det \Gamma_c[\tilde{\tilde{D}}(s)] \neq 0$. Then, $\tilde{\tilde{D}}(s)$ is a *spectral factor* of $H(s)$ that contains the zeros of $H(s)$ with negative real parts. It has been shown in [68] that the spectral factor $\tilde{\tilde{D}}(s)$ exists and that it is unique up to a multiplicative orthogonal matrix U, *i.e.*, $U^T U = I$.

The next theorem presents the frequency-domain solution of the LQR problem.

Theorem 8.1 (Frequency-domain solution of the LQR problem). *Consider the transfer behaviour of the system (8.1) and (8.2) in a right coprime MFD*

$$y(s) = C(sI - A)^{-1}Bu(s) = N(s)D^{-1}(s)u(s), \tag{8.13}$$

with $D(s)$ column reduced. Given are the $m \times m$ weighting matrix Q and the $p \times p$ weighting matrix R in the performance index (8.3), where $Q = Q_0^T Q_0$ is positive-semidefinite and $R = R^T$ is positive-definite. The linear quadratic regulator problem is solvable iff any greatest common right divisor of $D(s)$ and $Q_0 N(s)$ has no zeros on the imaginary axis.

If this condition is satisfied the unique polynomial matrix $\tilde{D}(s)$ characterizing the optimal and stabilizing state feedback in the frequency domain exists and can be obtained by determining the spectral factor $\tilde{\tilde{D}}(s)$ in

$$N^T(-s)QN(s) + D^T(-s)RD(s) = \tilde{\tilde{D}}^T(-s)\tilde{\tilde{D}}(s), \tag{8.14}$$

and a subsequent computation of

$$\tilde{D}(s) = \Gamma_c[D(s)]\Gamma_c^{-1}[\tilde{\tilde{D}}(s)]\tilde{\tilde{D}}(s). \tag{8.15}$$

Proof. The polynomial equation (8.14) can be obtained from the Riccati equation (8.6). First add and substract sP from (8.6) to obtain

$$P(sI - A) + (-sI - A^T)P + PBR^{-1}B^TP = C^TQC. \tag{8.16}$$

Pre-multiplying (8.16) by $B^T(-sI - A^T)^{-1}$ and postmultiplying the result by $(sI - A)^{-1}B$, while observing that $RK = B^TP$ (see (8.5)), leads to

$$\begin{aligned} &B^T(-sI - A^T)^{-1}K^TR + RK(sI - A)^{-1}B \\ &\quad + B^T(-sI - A^T)^{-1}K^TRK(sI - A)^{-1}B \\ &\qquad = B^T(-sI - A^T)^{-1}C^TQC(sI - A)^{-1}B. \end{aligned} \tag{8.17}$$

By adding R on both sides this can be rewritten as

$$\begin{aligned} &[I + B^T(-sI - A^T)^{-1}K^T]\, R\, [I + K(sI - A)^{-1}B] \\ &\qquad = B^T(-sI - A^T)^{-1}C^TQC(sI - A)^{-1}B + R. \end{aligned} \tag{8.18}$$

Using the connecting relation

$$I + K(sI - A)^{-1}B = \tilde{D}(s)D^{-1}(s), \tag{8.19}$$

(see (2.23)) and (8.13) this takes the form

$$[D^T(-s)]^{-1}\tilde{D}^T(-s)R\tilde{D}(s)D^{-1}(s) = [D^T(-s)]^{-1}N^T(-s)QN(s)D^{-1}(s) + R. \tag{8.20}$$

Pre-multiplying (8.20) by $D^T(-s)$ and postmultiplying the result by $D(s)$ one obtains the polynomial matrix equation

$$H(s) = \tilde{D}^T(-s)R\tilde{D}(s) = N^T(-s)QN(s) + D^T(-s)RD(s). \tag{8.21}$$

Since R is positive-definite and since any greatest common right divisor of $D(s)$ and $Q_0N(s)$ has no zero on the imaginary axis, which implies that the corresponding pair (Q_0C, A) has no unobservable eigenvalues on the imaginary axis (see Remark 1.5), there exists a stabilizing solution P of the ARE (8.6) and the polynomial matrix $H(s)$ has the properties (8.9) and (8.10). Conversely, if the pair (Q_0C, A) has no unobservable eigenvalues on the imaginary axis $Q_0N(s)$ and $D(s)$ do not have a greatest common right divisor that has a zero on the imaginary axis. In order to prove this assume that such a right divisor exists. This would imply that the corresponding pole is not controllable, because there are no unobservable eigenvalues on the imaginary axis. This, however, contradicts the fact that there exist a stabilizing solution of the ARE (8.6) and an optimal spectral factor $\tilde{D}(s)$ of (8.21) because the pre-requisites for the spectral factorization are not satisfied. Thus, the time-domain condition also implies the corresponding frequency-domain condition for the solution of the LQR problem. Therefore, the time-domain condition for the solution of the LQR problem also guarantees that the corresponding frequency-domain condition is satisfied.

As $R = R^T$ is a real-valued positive-definite matrix there exists a unique factorization $R = SS^T = S^2$, so that the *square root* $R^{1/2}$ of R is given by $R^{1/2} = S$ (see, *e.g.*, [37]). Consequently, there also exists a spectral factor $R^{1/2}\tilde{D}(s)$ of the polynomial matrix $H(s)$ in (8.21) because the ARE (8.6) has a stabilizing solution.

As the spectral factorization is unique up to an orthogonal matrix U the spectral factor $\tilde{\tilde{D}}(s)$ in (8.14) satisfies

$$U\tilde{\tilde{D}}(s) = R^{1/2}\tilde{D}(s), \tag{8.22}$$

or

$$\tilde{D}(s) = R^{-1/2}U\tilde{\tilde{D}}(s). \tag{8.23}$$

In view of

$$\Gamma_c[\tilde{D}(s)] = \Gamma_c[D(s)], \tag{8.24}$$

(see (2.29)) the orthogonal matrix U in (8.23) is given by

$$U = R^{1/2}\Gamma_c[D(s)]\Gamma_c^{-1}[\tilde{\tilde{D}}(s)], \tag{8.25}$$

because $\tilde{\tilde{D}}(s)$ is column reduced. Inserting (8.25) in (8.23) yields (8.15). □

Remark 8.1. The time-domain condition that the pair (Q_0C, A) has no unobservable eigenvalues on the imaginary axis, which corresponds to the frequency-domain condition that the polynomial matrices $D(s)$ and $Q_0N(s)$ do not have a greatest common right divisor with a zero on the imaginary axis, only assures that the eigenvalues of the closed-loop system are located in the open left-half complex plane. The location of all eigenvalues can only be influenced by the optimal control if the pair (Q_0C, A) is observable or equivalently, if the greatest common right divisor of $D(s)$ and $Q_0N(s)$ is a unimodular matrix. This becomes obvious by inspection of (8.14), because if all terms on the left-hand side contain a common factor, then this common factor also appears on the right-hand side.

Example 8.1. Frequency-domain solution of the LQR problem
Consider a system of the order 2 with two inputs and two outputs, whose transfer matrix is represented as

$$G(s) = N(s)D^{-1}(s) = \begin{bmatrix} 1 & 2 \\ 0 & 1 \end{bmatrix} \begin{bmatrix} s-3 & -2 \\ 3 & s+4 \end{bmatrix}^{-1}. \tag{8.26}$$

The system is unstable with one eigenvalue at $s = 2$ and the other at $s = -3$.

In a first step, consider the optimal control with minimal control effort characterized by the weighting matrices

$$Q = \begin{bmatrix} 0 & 0 \\ 0 & 0 \end{bmatrix} \quad \text{and} \quad R = \begin{bmatrix} 1 & 0 \\ 0 & 1 \end{bmatrix}, \tag{8.27}$$

i.e., only the input u is weighted in the functional (8.3). The left-hand side of (8.14) then takes the form

$$D^T(-s)RD(s) = \begin{bmatrix} -s^2+18 & 5s+18 \\ -5s+18 & -s^2+20 \end{bmatrix}. \tag{8.28}$$

Since $\det D(s)$ does not have zeros at the imaginary axis there exists a stable spectral factor of (8.28) and the spectral factorization of this polynomial matrix yields

$$\tilde{\tilde{D}}(s) = \begin{bmatrix} s+0.6 & -0.8 \\ 4.2 & s+4.4 \end{bmatrix}. \tag{8.29}$$

Since

$$\Gamma_c[\tilde{\tilde{D}}(s)] = \begin{bmatrix} 1 & 0 \\ 0 & 1 \end{bmatrix} \quad \text{and} \quad \Gamma_c[D(s)] = \begin{bmatrix} 1 & 0 \\ 0 & 1 \end{bmatrix} \tag{8.30}$$

(8.15) leads to the optimal parameterizing polynomial matrix $\tilde{D}(s) = \tilde{\tilde{D}}(s)$ with $\det \tilde{D}(s) = (s+2)(s+3)$. When using an optimal linear quadratic control with minimum control effort, the eigenvalues of the optimal closed-loop system are the stable eigenvalues of the system plus the stable mirror images of the unstable eigenvalues of the system. If the system had been stable, the optimal parameterizing polynomial matrix would have been $\tilde{D}(s) = D(s)$, *i.e.*, in a time-domain approach, the optimal gain would have been $K = 0$.

Now consider an LQR control with

$$Q = \begin{bmatrix} 32 & 0 \\ 0 & 32 \end{bmatrix} \quad \text{and} \quad R = \begin{bmatrix} 2 & 0 \\ 0 & 2 \end{bmatrix}, \tag{8.31}$$

i.e., this time the output y is also weighted in the performace index (8.3), so that one can expect faster decaying transient responses of the closed-loop system. The left-hand side of (8.14) then takes the form

$$N^T(-s)QN(s) + D^T(-s)RD(s) = \begin{bmatrix} -2s^2+68 & 10s+100 \\ -10s+100 & -2s^2+200 \end{bmatrix}. \tag{8.32}$$

Since the pair $(D(s), Q_0N(s))$ does not have a greatest common right divisor with zeros on the imaginary axis, a stable spectral factor of (8.32) exists and the spectral factorization of this polynomial matrix yields

$$\tilde{\tilde{D}}(s) = \sqrt{2}\begin{bmatrix} s+3 & 0 \\ 5 & s+10 \end{bmatrix}. \tag{8.33}$$

Here, the highest column-degree-coefficient matrix $\Gamma_c[\tilde{\tilde{D}}(s)]$ of $\tilde{\tilde{D}}(s)$ is $\sqrt{2}I$, so that (8.15) finally yields

$$\tilde{D}(s) = \Gamma_c[D(s)]\Gamma_c^{-1}[\tilde{\tilde{D}}(s)]\tilde{\tilde{D}}(s) = \begin{bmatrix} s+3 & 0 \\ 5 & s+10 \end{bmatrix}. \tag{8.34}$$

The eigenvalues of the closed-loop system are now $\tilde{s}_1 = -3$ and $\tilde{s}_2 = -10$, *i.e.*, they have been shifted further to the left.

8.3 The Stationary Kalman Filter in the Time Domain

Considered are linear, time-invariant systems with the state equation

$$\dot{x}(t) = Ax(t) + Bu(t) + Gw(t), \tag{8.35}$$

where $x \in \mathbb{R}^n$ is the state, $u \in \mathbb{R}^p$ is the input and $w \in \mathbb{R}^q$ is a stochastic disturbance. It is assumed that the pairs (A, B) and (A, G) are controllable. The measured output $y \in \mathbb{R}^m$ is subdivided as

$$y_1(t) = C_1x(t) + v_1(t), \tag{8.36}$$

$$y_2(t) = C_2x(t), \tag{8.37}$$

where $y_2 \in \mathbb{R}^\kappa$ are the measurements not corrupted by noise (ideal measurements) and $y_1 \in \mathbb{R}^{m-\kappa}$ are the disturbed measurements. It is assumed that the system, *i.e.*, the pair (C, A) is observable, where the $m \times n$ matrix C is defined as

$$C = \begin{bmatrix} C_1 \\ C_2 \end{bmatrix}. \tag{8.38}$$

If all measurements are corrupted by noise, one has $\kappa = 0$, $C_1 = C$ and C_2 vanishes, and for completely undisturbed measurements one has $\kappa = m$, $C_2 = C$ and C_1 vanishes. The stochastic inputs $w \in \mathbb{R}^q$ and $v_1 \in \mathbb{R}^{m-\kappa}$ are independent, zero-mean, stationary Gaussian white noise with

$$\mathrm{E}\{w(t)w^T(\tau)\} = \bar{Q}\delta(t-\tau), \tag{8.39}$$

$$\mathrm{E}\{v_1(t)v_1^T(\tau)\} = \bar{R}_1\delta(t-\tau), \tag{8.40}$$

where $\mathrm{E}\{\cdot\}$ denotes the mathematical expectation and $\delta(t)$ is the Dirac delta function. The covariance matrices $\bar{Q}$ and $\bar{R}_1$ are real and symmetric, where $\bar{Q}$ is positive-semidefinite and $\bar{R}_1$ is positive-definite. The initial state $x(0) = x_0$ is not correlated with the disturbances, *i.e.*, $\mathrm{E}\{x_0w^T(t)\} = 0$ and $\mathrm{E}\{x_0v_1^T(t)\} = 0$ for all $t \geq 0$.

The *stationary Kalman filtering problem* (see, *e.g.*, [3]) amounts to determining an estimate $\hat{x}$ for the state of (8.35) such that the stationary variance $\mathrm{E}\{(x-\hat{x})^T(x-\hat{x})\}$ of the estimation error $x - \hat{x}$ becomes minimal. If all measurements are corrupted by noise a full-order optimal observer, the so-called (stationary) full-order *Kalman filter* can be used to solve the filtering problem. Whenever part of the measurements are ideal, *i.e.*, there exist measurements not corrupted by noise, eigenvalues of the full-order Kalman filter tend to infinity. A regular filter results, if the order of the filter is reduced by the number κ of ideal measurements. To derive the (stationary) reduced-order Kalman filter consider $n - \kappa$ linear combinations

$$\zeta(t) = Tx(t) \tag{8.41}$$

and the κ ideal measurements y_2, which can be used to represent the state x of the system as

$$x(t) = \begin{bmatrix} C_2 \\ T \end{bmatrix}^{-1} \begin{bmatrix} y_2(t) \\ \zeta(t) \end{bmatrix} = \Psi_2 y_2(t) + \Theta\zeta(t). \tag{8.42}$$

In the solution of this filtering problem it is assumed that the covariance matrix

$$\Phi = C_2 G\bar{Q}G^T C_2^T \tag{8.43}$$

characterizing the measurement noise with respect to the time derivative $\dot{y}_2$ is positive-definite. Thus, the covariance matrix

$$\tilde{R} = \begin{bmatrix} \bar{R}_1 & 0 \\ 0 & \Phi \end{bmatrix} \tag{8.44}$$

of the measurement noise with respect to y_1 and $\dot{y}_2$ is also positive-definite, which is a standing assumption in the design of reduced-order optimal estimators (see, *e.g.*, [7,20,23,26,51]). The stationary minimum variance estimate $\hat{x}$ results from (8.42) when the estimate $\hat{\zeta}$ for ζ is obtained from a *reduced-order optimal estimator* (*reduced-order Kalman filter*)

$$\dot{\hat{\zeta}}(t) = T(A - L_1C_1)\Theta\hat{\zeta}(t) + [TL_1 \quad T(A - L_1C_1)\Psi_2]\begin{bmatrix} y_1(t) \\ y_2(t) \end{bmatrix} + TBu(t), \tag{8.45}$$

$$\hat{x}(t) = \Theta\hat{\zeta}(t) + \Psi_2 y_2(t) \tag{8.46}$$

(see [21]). The optimal estimate $\hat{\zeta}$ results if the matrices L_1 and Ψ_2 are chosen such that

$$L_1 = \bar{P}C_1^T \bar{R}_1^{-1}, \tag{8.47}$$

and

$$\Psi_2 = (\bar{P}A^T C_2^T + G\bar{Q}G^T C_2^T)\,\Phi^{-1} \tag{8.48}$$

with Φ as defined in (8.43). The stationary error covariance $\bar{P} = \bar{P}(\infty)$ appearing in (8.47) and (8.48), where

$$\bar{P}(t) = \mathrm{E}\{(x(t) - \hat{x}(t))(x(t) - \hat{x}(t))^T\} \tag{8.49}$$

can be obtained as the solution of an ARE. In [21,62] this ARE is presented as

$$\begin{aligned}(A - \Psi_2C_2A - L_1C_1)\bar{P} + \bar{P}(A^T - A^TC_2^T\Psi_2^T - C_1^TL_1^T)\\ +L_1\bar{R}_1L_1^T + (I - \Psi_2C_2)G\bar{Q}G^T(I - C_2^T\Psi_2^T) = 0.\end{aligned} \tag{8.50}$$

Observing (8.43) and

$$L_1\bar{R}_1 = \bar{P}C_1^T, \tag{8.51}$$

and

$$\bar{P}A^TC_2^T = \Psi_2\Phi - G\bar{Q}G^TC_2^T, \tag{8.52}$$

(which result from (8.47) and (8.48)) the ARE (8.50) can also be represented in the following form

$$A\bar{P} + \bar{P}A^T - \begin{bmatrix} L_1 & \Psi_2 \end{bmatrix} \begin{bmatrix} \bar{R}_1 & 0 \\ 0 & \Phi \end{bmatrix} \begin{bmatrix} L_1^T \\ \Psi_2^T \end{bmatrix} + G\bar{Q}G^T = 0, \tag{8.53}$$

which will be used to derive the equivalent frequency-domain solution. By inserting the optimal solutions (8.47) and (8.48) in (8.53) one obtains a form of the ARE that can be solved for $\bar{P}$ and that then leads to the optimal solutions L_1 and Ψ_2. Though the optimal estimator (8.45) is of the order $n - \kappa$, the corresponding Riccati equation is of the order n, *i.e.*, $\bar{P}$ is an $n \times n$ matrix. This causes numerical problems in some algorithms for the solution of such equations (see, *e.g.*, [4]) because of multiple eigenvalues at $s = 0$. This may be one of the reasons why the above solution to the Kalman filtering problem has received little attention so far. How to remove the problems arizing from the eigenvalues at $s = 0$ is, *e.g.*, discussed in [33].

The matrices L_1 and Ψ_2 completely characterize the optimal Kalman filter, because the remaining parameters T and Θ are specified by a solution of

$$T\Psi_2 = 0, \tag{8.54}$$

with $\operatorname{rank} T = n - \kappa$ and

$$\Theta = \begin{bmatrix} C_2 \\ T \end{bmatrix}^{-1} \begin{bmatrix} 0 \\ I_{n-\kappa} \end{bmatrix} \tag{8.55}$$

(see also (3.21) and (3.22)).

In Section 3.1 it is shown that the matrices L_1 and Ψ_2 in the reduced-order observer (8.45) have to satisfy the conditions

$$C_2\Psi_2 = I \quad \text{and} \quad C_2L_1 = 0. \tag{8.56}$$

To verify that (8.47) and (8.48) satisfy (8.56) consider

$$\begin{aligned} C_2\bar{P}(t) &= C_2\mathrm{E}\{(x(t) - \hat{x}(t))(x(t) - \hat{x}(t))^T\} \\ &= \mathrm{E}\{(y_2(t) - \hat{y}_2(t))(x(t) - \hat{x}(t))^T\}. \end{aligned} \tag{8.57}$$

Since $y_2 = \hat{y}_2$ (see (3.12)) the result

$$C_2\bar{P}(\infty) = C_2\bar{P} = 0 \tag{8.58}$$

is obtained from (8.57). Observing this and (8.43) it is easy to show that the optimal matrices L_1 and Ψ_2 satisfy (8.56).

The above optimal estimation scheme contains the full-order Kalman filter of the order n (resulting when all measurements are corrupted by noise, *i.e.*, the case $\kappa = 0$) and the completely reduced-order optimal estimator of the order $n - m$ (resulting when all measurements are free of noise, *i.e.*, the case $\kappa = m$) as special cases.

8.4 The Stationary Kalman Filter in the Frequency Domain

The system (8.35)–(8.37) can be described by

$$y_1(s) = G_1(s)u(s) + G_{w1}(s)w(s) + v_1(s), \tag{8.59}$$
$$y_2(s) = G_2(s)u(s) + G_{w2}(s)w(s) \tag{8.60}$$

in the frequency domain. Its transfer behaviour between the input u and the measurement y is described by the left coprime MFD

$$G(s) = \begin{bmatrix} G_1(s) \\ G_2(s) \end{bmatrix} = \begin{bmatrix} C_1 \\ C_2 \end{bmatrix} (sI - A)^{-1}B = \bar{D}^{-1}(s)\bar{N}(s), \tag{8.61}$$

and its transfer behaviour between the disturbance input w and the measurement y by the left coprime MFD

$$G_w(s) = \begin{bmatrix} G_{w1}(s) \\ G_{w2}(s) \end{bmatrix} = \begin{bmatrix} C_1 \\ C_2 \end{bmatrix} (sI - A)^{-1}G = \bar{D}^{-1}(s)\bar{N}_w(s), \tag{8.62}$$

where $\bar{D}(s)$ in (8.61) and (8.62) is such that

$$\bar{D}_\kappa(s) = \Pi\left\{\bar{D}(s)\begin{bmatrix} I_{m-\kappa} & 0 \\ 0 & s^{-1}I_\kappa \end{bmatrix}\right\} \tag{8.63}$$

is row reduced (see also Theorem 3.1 and Section A.1).

In the frequency domain the (reduced-order) estimator is characterized by the denominator matrix $\bar{\bar{D}}(s)$ that has the properties (3.79) and (3.80). Therefore, a polynomial matrix equation has to be determined yielding the optimal solution $\bar{\bar{D}}(s)$ of the Kalman filtering problem in the frequency domain. This polynomial matrix equation directly results from the ARE (8.53) when using the connecting relations between the time- and the frequency-domain representations of the estimation problem. It leads to an $m \times m$ polynomial matrix $\bar{H}(s)$ in the complex variable $s = \sigma + j\omega$ that is symmetric and positive. It can be solved by a dual version of the spectral factorization described above that yields an $m \times m$ polynomial matrix $\bar{\bar{D}}(s)$ satisfying

$$\bar{H}(s) = \bar{\bar{D}}(s)\bar{\bar{D}}^T(-s), \tag{8.64}$$

with

$$\det \bar{\bar{D}}(s) \neq 0 \text{ for all } \mathrm{Re}(s) \geq 0, \tag{8.65}$$

i.e., the determinant $\det \bar{\bar{D}}(s)$ is a Hurwitz polynomial and $\det \Gamma_r[\bar{\bar{D}}(s)] \neq 0$. Then, $\bar{\bar{D}}(s)$ is a *spectral factor* of $\bar{H}(s)$ that contains the zeros of $\bar{H}(s)$ with negative real parts. It has been shown in [68] that the spectral factor $\bar{\bar{D}}(s)$ exists and that it is unique up to a multiplicative orthogonal matrix

$\bar{U}$, *i.e.*, $\bar{U}\bar{U}^T = I$. This dual factorization can be obtained by applying to the transpose of $\bar{H}(s)$ an algorithm to solve (8.11) and (8.12) and taking the transpose of the result.

The next theorem shows how the optimal polynomial matrix $\tilde{\bar{D}}(s)$ can be obtained in the frequency domain.

Theorem 8.2 (Design of the reduced-order optimal estimator in the frequency domain). *Given is a system (8.59) and (8.60) driven by white noise with $m - \kappa$, $0 \leq \kappa \leq m$, measurements y_1 also corrupted by white noise. The transfer matrix between the input noise and the measurements is represented in a left coprime MFD $G_w(s) = \bar{D}^{-1}(s)\bar{N}_w(s)$ (see (8.62)). Also given are the symmetric covariance matrices $\bar{Q} = \bar{Q}_0\bar{Q}_0^T$ of the noise at the input of the system and $\bar{R}_1 = \bar{R}_{10}\bar{R}_{10}^T$ of the measurement noise (see (8.39) and (8.40)), where the $q \times q$ matrix $\bar{Q}$ is positive-semidefinite and the $(m-\kappa) \times (m-\kappa)$ matrix $\bar{R}_1$ is positive-definite. It is further assumed that the $\kappa \times \kappa$ constant matrix*

$$\Phi = \Pi\{sG_{w2}(s)\}\bar{Q}\left(\Pi\{sG_{w2}(s)\}\right)^T \tag{8.66}$$

is positive-definite (see (8.62)) and that any greatest common left divisor of

$$\bar{D}(s)\begin{bmatrix} \bar{R}_{10} & 0 \\ 0 & 0 \end{bmatrix} \quad \text{and} \quad \bar{N}_w(s)\bar{Q}_0 \tag{8.67}$$

has no zeros on the imaginary axis. If these conditions are satisfied, then the spectral factor $\bar{\bar{D}}(s)$ in

$$\bar{D}(s)\begin{bmatrix} \bar{R}_1 & 0 \\ 0 & 0 \end{bmatrix}\bar{D}^T(-s) + \bar{N}_w(s)\bar{Q}\bar{N}_w^T(-s) = \bar{\bar{D}}(s)\bar{\bar{D}}^T(-s) \tag{8.68}$$

exists. If the ARE (8.53) has a stabilizing solution, then

$$\tilde{\bar{D}}(s) = \bar{\bar{D}}(s)\Gamma_r^{-1}[\bar{\bar{D}}(s)]\Gamma_r[\bar{D}_\kappa(s)] \tag{8.69}$$

characterizes the optimal and stable estimator in the frequency domain.

Proof. Using the expansion

$$(sI - A)^{-1} = Is^{-1} + As^{-2} + \ldots \tag{8.70}$$

it is straightforward to show that

$$sC_2(sI - A)^{-1}G = C_2G + C_2A(sI - A)^{-1}G, \tag{8.71}$$

so that

$$\Pi\{sC_2(sI - A)^{-1}G\} = \Pi\{sG_{w2}(s)\} = C_2G, \tag{8.72}$$

which shows that (8.66) is equivalent to (8.43).

The polynomial matrix equation (8.68) can be obtained from the Riccati equation (8.53). First, add and substract $s\bar{P}$ from (8.53) to obtain

$$(sI - A)\bar{P} + \bar{P}(-sI - A^T) + L_1\bar{R}_1L_1^T + \Psi_2\Phi\Psi_2^T = G\bar{Q}G^T. \tag{8.73}$$

Pre-multiplying (8.73) by $\begin{bmatrix} C_1 \\ C_2 \end{bmatrix}(sI - A)^{-1}$ and postmultiplying the result by $(-sI - A^T)^{-1}\begin{bmatrix} C_1^T & C_2^T \end{bmatrix}$ leads to

$$\begin{aligned}
&\begin{bmatrix} C_1 \\ C_2 \end{bmatrix}\bar{P}(-sI - A^T)^{-1}\begin{bmatrix} C_1^T & C_2^T \end{bmatrix} + \begin{bmatrix} C_1 \\ C_2 \end{bmatrix}(sI - A)^{-1}\bar{P}\begin{bmatrix} C_1^T & C_2^T \end{bmatrix} \\
&+ \begin{bmatrix} C_1 \\ C_2 \end{bmatrix}(sI - A)^{-1}[L_1\bar{R}_1L_1^T + \Psi_2\Phi\Psi_2^T](-sI - A^T)^{-1}\begin{bmatrix} C_1^T & C_2^T \end{bmatrix} \quad (8.74) \\
&= \begin{bmatrix} C_1 \\ C_2 \end{bmatrix}(sI - A)^{-1}G\bar{Q}G^T(-sI - A^T)^{-1}\begin{bmatrix} C_1^T & C_2^T \end{bmatrix}.
\end{aligned}$$

Because of $\bar{P}C_1^T = L_1\bar{R}_1$ and $C_2\bar{P} = 0$ (see (8.47) and (8.58)) this can be written as

$$\begin{aligned}
&\left\{\begin{bmatrix} C_1 \\ C_2 \end{bmatrix}(sI - A)^{-1}\begin{bmatrix} L_1 & \Psi_2 \end{bmatrix} + \begin{bmatrix} I & 0 \\ 0 & 0_\kappa \end{bmatrix}\right\}\begin{bmatrix} \bar{R}_1 & 0 \\ 0 & \Phi \end{bmatrix} \\
&\cdot\left\{\begin{bmatrix} L_1^T \\ \Psi_2^T \end{bmatrix}(-sI - A^T)^{-1}\begin{bmatrix} C_1^T & C_2^T \end{bmatrix} + \begin{bmatrix} I & 0 \\ 0 & 0_\kappa \end{bmatrix}\right\} \quad (8.75) \\
&= \begin{bmatrix} \bar{R}_1 & 0 \\ 0 & 0 \end{bmatrix} + \begin{bmatrix} C_1 \\ C_2 \end{bmatrix}(sI - A)^{-1}G\bar{Q}G^T(-sI - A^T)^{-1}\begin{bmatrix} C_1^T & C_2^T \end{bmatrix}.
\end{aligned}$$

Using the connecting relation

$$\bar{D}^{-1}(s)\tilde{\bar{D}}(s) = \begin{bmatrix} C_1 \\ C_2 \end{bmatrix}(sI - A)^{-1}[L_1 \quad \Psi_2] + \begin{bmatrix} I_{m-\kappa} & 0 \\ 0 & 0_\kappa \end{bmatrix} \tag{8.76}$$

(see (3.65)) and (8.62) this takes the form

$$\begin{aligned}
&\bar{D}^{-1}(s)\tilde{\bar{D}}(s)\begin{bmatrix} \bar{R}_1 & 0 \\ 0 & \Phi \end{bmatrix}\tilde{\bar{D}}^T(-s)[\bar{D}^T(-s)]^{-1} \\
&= \begin{bmatrix} \bar{R}_1 & 0 \\ 0 & 0 \end{bmatrix} + \bar{D}^{-1}(s)\bar{N}_w(s)\bar{Q}\bar{N}_w^T(-s)[\bar{D}^T(-s)]^{-1}. \quad (8.77)
\end{aligned}$$

Finally, pre-multiplying (8.77) by $\bar{D}(s)$ and postmultiplying it by $\bar{D}^T(-s)$ leads to the polynomial matrix equation

$$\tilde{\bar{D}}(s)\tilde{R}^{1/2}\tilde{R}^{1/2}\tilde{\bar{D}}^T(-s) = \bar{D}(s)\begin{bmatrix} \bar{R}_1 & 0 \\ 0 & 0 \end{bmatrix}\bar{D}^T(-s) + \bar{N}_w(s)\bar{Q}\bar{N}_w^T(-s), \tag{8.78}$$

where $\tilde{R}$ is defined in (8.44). On the one hand, from the right-hand side of

(8.78) the spectral factor $\bar{\bar{\bar{D}}}(s)$ (see (8.68)) with $\det \bar{\bar{\bar{D}}}(s)$ a Hurwitz polynomial can be obtained. It exists because the greatest common left divisor of the pair (8.67) has no zeros on the imaginary axis (see [68]). On the other hand, as the ARE (8.53) has a stabilizing solution by assumption, there exists a corresponding $\bar{\tilde{\bar{D}}}(s)$ with $\det \bar{\tilde{\bar{D}}}(s)$ a Hurwitz polynomial and that satisfies (8.78). Thus,

$$\bar{\bar{\bar{D}}}'(s) = \bar{\tilde{\bar{D}}}(s)\tilde{R}^{1/2} \tag{8.79}$$

is also a spectral factor with respect to the right-hand side of (8.78). Since a spectral factorization is unique up to an orthogonal matrix (see [68]) there must exist an orthogonal matrix $\bar{U}$, such that

$$\bar{\bar{\bar{D}}}(s)\bar{U} = \bar{\tilde{\bar{D}}}(s)\tilde{R}^{1/2}, \tag{8.80}$$

or equivalently

$$\bar{\tilde{\bar{D}}}(s) = \bar{\bar{\bar{D}}}(s)\bar{U}\tilde{R}^{-1/2}. \tag{8.81}$$

Since

$$\Gamma_r[\bar{\tilde{\bar{D}}}(s)] = \Gamma_r[\bar{D}_\kappa(s)], \tag{8.82}$$

(see (3.80)) the orthogonal matrix $\bar{U}$ in (8.81) has the form

$$\bar{U} = \Gamma_r^{-1}[\bar{\bar{\bar{D}}}(s)]\Gamma_r[\bar{D}_\kappa(s)]\tilde{R}^{1/2}, \tag{8.83}$$

which shows (8.69). □

Unfortunately, no condition is known for the solvability of the ARE (8.53), so that the optimality of the spectral factor (8.69) can only be checked by computing the equivalent time-domain solution. In all examples it turns out that the spectral factor of (8.68) yields an optimal solution, so that it is most likely that the conditions for the existence of the spectral factorization are also sufficient for the optimality of the filter. This is always true for the full-order filter, where the optimality directly follows from the frequency-domain results. The proof of this is dual to the proof of Theorem 8.1 when using the results in [22].

To check the optimality of the spectral factor in the time domain consider the ARE (8.53). Since $\bar{R}_1$ and $\bar{Q}$ are given and the observer gains L_1 and Ψ_2 related to a minimal realization (C, A, G) of (8.62) can be obtained by solving the Diophantine equation (3.91) the ARE (8.53) leads to the Lyapunov equation

$$A\bar{P} + \bar{P}A^T = \begin{bmatrix} L_1 & \Psi_2 \end{bmatrix} \begin{bmatrix} \bar{R}_1 & 0 \\ 0 & \Phi \end{bmatrix} \begin{bmatrix} L_1^T \\ \Psi_2^T \end{bmatrix} - G\bar{Q}G^T \tag{8.84}$$

for $\bar{P}$. If A is a Hurwitz matrix this equation has a unique solution for arbitrary right-hand sides (see [9]). In general, A will not be Hurwitz, so that (8.84) cannot be used to compute $\bar{P}$. However, a solution for an arbitrary A can be obtained in the following way.

First, observe that

$$(A - L_1C_1)\bar{P} + \bar{P}(A^T - C_1^T L_1^T) = A\bar{P} + \bar{P}A^T - L_1C_1\bar{P} - \bar{P}C_1^T L_1^T$$
$$= A\bar{P} + \bar{P}A^T - L_1\bar{R}_1L_1^T - L_1\bar{R}_1L_1^T, \quad (8.85)$$

because of $L_1\bar{R}_1 = \bar{P}C_1^T$ (see (8.47)). Therefore,

$$A\bar{P} + \bar{P}A^T - L_1\bar{R}_1L_1^T = (A - L_1C_1)\bar{P} + \bar{P}(A^T - C_1^T L_1^T) + L_1\bar{R}_1L_1^T. \quad (8.86)$$

Using this result (8.84) becomes

$$(A - L_1C_1)\bar{P} + \bar{P}(A^T - C_1^T L_1^T) = -L_1\bar{R}_1L_1^T + \Psi_2\Phi\Psi_2^T - G\bar{Q}G^T. \quad (8.87)$$

Therefore, pre-multiplying (8.87) by ΘT and postmultiplying it by $T^T\Theta^T$ leads to

$$\Theta T(A - L_1C_1)\bar{P}T^T\Theta^T + \Theta T\bar{P}(A^T - C_1^T L_1^T)T^T\Theta^T \quad (8.88)$$
$$= -\Theta TL_1\bar{R}_1L_1^TT^T\Theta^T - \Theta TG\bar{Q}G^TT^T\Theta^T,$$

when using $T\Psi_2 = 0$ (see (3.6)). In view of

$$\Theta T\bar{P} = (I - \Psi_2C_2)\bar{P} = \bar{P} \quad (8.89)$$

(see (3.7) and (8.58)) the result (8.88) becomes

$$\Theta T(A - L_1C_1)\Theta T\bar{P}T^T\Theta^T + \Theta T\bar{P}T^T\Theta^T(A^T - C_1^T L_1^T)T^T\Theta^T \quad (8.90)$$
$$= -\Theta TL_1\bar{R}_1L_1^TT^T\Theta^T - \Theta TG\bar{Q}G^TT^T\Theta^T.$$

Since L_1 and Ψ_2 are obtained from a spectral factor $\bar{\tilde{D}}(s)$ the corresponding matrix $F = T(A - L_1C_1)\Theta$ is a Hurwitz matrix (see (8.45)). Therefore, (8.90) can also be expressed as

$$\Theta FT\bar{P}T^T\Theta^T + \Theta T\bar{P}T^TF^T\Theta^T = -\Theta TL_1\bar{R}_1L_1^TT^T\Theta^T - \Theta TG\bar{Q}G^TT^T\Theta^T. \quad (8.91)$$

Pre-multiplying this by T and postmultiplying it by T^T leads to

$$F\bar{P}_{red} + \bar{P}_{red}^TF^T = -TL_1\bar{R}_1L_1^TT^T - TG\bar{Q}G^TT^T, \quad (8.92)$$

with $\bar{P}_{red} = T\bar{P}T^T$ because of $T\Theta = I$ (see (3.6)). This is a Lyapunov equation for $\bar{P}_{red}$ that has a unique positive-definite solution because F is Hurwitz and the right-hand side of (8.92) is negative-definite (see, *e.g.*, [37]). Finally, observing that

$$\begin{bmatrix} C_2 \\ T \end{bmatrix} \bar{P} \begin{bmatrix} C_2^T & T^T \end{bmatrix} = \begin{bmatrix} 0_\kappa & 0 \\ 0 & T\bar{P}T^T \end{bmatrix} = \begin{bmatrix} 0_\kappa & 0 \\ 0 & \bar{P}_{red} \end{bmatrix}, \quad (8.93)$$

in view of (8.58) the matrix $\bar{P}$ is related to $\bar{P}_{red}$ by

$$\bar{P} = \begin{bmatrix} \Psi_2 & \Theta \end{bmatrix} \begin{bmatrix} 0_\kappa & 0 \\ 0 & T\bar{P}T^T \end{bmatrix} \begin{bmatrix} \Psi_2^T \\ \Theta^T \end{bmatrix} = \Theta \bar{P}_{red} \Theta^T \tag{8.94}$$

(see (3.5)) and yields a positive-semidefinite matrix $\bar{P}$ since $\bar{P}_{red}$ is positive-definite. By inserting (8.94) in (8.84) one has to verify that (8.94) solves the Lyapunov equation (8.84). Then, if the matrices L_1 and Ψ_2 that follow from (8.47) and (8.48) coincide with L_1 and Ψ_2 resulting from the Diophantine equation (3.91) the polynomial matrix $\bar{\tilde{D}}(s)$ characterizes the optimal solution.

Remark 8.2. In the design of full-order Kalman filters one has to solve the polynomial matrix equation

$$\bar{D}(s)\bar{R}_1\bar{D}^T(-s) + \bar{N}_w(s)\bar{Q}\bar{N}_w^T(-s) = \bar{\bar{D}}(s)\bar{\bar{D}}^T(-s), \tag{8.95}$$

where $\bar{R}_1$ is the $m \times m$ covariance matrix of the measurement noise that affects all measurements $y_1 = y$. A comparison of (8.68) and (8.95) shows that the polynomial matrix equation for the design of the reduced-order Kalman filter is obtained from (8.95) by inserting the covariance matrix

$$\mathrm{E}\left\{ \begin{bmatrix} v_1(t) \\ 0 \end{bmatrix} \begin{bmatrix} v_1^T(\tau) & 0^T \end{bmatrix} \right\} = \begin{bmatrix} \bar{R}_1 & 0 \\ 0 & 0 \end{bmatrix} \delta(t-\tau) \tag{8.96}$$

of the measurement noise which only affects the first $m - \kappa$ measurements y_1. This procedure is not feasible in the time domain since the inverse of this rank deficient covariance matrix does not exist in the ARE of the filter.

Example 8.2. Frequency-domain design of a reduced-order Kalman filter
Considered is a system of the order 3 with two input disturbances w and two measurements, where only the first measurement is corrupted by a noise v_1, *i.e.*, one has the case $\kappa = 1$. The transfer behaviour of the system is characterized by the matrix

$$G_w(s) = \begin{bmatrix} G_{w1}(s) \\ G_{w2}(s) \end{bmatrix} = \begin{bmatrix} s^2+6s+12 & 2s^2+6s+6 \\ -s^2-6s-8 & s^2+3s+2 \end{bmatrix} \frac{1}{(s+2)^3}, \tag{8.97}$$

(see (8.62)) and a left coprime MFD of this transfer matrix is, *e.g.*,

$$G_w(s) = \bar{D}^{-1}(s)\bar{N}_w(s) = \begin{bmatrix} s+2 & 2 \\ s+2 & s^2+4s+6 \end{bmatrix}^{-1} \begin{bmatrix} 1 & 2 \\ -s-3 & s+3 \end{bmatrix}. \tag{8.98}$$

This factorization leads to a row-reduced

$$\bar{D}_1(s) = \begin{bmatrix} s+2 & 0 \\ s+2 & s+4 \end{bmatrix} \tag{8.99}$$

(see (8.63)). It is assumed that the covariance matrix of the input disturbance w is

$$\bar{Q} = \begin{bmatrix} \alpha & 0 \\ 0 & \beta \end{bmatrix}, \tag{8.100}$$

with $\alpha = 25/21$ and $\beta = 175/168$, and that the covariance of the measurement noise v_1 is $\bar{R}_1 = 1$. With

$$C_2G = \Pi\{sG_{w2}(s)\} = [-1 \quad 1], \tag{8.101}$$

(see (8.72)) the quantity

$$\Phi = C_2G\bar{Q}G^TC_2^T = \alpha + \beta = 375/168 \tag{8.102}$$

is positive and therefore, a reduced-order Kalman filter of the order $n - \kappa = 2$ can be designed along the lines of Section 8.4.

When substituting the above quantities in (8.68) one obtains

$$H(s) = \bar{\bar{D}}(s)\bar{\bar{D}}^T(-s) = \bar{D}(s)\begin{bmatrix} \bar{R}_1 & 0 \\ 0 & 0 \end{bmatrix}\bar{D}^T(-s) + \bar{N}_w(s)\bar{Q}\bar{N}_w^T(-s) \tag{8.103}$$

$$= \begin{bmatrix} -s^2 + 4 + \alpha + 4\beta & -s^2 + (\alpha - 2\beta)s + 4 - 3\alpha + 6\beta \\ -s^2 - (\alpha - 2\beta)s + 4 - 3\alpha + 6\beta & -(1 + \alpha + \beta)s^2 + 4 + 9\alpha + 9\beta \end{bmatrix}.$$

If the pair (8.67) does not have a greatest common left divisor with zeros on the imaginary axis it is assured that $\det H(s)$ in (8.103) does not have zeros on the imaginary axis and consequently that a spectral factor exists. With

$$\bar{Q}_0 = \begin{bmatrix} \sqrt{\alpha} & 0 \\ 0 & \sqrt{\beta} \end{bmatrix} \text{ and } \bar{R}_{10} = 1, \tag{8.104}$$

the greatest common left divisor of the pair (8.67), namely

$$\bar{D}(s)\begin{bmatrix} \bar{R}_{10} & 0 \\ 0 & 0 \end{bmatrix} = \begin{bmatrix} s+2 & 0 \\ s+2 & 0 \end{bmatrix} \text{ and } \bar{N}_w(z)\bar{Q}_0 = \begin{bmatrix} \sqrt{\alpha} & 2\sqrt{\beta} \\ -\sqrt{\alpha}(s+3) & \sqrt{\beta}(s+3) \end{bmatrix} \tag{8.105}$$

is a unimodular matrix, so that the conditions for the existence of a spectral factor are satisfied. Spectral factorization yields the polynomial matrix

$$\bar{\bar{D}}(s) = \frac{1}{509}\begin{bmatrix} 509s + 1557 & 2\sqrt{\Phi} \\ 509s + 1107 & (509s + 1499)\sqrt{\Phi} \end{bmatrix}. \tag{8.106}$$

With

$$\Gamma_r[\bar{\bar{D}}(s)] = \begin{bmatrix} 1 & 0 \\ 1 & \sqrt{\Phi} \end{bmatrix} \quad \text{and} \quad \Gamma_r[\bar{D}_1(s)] = \begin{bmatrix} 1 & 0 \\ 1 & 1 \end{bmatrix}, \tag{8.107}$$

Equation (8.69) finally leads to

$$\tilde{\bar{\bar{D}}}(s) = \frac{1}{509}\begin{bmatrix} 509s + 1557 & 2 \\ 509s + 1107 & 509s + 1499 \end{bmatrix}. \tag{8.108}$$

In order to show that (8.108) characterizes the optimal solution in the frequency domain consider a minimal realization

$$A = \begin{bmatrix} -2 & -2 & 0 \\ -1 & -4 & 1 \\ -2 & -6 & 0 \end{bmatrix}, \quad G = \begin{bmatrix} 1 & 2 \\ -1 & 1 \\ -3 & 3 \end{bmatrix}, \quad C = \begin{bmatrix} 1 & 0 & 0 \\ 0 & 1 & 0 \end{bmatrix} \tag{8.109}$$

of (8.97), so that the Diophantine equation (3.91) leads to

$$L_1 = \frac{1}{509}\begin{bmatrix} 539 \\ 0 \\ 89 \end{bmatrix} \quad \text{and} \quad \Psi_2 = \frac{1}{509}\begin{bmatrix} 2 \\ 509 \\ 1499 \end{bmatrix}. \tag{8.110}$$

Equation (8.54) is, *e.g.*, solved by

$$T = \begin{bmatrix} -254.5 & 1 & 0 \\ -749.5 & 0 & 1 \end{bmatrix}, \tag{8.111}$$

and with this result (8.55) leads to

$$\Theta = \frac{1}{509}\begin{bmatrix} -2 & 0 \\ 0 & 0 \\ -1499 & 509 \end{bmatrix}. \tag{8.112}$$

With these results and the above $\bar{R}_1$ and $\bar{Q}$ the right-hand side of the Lyapunov equation (8.92) obtains the form

$$-TL_1\bar{R}_1L_1^TT^T - TG\bar{Q}G^TT^T = \begin{bmatrix} -419161.5 & -1234367 \\ 0 & 0 \\ -1234367 & -3635023.5 \end{bmatrix}, \tag{8.113}$$

so that (8.92) can be solved for

$$\bar{P}_{red} = \begin{bmatrix} 68587.75 & 201945.75 \\ 201945.75 & 594597.25 \end{bmatrix}. \tag{8.114}$$

By using (8.94) it can be verified that

$$\bar{P} = \Theta\bar{P}_{red}\Theta^T = \frac{1}{509}\begin{bmatrix} 539 & 0 & 89 \\ 0 & 0 & 0 \\ 89 & 0 & 26.5 \end{bmatrix} \tag{8.115}$$

is a solution of (8.84). With this $\bar{P}$ the above gains L_1 and Ψ_2 result from (8.47) and (8.48). This verifies that (8.108) is in fact the optimal solution, *i.e.*, it characterizes the stationary Kalman filter in the frequency domain.

Together with a linear quadratic regulator parameterized by an optimal polynomial matrix $\tilde{D}(s)$ (see Section 8.2) the polynomial matrix (8.108) can be used to solve the LQG problem in the frequency domain. The design of the corresponding observer-based compensator is described in Chapter 4.

9

Model-matching Control with Two Degrees of Freedom

In the preceding chapters the reference behaviour is specified by assigning a desired reference transfer matrix to the closed-loop system. This, however, does not assure a robust tracking in the presence of modelling errors because the reference input matrix M and the observer-based compensator depend on the parameters of the plant. Furthermore, the disturbance behaviour cannot be assigned independently from the reference behaviour since both depend on the dynamics assigned by the state feedback control. These drawbacks can be circumvented by using *model matching* that is discussed in this chapter. In the model-matching problem a controller is determined that generates an input to the system, so that the output of the system exactly tracks the output of a given reference model. In this chapter the model-matching problem is solved in a two-step approach:

1. Design of a reference model that consists of a model of the system to be controlled and a state feedback control that assigns a desired reference behaviour to the resulting closed-loop system.
2. Design of a tracking controller that achieves an asymptotic tracking of the output generated by the reference signal model also in the presence of modelling errors and of persistently acting disturbances that can be modelled in a signal process.

If no modelling errors and no disturbances are present, and provided that the initial values of the system to be controlled and of the reference signal model coincide, an input $u = u_d$ to the plant assures a reference trajectory $y_c = y_{cd}$. Thus, in a first step a *model-based feedforward controller* is designed that is a model of the closed-loop system having a desired reference transfer behaviour. Since in practical applications initial errors $y_c(0) - y_{cd}(0)$ are present a feedback controller is determined in the second step to stabilize the tracking error $y_c - y_{cd}$. If disturbances have to be rejected that can be modelled by a signal process, the feedback controller is designed by an application of the internal model principle (see Chapter 7). This assures robust disturbance re-

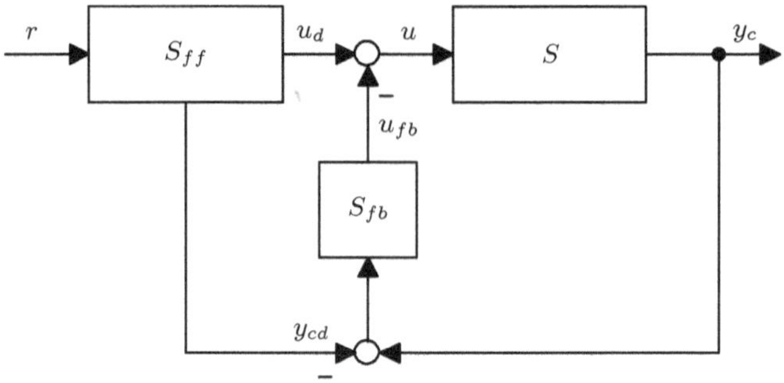

Figure 9.1. Model-matching control for the system S with two degrees of freedom consisting of a model-based feedforward controller S_{ff} and a feedback controller S_{fb}

jection provided that the existing modelling errors do not lead to an unstable closed-loop system.

Different from the control scheme in the previous chapters the feedforward controller achieves the nominal reference transfer behaviour specified by the reference signal model even in the presence of modelling errors. Since the disturbance rejecting feedback controller contains an unstable model of the disturbances and the input signal can saturate, controller windup needs to be prevented. This can be accomplished by designing the driven model of the disturbance signals as an observer (see Chapter 7), so that the methods for the prevention of windup in Section 4.5 are applicable. The stabilizing feedback controller is only active if tracking errors $y_c - y_{cd}$ occur. Therefore, the reference behaviour specified by the feedforward controller S_{ff} and the disturbance behaviour specified by the feedback controller S_{fb} can be designed independently. This leads to the *model-matching control with two degrees of freedom* shown in Figure 9.1.

In the next section the design of the model-based feedforward controller is presented in the time domain. In Section 9.2 a frequency-domain formulation of this model-based feedforward controller is derived. The tracking error is stabilized by a state feedback controller and an application of the internal model principle. The time-domain design of this control scheme is presented in Section 9.3 and its frequency-domain design in Section 9.4. In general, the feedback controller can only be implemented with the aid of an observer. The time-domain design of these observer-based compensators is presented in Section 9.5 and Section 9.6 contains the frequency-domain design of observer-based tracking control.

9.1 Model-based Feedforward Control in the Time Domain

Consider the system

$$\dot{x}(t) = Ax(t) + Bu(t), \quad x(0) = x_0, \tag{9.1}$$
$$y(t) = Cx(t), \tag{9.2}$$

with the state $x \in \mathbb{R}^n$, the input $u \in \mathbb{R}^p$ and the measurement $y \in \mathbb{R}^m$ with $m \geq p$. It is assumed that (A, B) is controllable, (C, A) is observable and that the controlled output $y_c \in \mathbb{R}^p$ is measurable, so that it can be represented by

$$y_c(t) = \left[y_c^1(t)\ y_c^2(t) \ldots y_c^p(t)\right]^T = \Xi y(t) = \Xi Cx(t) = C_c x(t), \tag{9.3}$$

where Ξ is a selection matrix. The relative degree of the ith output y_c^i is given by δ_i, $i = 1, 2, \ldots, p$, and the system is assumed to be decouplable and minimum phase (see Chapter 6).

The desired reference transfer behaviour of the closed-loop system is specified by designing a state feedback controller

$$u_d(t) = -K_d x_d(t) + M_d r(t) \tag{9.4}$$

for a model

$$\dot{x}_d(t) = Ax_d(t) + Bu_d(t), \quad x_d(0) = x_{d,0}, \tag{9.5}$$
$$y_{cd}(t) = C_c x_d(t) \tag{9.6}$$

of the system (9.1) and (9.3). In general, the exact model of the system is not known, so that the design can only be based on a nominal model. In what follows modelling errors are neglected. They will be discussed when designing the feedback controller. In order to specify the tracking behaviour independently for every controlled output the state feedback controller (9.4) is designed to decouple the reference transfer behaviour. In Chapter 6 it was shown that there exists a decoupling controller (9.4) for the model (9.5) and (9.6) that achieves the reference transfer behaviour

$$y_{cd}(s) = \text{diag}(g_{ii}(s))r(s), \tag{9.7}$$

with

$$g_{ii}(s) = \frac{\tilde{d}_{ii}(0)}{\tilde{d}_{ii}(s)}, \quad i = 1, 2, \ldots, p, \tag{9.8}$$

and

$$\tilde{d}_{ii}(s) = (s - \tilde{s}_{i1}) \cdot \ldots \cdot (s - \tilde{s}_{i\delta_i}). \tag{9.9}$$

Since the states x_d of the model (9.5) are measurable the feedback controller (9.4) is directly implementable. By applying the control (9.4) to the model (9.5) one obtains the *model-based feedforward controller*

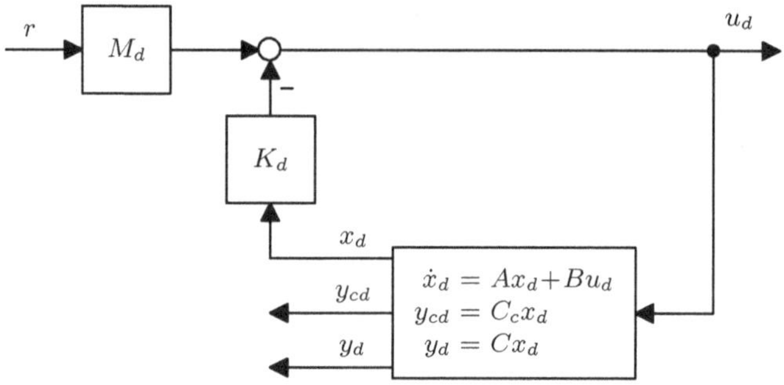

Figure 9.2. Model-based feedforward control in the time domain

$$\dot{x}_d(t) = Ax_d(t) + Bu_d(t), \quad x_d(0) = x_{d,0}, \tag{9.10}$$
$$u_d(t) = -K_d x_d(t) + M_d r(t), \tag{9.11}$$
$$y_{cd}(t) = C_c x_d(t), \tag{9.12}$$
$$y_d(t) = Cx_d(t), \tag{9.13}$$

with the state $x_d \in \mathbb{R}^n$, the input $u_d \in \mathbb{R}^p$ and the outputs $y_{cd} \in \mathbb{R}^p$ and $y_d \in \mathbb{R}^m$. By using the output signal u_d of the state feedback controller (9.4) as a feedforward control $u = u_d$ for the system (9.1) and (9.3) the controlled output y_c of the system exactly tracks the trajectory y_{cd} generated by the model-based feedforward controller. In order to verify this consider the dynamics

$$\begin{aligned} \dot{x}(t) - \dot{x}_d(t) &= Ax(t) + Bu_d(t) - Ax_d(t) - Bu_d(t) \\ &= A(x(t) - x_d(t)), \end{aligned} \tag{9.14}$$
$$x(0) - x_d(0) = x_0 - x_{d,0} \tag{9.15}$$

of the tracking error $e_x = x - x_d$ that result by substracting (9.10) from (9.1) with $u = u_d$. If the initial tracking error vanishes, *i.e.*, $x(0) - x_d(0) = 0$, the unique solution of (9.14) is $x(t) - x_d(t) \equiv 0$, so that the tracking error $e_{y_c} = y_c - y_{cd} = C_c(x - x_d)$ (see (9.3) and (9.12)) is zero at all times. This shows that the model-based feedforward controller (9.10)–(9.13) assures that the output y_c of the system (9.1)–(9.3) exactly tracks the reference trajectory y_{cd}. The resulting model-based feedforward controller is shown in Figure 9.2.

Remark 9.1. If the system (9.1) and (9.3) is not decouplable and/or non-minimum phase a dynamic state feedback can be used to obtain a diagonal decoupling with internal stability (see, *e.g.*, [47,49]). Alternatively, one can use the results of Chapter 6 to achieve a reference transfer matrix with coupled rows.

9.2 Model-based Feedforward Control in the Frequency Domain

In order to represent the model-based feedforward controller (9.10)–(9.13) in the frequency domain introduce the right coprime MFD

$$(sI - A)^{-1}B = N_x(s)D^{-1}(s) \tag{9.16}$$

with $D(s)$ column reduced, so that the transfer behaviours

$$y_{cd}(s) = C_c(sI - A)^{-1}Bu_d(s) = G_c(s)u_d(s), \tag{9.17}$$
$$y_d(s) = C(sI - A)^{-1}Bu_d(s) = G(s)u_d(s) \tag{9.18}$$

of the model (9.10)–(9.13) can be characterized by the right coprime MFD

$$G(s) = N(s)D^{-1}(s), \tag{9.19}$$

and the right MFD

$$G_c(s) = N_c(s)D^{-1}(s), \tag{9.20}$$

where

$$N_c(s) = C_cN_x(s), \tag{9.21}$$
$$N(s) = CN_x(s). \tag{9.22}$$

The transfer behaviour of the model-based feedback controller between the reference input r and the model state x_d is

$$x_d(s) = (sI - A + BK_d)^{-1}BM_dr(s) = N_x(s)\tilde{D}_d^{-1}(s)M_dr(s) \tag{9.23}$$

(see (2.16) and (2.21) for $C_c = I$). With (9.12), (9.13), (9.21) and (9.22) the transfer behaviours between the reference input and the outputs of the feedforward controller are

$$y_{cd}(s) = C_c(sI - A + BK_d)^{-1}BM_dr(s) = N_c(s)\tilde{D}_d^{-1}(s)M_dr(s), \tag{9.24}$$
$$y_d(s) = C(sI - A + BK_d)^{-1}BM_dr(s) = N(s)\tilde{D}_d^{-1}(s)M_dr(s). \tag{9.25}$$

Taking the Laplace transform of (9.11) and inserting (9.23) yields

$$u_d(s) = -K_dx_d(s) + M_dr(s) = (\tilde{D}_d(s) - K_dN_x(s))\tilde{D}_d^{-1}(s)M_dr(s), \tag{9.26}$$

after a simple rearrangement. Solving the relation

$$D(s) + K_dN_x(s) = \tilde{D}_d(s) \tag{9.27}$$

(see (2.18)) for $D(s)$ leads to

$$D(s) = \tilde{D}_d(s) - K_dN_x(s). \tag{9.28}$$

By inserting this in (9.26) one obtains

$$u_d(s) = D(s)\tilde{D}_d^{-1}(s)M_d r(s). \tag{9.29}$$

Consequently, a frequency-domain representation of the model-based feedforward controller (9.10)–(9.13) has the form

$$\begin{bmatrix} u_d(s) \\ y_{cd}(s) \\ y_d(s) \end{bmatrix} = \begin{bmatrix} D(s) \\ N_c(s) \\ N(s) \end{bmatrix} \tilde{D}_d^{-1}(s)M_d r(s) \tag{9.30}$$

(see (9.24), (9.25) and (9.29)). In order to determine the parameterizing matrices $\tilde{D}_d(s)$ and M_d of the feedforward controller (9.30) a frequency-domain parameterization of the decoupling controller (9.4) is needed. It was shown in Theorem 6.3 that a decoupling control for the model-based feedforward controller leads to the reference transfer behaviour

$$y_{cd}(s) = N_c(s)\tilde{D}_d^{-1}(s)M_d r(s) = \Lambda_d^{-1}(s)r(s). \tag{9.31}$$

Consequently, the reference transfer behaviour (9.7) is achieved by the decoupling controller if

$$\Lambda_d(s) = \operatorname{diag}\left(\frac{\tilde{d}_{ii}(s)}{\tilde{d}_{ii}(0)}\right). \tag{9.32}$$

Thus, by using the results of Theorem 6.3 the frequency-domain parameterization of the controller is given by

$$\tilde{D}_d(s) = M_d\Lambda_d(s)N_c(s), \tag{9.33}$$

with the constant matrix

$$M_d = \Gamma_c[D(s)]\Gamma_c^{-1}[\Lambda_d(s)N_c(s)]. \tag{9.34}$$

9.3 Tracking Control by State Feedback in the Time Domain

In practical applications initial errors, external disturbances or modelling errors are present, so that a feedback controller is needed for the stabilization of the tracking error and for the rejection of disturbances. In this section state feedback controllers are designed for the tracking control in the time domain.

9.3.1 Tracking Controller without Disturbance Rejection

Consider the control signal

$$u(t) = u_d(t) + u_{fb}(t), \tag{9.35}$$

where u_d is the feedforward control provided by the model-based feedforward controller (9.10) and (9.11), and u_{fb} is an additional feedback control to stabilize the tracking error. If (9.35) is applied to the system, the dynamics of the tracking error $e_x = x - x_d$ are described by

$$\begin{aligned}\dot{e}_x(t) &= \dot{x}(t) - \dot{x}_d(t) = Ax(t) + B(u_d(t) + u_{fb}(t)) - Ax_d(t) - Bu_d(t) \\ &= Ae_x(t) + Bu_{fb}(t)\end{aligned} \tag{9.36}$$

(see (9.1) and (9.10)). By introducing (9.3) and (9.12) in $e_{y_c} = y_c - y_{cd}$ the *error system*

$$\dot{e}_x(t) = Ae_x(t) + Bu_{fb}(t), \quad e_x(0) = x_0 - x_d(0), \tag{9.37}$$

$$e_{y_c}(t) = C_c e_x(t) \tag{9.38}$$

is obtained from (9.36). Since the pair (A, B) is controllable the error dynamics (9.37) are stabilizable by the state feedback

$$u_{fb}(t) = -Ke_x(t). \tag{9.39}$$

This yields the asymptotically stable closed-loop dynamics

$$\dot{e}_x(t) = (A - BK)e_x(t), \quad e_x(0) = x_0 - x_d(0). \tag{9.40}$$

If initial errors $e_x(0) \neq 0$ or asymptotically decaying external disturbances are present, the tracking error $e_{y_c} = y_c - y_d = C_c e_x$ vanishes asymptotically. Finally, by introducing (9.39) in (9.35) the stabilizing tracking controller

$$u(t) = u_d(t) - Ke_x(t) \tag{9.41}$$

is obtained. The feedforward control u_d in (9.41) assures the desired tracking for the controlled output y_c (see Section 9.1). Only in the presence of initial errors or if disturbances act on the system (9.1) and (9.3), is the feedback controller (9.39) active and assures asymptotic tracking. Therefore, the reference behaviour and the disturbance behaviour can be assigned independently, so that the presented tracking-control scheme has two degrees of freedom.

9.3.2 Tracking Controller with Disturbance Rejection

If non-vanishing disturbances affect the system (9.1) and (9.2) and/or modelling errors are present, the tracking controller (9.41) cannot assure an asymptotic tracking of the desired reference trajectory y_{cd}. The influence of constant modelling errors can be compensated by an integral action in the feedback controller and if the disturbances can be modelled by signal processes, asymptotic disturbance rejection can be achieved using the internal model principle (see Section 7.1). In what follows it is shown how this measure can be incorporated in the design of the tracking controller (9.41). Since the resulting controller contains the unstable signal model of the disturbances, a prevention of windup is needed to assure a desired control performance also in the presence of restricted input signals. This becomes possible by formulating the controlled signal model of the disturbances as an observer, so that the prevention of controller windup for observer-based controllers can be applied (see also Section 7.1).

When disturbances are taken into account, the system (9.1) and (9.3) has the form

$$\dot{x}(t) = Ax(t) + Bu(t) + B_d d(t), \tag{9.42}$$

$$y_c(t) = C_c x(t) + D_{cd} d(t). \tag{9.43}$$

It is assumed that the unmeasurable disturbance $d \in \mathbb{R}^p$ can be modelled by the signal process

$$\dot{v}^*(t) = S^* v^*(t), \tag{9.44}$$

$$d(t) = H v^*(t), \tag{9.45}$$

where $v^* \in \mathbb{R}^q$ is the state of the disturbance process with $v^*(0) = v_0^*$ unknown, and the pair (H, S^*) is observable (see Section 7.1).

When using the internal model principle as introduced by Davison (see [10, 11]) for the asymptotic rejection of disturbances modelled by (9.44) and (9.45), the controller is extended by the driven signal model

$$\dot{v}(t) = Sv(t) + B_\varepsilon e_{y_c}(t), \quad v(0) = v_0, \tag{9.46}$$

with the state $v \in \mathbb{R}^{pq}$ and the matrices S and B_ε defined in (7.13) and (7.15). A drawback of this approach is that it does not allow to use the observer technique for the prevention of controller windup (see Section 4.5). Therefore, the driven signal model (9.46) is formulated as an observer. To this end, consider the modified driven signal model

$$\dot{v}(t) = Sv(t) + B_\varepsilon e_{y_c}(t) - B_\sigma \sigma(t) \tag{9.47}$$

with a $pq \times p$ constant matrix B_σ. In (9.47) σ is defined by

$$\sigma(t) = u(t) - u_d(t) + \tilde{u}_{fb}(t), \tag{9.48}$$

where

$$u_{fb}(t) = u(t) - u_d(t) = -\tilde{u}_{fb}(t) \tag{9.49}$$

(see (9.35)). Equation (9.49) implies that $\sigma(t) \equiv 0$, so that the driven signal models (9.46) and (9.47) coincide. In what follows the additional degrees of freedom contained in B_σ are used to assure that the state v of the driven signal model (9.47) reconstructs Σe_x, where Σ is a suitable constant $pq \times n$ matrix and e_x is the state of the error system (9.37). The error dynamics of this observer and the dynamics of the tracking error e_x are stabilized by the feedback

$$\tilde{u}_{fb}(t) = K_x e_x(t) + K_v v(t). \tag{9.50}$$

By substituting (9.48) and (9.50) in (9.47) one obtains the observer

$$\dot{v}(t) = (S - B_\sigma K_v)v(t) + B_\varepsilon e_{y_c}(t) - B_\sigma(u(t) - u_d(t)) - B_\sigma K_x e_x(t), \quad (9.51)$$

with the input $u - u_d$ and the measurements e_{y_c} and e_x. In order to determine the matrix B_σ in (9.51) and the state feedback gains K_x and K_v in (9.50), consider the error dynamics of the observer (9.51)

$$\begin{aligned} \Sigma\dot{e}_x(t) - \dot{v}(t) &= (S - B_\sigma K_v)(\Sigma e_x(t) - v(t)) \\ &\quad + (-(S - B_\sigma K_v)\Sigma + \Sigma A + B_\sigma K_x - B_\varepsilon C_c)e_x(t) \\ &\quad + (\Sigma B + B_\sigma)(u(t) - u_d(t)), \end{aligned} \quad (9.52)$$

where (9.35), (9.37), (9.38) and (9.51) have been used. Consequently, with

$$B_\sigma = -\Sigma B \quad (9.53)$$

the homogeneous error dynamics

$$\Sigma\dot{e}_x(t) - \dot{v}(t) = (S - B_\sigma K_v)(\Sigma e_x(t) - v(t)), \quad \Sigma e_x(0) - v(0) = \Sigma e_x(0) - v_0 \quad (9.54)$$

result if there exists a solution Σ of

$$\Sigma(A - BK_x) - (S + \Sigma BK_v)\Sigma = B_\varepsilon C_c. \quad (9.55)$$

In the resulting closed-loop system, which is shown in Figure 9.3, the saturated input $u - u_d$ is used in (9.51), so that controller windup is prevented (see Section 4.5).

When the disturbances are neglected, the error dynamics of the closed-loop system are described by

$$\begin{bmatrix} \dot{e}_x(t) \\ \Sigma\dot{e}_x(t) - \dot{v}(t) \end{bmatrix} = \begin{bmatrix} A - B(K_x + K_v\Sigma) & BK_v \\ 0 & S - B_\sigma K_v \end{bmatrix} \begin{bmatrix} e_x(t) \\ \Sigma e_x(t) - v(t) \end{bmatrix}. \quad (9.56)$$

These equations result from (9.54) and by inserting (9.49) and (9.50) in (9.37) after a simple rearrangement. Due to the triangular structure of (9.56) the error dynamics can be stabilized by determining the state feedback gains K_x and K_v such that $A - B(K_x + K_v\Sigma)$ and $S - B_\sigma K_v$ are Hurwitz matrices. To this end, the state feedback gain

$$K = K_x + K_v\Sigma \quad (9.57)$$

is computed such that $A - B(K_x + K_v\Sigma) = A - BK$ has all its eigenvalues in the open left-half complex plane. This is always possible since (A, B) is assumed to be controllable. Then, the matrix equation (9.55) can be rewritten as

$$\Sigma(A - BK) - S\Sigma = B_\varepsilon C_c, \quad (9.58)$$

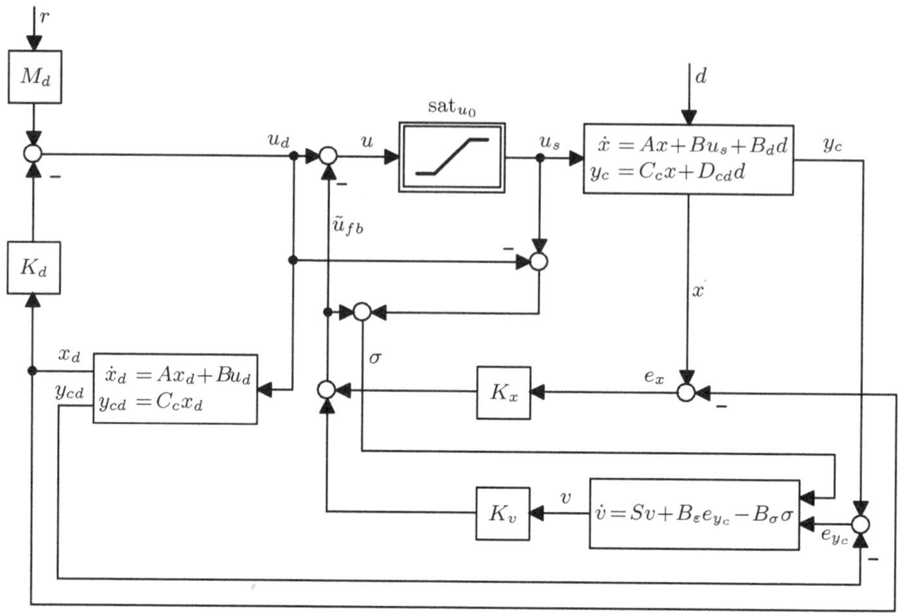

Figure 9.3. Tracking control with disturbance rejection and windup prevention using state feedback in the time domain

where Σ is the only unknown matrix. As (9.58) is a Sylvester equation for Σ, there exists a unique solution if the eigenvalues of $A - BK$ and S do not coincide (see [9]). This is always satisfied, because $A - BK$ is a Hurwitz matrix and the eigenvalues of S are located in the closed right-half complex plane (see Table 7.1). With Σ obtained by solving (9.58) the matrix B_σ follows from (9.53). If the pair (S, B_σ) is controllable, then the eigenvalues of $S - B_\sigma K_v$ are arbitrarily assignable by choosing K_v. By comparing the system (9.42) and (9.43) with (7.37) and (7.38) the results of Theorem 7.1 apply. Consequently, the pair (S, B_σ) is controllable if

- the pairs (A, B) and (S^*, b_ε) are controllable (see (7.13)) and
- rank $\begin{bmatrix} \lambda_i I - A & B \\ C_c & 0 \end{bmatrix} = n + p$ for all eigenvalues λ_i, $i = 1, 2, \ldots, q$, of S^* (see (7.13)).

Finally, the feedback gain K_x can be obtained from

$$K_x = K - K_v \Sigma \tag{9.59}$$

(see (9.57)).

9.4 Tracking Control by State Feedback in the Frequency Domain

In this section, the frequency-domain design of the tracking controller with disturbance rejection is presented. The design of tracking controllers without disturbance rejection is omitted since these controllers are easily derivable from the presented results.

In what follows the frequency-domain parameterization is considered of the stabilization of the extended system

$$\dot{e}_x(t) = Ae_x(t) + Bu_{fb}(t), \tag{9.60}$$
$$\dot{v}(t) = Sv(t) + B_\varepsilon e_{y_c}(t) - B_\sigma(u_{fb}(t) + \tilde{u}_{fb}(t)), \tag{9.61}$$
$$e_{y_c}(t) = C_c e_x(t) \tag{9.62}$$

(see (9.37), (9.38) and (9.47)–(9.49)) by the state feedback

$$u_{fb}(t) = -\tilde{u}_{fb}(t) = -K_x e_x(t) - K_v v(t). \tag{9.63}$$

In Section 7.1 the extended system

$$\dot{x}(t) = Ax(t) + Bu(t), \tag{9.64}$$
$$\dot{v}(t) = Sv(t) + B_\varepsilon y_c(t) - B_\sigma(u(t) + \tilde{u}_{aug}(t)), \tag{9.65}$$
$$y_c(t) = C_c x(t), \tag{9.66}$$

(see (7.17)–(7.21)) is stabilized by the state feedback

$$u(t) = -\tilde{u}_{aug}(t) = -K_x x(t) - K_v v(t). \tag{9.67}$$

Obviously, the above stabilization problems coincide. Therefore, the results of Section 7.2 can be directly used to solve the stabilization problem for the tracking control in the frequency domain, so that only a summary of the frequency-domain parameterization of the state feedback (9.63) is presented.

Before introducing the parameterizing polynomial matrices a frequency-domain description of the error system and of the driven signal model is needed. The transfer behaviour

$$e_{y_c}(s) = C_c(sI - A)^{-1}Bu_{fb}(s) \tag{9.68}$$

of the error system (9.60) and (9.62) can be represented in the frequency domain by using the right coprime MFD

$$(sI - A)^{-1}B = N_x(s)D^{-1}(s), \tag{9.69}$$

with $D(s)$ column reduced. Then, the transfer matrix in (9.68) is characterized by the MFD

$$C_c(sI - A)^{-1}B = N_c(s)D^{-1}(s), \tag{9.70}$$

with

$$N_c(s) = C_c N_x(s). \tag{9.71}$$

The transfer behaviour

$$v(s) = (sI - S)^{-1}B_\varepsilon e_{y_c}(s) - (sI - S)^{-1}B_\sigma(u_{fb}(s) + \tilde{u}_{fb}(s)) \tag{9.72}$$

of the driven signal model (9.61) can be described by polynomial matrices by introducing the right coprime MFDs

$$(sI - S)^{-1}B_\varepsilon = N_S(s)D_S^{-1}(s), \tag{9.73}$$

$$(sI - S)^{-1}B_\sigma = N_{Sr}(s)D_S^{-1}(s), \tag{9.74}$$

where

$$D_S(s) = \det(sI - S^*)I_p \tag{9.75}$$

(see (7.70)). In the frequency domain the dynamics of the closed-loop system (9.60), (9.61) and (9.63) are parameterized by the $p \times p$ polynomial matrix

$$\tilde{D}_{aug}(s) = D_S(s)D(s) + K_x N_x(s)D_S(s) + K_v N_S(s)N_c(s) \tag{9.76}$$

(see (7.79)–(7.81)). If

- the pairs $(N_x(s), D(s))$ and $(N_S(s), D_S(s))$ are right coprime and
- the polynomials $\det N_c(s)$ and $\det D_S(s) = \det(sI - S)$ do not have a common zero,

the stabilization problem is solvable (see Section 7.2). In Section 7.2 it was shown that due to the observer property of the driven signal model the parameterizing polynomial matrix $\tilde{D}_{aug}(s)$ can be specified by

$$\tilde{D}_{aug}(s) = \tilde{D}_S(s)\tilde{D}(s). \tag{9.77}$$

In (9.77) the $p \times p$ polynomial matrix

$$\tilde{D}_S(s) = D_S(s) + K_v N_{Sr}(s) \tag{9.78}$$

(see (7.72)) characterizes the error dynamics (9.54) of the observer formulation of the driven signal model (9.51) and the $p \times p$ polynomial matrix

$$\tilde{D}(s) = D(s) + K N_x(s) \tag{9.79}$$

(see (7.66)) assigns the dynamics of the closed-loop system

$$\dot{e}_x(t) = (A - BK)e_x(t) \tag{9.80}$$

(see (9.56) and (9.57)). As a consequence of (9.75), (9.78) and (9.79) the polynomial matrices $\tilde{D}_S(s)$ and $\tilde{D}(s)$ must satisfy

$$\delta_{ci}[\tilde{D}_S(s)] = q, \quad i = 1, 2, \ldots, p, \tag{9.81}$$

$$\Gamma_c[\tilde{D}_S(s)] = I_p, \tag{9.82}$$

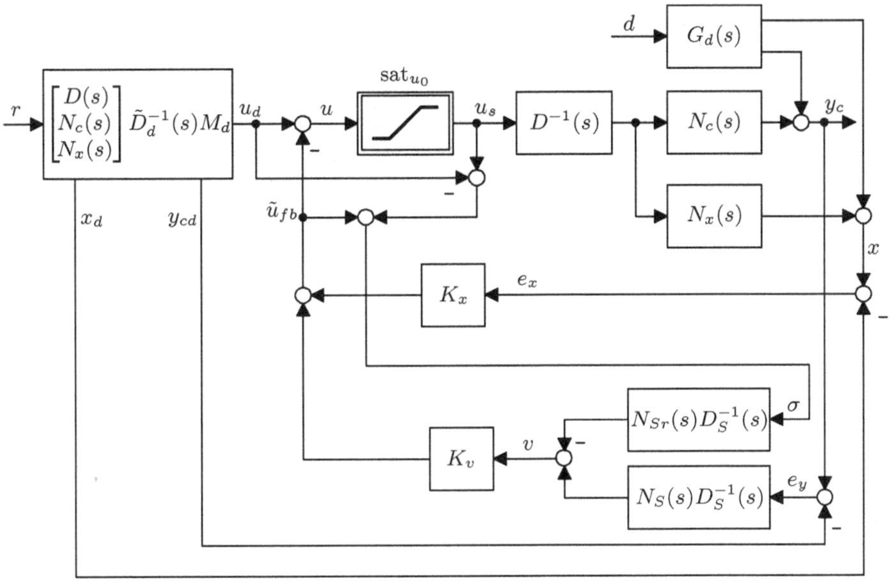

Figure 9.4. Tracking control with disturbance rejection and windup prevention using state feedback in frequency domain

and

$$\delta_{ci}[\tilde{D}(s)] = \delta_{ci}[D(s)], \quad i = 1, 2, \ldots, p, \tag{9.83}$$
$$\Gamma_c[\tilde{D}(s)] = \Gamma_c[D(s)]. \tag{9.84}$$

The resulting closed-loop system in Figure 9.4 shows that one needs to know the transfer behaviour (9.74) of the driven signal model and the state feedback gains K_x and K_v in order to implement the tracking control scheme. The transfer behaviour (9.74) results from the solution $(N_{Sr}(s), \Sigma N_x(s))$ of the polynomial matrix equation

$$N_{Sr}(s)\tilde{D}(s) + \Sigma N_x(s)D_S(s) = N_S(s)N_c(s) \tag{9.85}$$

(see (7.104)). The feedback gain K_v is obtained from determining the solution $(\tilde{D}_x(s), K_v)$ of

$$\tilde{D}_x(s)D_S(s) + K_vN_S(s)N_c(s) = \tilde{D}_S(s)\tilde{D}(s) \tag{9.86}$$

(see (7.100)). Then, K_x can be computed by determining the solution $P(s) = K_x$ and $Q(s) = I$ of the Diophantine equation

$$P(s)N_x(s) + Q(s)D(s) = \tilde{D}_x(s) \tag{9.87}$$

(see (7.97)).

9.5 Observer-based Tracking Control in the Time Domain

A drawback of the tracking controllers presented in Section 9.3 is that the tracking error e_x is assumed to be measurable. In most applications this assumption is not satisfied, so that an observer has to estimate the tracking error e_x. This leads to an observer-based tracking controller where a reduced-order observer is introduced to implement the tracking controller with disturbance rejection.

The observer-based tracking controller is designed for the error system

$$\dot{e}_x(t) = Ae_x(t) + Bu_{fb}(t), \quad e_x(0) = x_0 - x_d(0), \tag{9.88}$$

$$e_y(t) = Ce_x(t), \tag{9.89}$$

(see (9.37) and (9.38)) with the state $e_x \in \mathbb{R}^n$, the input $u_{fb} \in \mathbb{R}^p$ and the measurement $e_y \in \mathbb{R}^m$ with $m \geq p$. It is assumed that the pair (A, B) is controllable and the pair (C, A) is observable. The measurement is partitioned in the output $e_{y_1} \in \mathbb{R}^{m-\kappa}$ and the output $e_{y_2} \in \mathbb{R}^{\kappa}$, *i.e.*,

$$e_y(t) = \begin{bmatrix} e_{y_1}(t) \\ e_{y_2}(t) \end{bmatrix} = \begin{bmatrix} C_1 \\ C_2 \end{bmatrix} e_x(t), \tag{9.90}$$

where e_{y_2} is directly used to reconstruct the state. On the basis of these outputs a reduced-order observer of the order $n_O = n - \kappa$, $0 \leq \kappa \leq m$, can be designed (see Section 3.1). The tracking error e_{y_c} of the controlled output y_c is measurable, such that there exists a representation

$$e_{y_c}(t) = \Xi e_y(t) = \Xi Ce_x(t) = C_c e_x(t), \tag{9.91}$$

with a selection matrix Ξ. In order to implement the tracking controller of Section 9.3.2 the reduced-order observer

$$\dot{\hat{\zeta}}(t) = T(A - L_1C_1)\Theta\hat{\zeta}(t) + \begin{bmatrix} TL_1 & T(A - L_1C_1)\Psi_2 \end{bmatrix} \begin{bmatrix} e_{y_1}(t) \\ e_{y_2}(t) \end{bmatrix}$$
$$+TBu_{fb}(t), \quad \hat{\zeta}(0) = \hat{\zeta}_0, \tag{9.92}$$

$$\hat{e}_x(t) = \Theta\hat{\zeta}(t) + \Psi_2 e_{y_2}(t), \tag{9.93}$$

with $\hat{\zeta} \in \mathbb{R}^{n-\kappa}$, $0 \leq \kappa \leq m$, is designed for the error system (9.88) and (9.89). The observer gains L_1 and Ψ_2 can, *e.g.*, be obtained by applying the parametric approach presented in Section 5.5. Then, the remaining matrices T and Θ follow from (3.21) and (3.22). By inserting the estimate $\hat{e}_x$ (see (9.93)) in (9.50) and by using (9.49) the feedback

$$u_{fb}(t) = -K_x\hat{e}_x(t) - K_v v(t) = -K_x\Theta\hat{\zeta}(t) - K_x\Psi_2 e_{y_2}(t) - K_v v(t) \tag{9.94}$$

is obtained. In order to use the results for preventing controller windup (see Section 4.5) the input u_{fb} of the observer (9.92) and (9.93) is substituted by

$$\bar{\sigma}(t) = u_s(t) - u_d(t), \tag{9.95}$$

(see (9.49)) where u_s denotes the saturated input. By joining the observer (9.92) and (9.93) and the observer formulation (9.51) of the driven signal model to obtain the overall observer (see also Section 7.1) the observer-based tracking controller

$$\begin{aligned}\dot{\hat{\zeta}}(t) &= T(A - L_1C_1)\Theta\hat{\zeta}(t) + \begin{bmatrix}TL_1 & T(A-L_1C_1)\Psi_2\end{bmatrix}\begin{bmatrix}e_{y_1}(t)\\ e_{y_2}(t)\end{bmatrix}\\ &\quad +TB(u_s(t) - u_d(t)), \quad \hat{\zeta}(0) = \hat{\zeta}_0, &(9.96)\\ \dot{v}(t) &= (S - B_\sigma K_v)v(t) + B_\varepsilon e_{y_c}(t) - B_\sigma(u_s(t) - u_d(t))\\ &\quad -B_\sigma(K_x\Theta\hat{\zeta}(t) + K_x\Psi_2 e_{y_2}(t)), \quad v(0) = v_0, &(9.97)\\ \hat{\tilde{u}}_{fb}(t) &= K_x\Theta\hat{\zeta}(t) + K_x\Psi_2 e_{y_2}(t) + K_v v(t) &(9.98)\end{aligned}$$

results, which has the order $n - \kappa + pq$. The resulting tracking control scheme is shown in Figure 9.5.

In Figure 9.5 the abbreviations

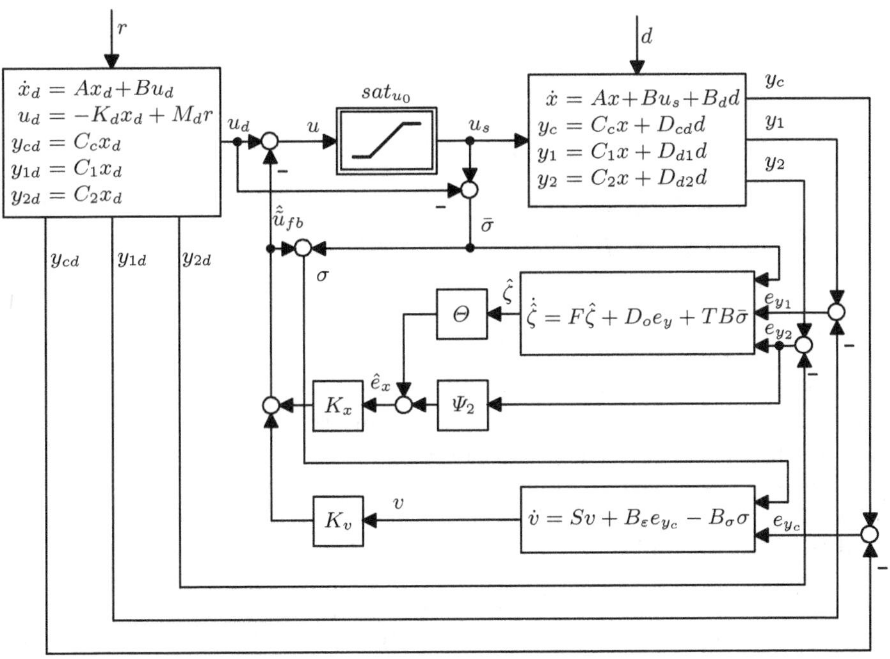

Figure 9.5. Tracking control with disturbance rejection and windup prevention using observer-based state feedback in the time domain

$$F = T(A - L_1C_1)\Theta, \tag{9.99}$$

$$D_0 = \begin{bmatrix} TL_1 & T(A - L_1C_1)\Psi_2 \end{bmatrix} \tag{9.100}$$

have been used.

Remark 9.2. Different from the usual observer-based compensator (see Chapter 4) all degrees of freedom contained in the state feedback gains K_x and K_v and in the observer gains L_1 and Ψ_2 can be used to shape the disturbance behaviour of the closed-loop system, because the feedback controller is only active if disturbances are present. Thus, more degrees of freedom are available to obtain a desired disturbance behaviour because in the design of the usual observer-based compensator the state feedback gain K is often determined to achieve a desired reference behaviour.

9.6 Observer-based Tracking Control in the Frequency Domain

In this section, the frequency-domain design of the tracking controller with disturbance rejection is presented. Since the time-domain representation (9.96)–(9.98) of the observer-based tracking controller coincides with the time-domain representation

$$\begin{aligned} \dot{\hat{\zeta}}(t) &= T(A - L_1C_1)\Theta\hat{\zeta}(t) + \begin{bmatrix} TL_1 & T(A - L_1C_1)\Psi_2 \end{bmatrix} \begin{bmatrix} y_1(t) \\ y_2(t) \end{bmatrix} \\ &\quad + TBu_s(t), \quad \hat{\zeta}(0) = \hat{\zeta}_0, \end{aligned} \tag{9.101}$$

$$\begin{aligned} \dot{v}(t) &= (S - B_\sigma K_v)v(t) + B_\varepsilon y_c(t) - B_\sigma u_s(t) \\ &\quad - B_\sigma(K_x\Theta\hat{\zeta}(t) + K_x\Psi_2 y_2(t)), \quad v(0) = v_0, \end{aligned} \tag{9.102}$$

$$\hat{\bar{u}}_{aug}(t) = K_x\Theta\hat{\zeta}(t) + K_x\Psi_2 y_2(t) + K_v v(t) \tag{9.103}$$

of the observer-based compensator with internal signal model (see (7.60) and (7.61)) the frequency-domain design goes along the same lines as in Section 7.4. Therefore, the design procedure of Section 7.4 can also be applied to the frequency-domain design of tracking controllers.

The next theorem summarizes the frequency-domain design of observer-based tracking controllers.

Theorem 9.1 (Design of observer-based tracking controllers with disturbance rejection in the frequency domain). *Given is the transfer behaviour $y(s) = G(s)u(s)$ of the system (9.42) and (9.43) with the $m \times p$ transfer matrix $G(s)$ represented in a right coprime and a left coprime MFD*

$$G(s) = N(s)D^{-1}(s) = \bar{D}^{-1}(s)\bar{N}(s). \tag{9.104}$$

In (9.104) the polynomial matrix $D(s)$ is column reduced and

$$\bar{D}_{\kappa}(s) = \Pi\left\{\bar{D}(s)\begin{bmatrix} I_{m-\kappa} & 0 \\ 0 & s^{-1}I_{\kappa}\end{bmatrix}\right\} \tag{9.105}$$

is row reduced. The transfer behaviour between $e_{y_c} \in \mathbb{R}^p$ and $v \in \mathbb{R}^{pq}$ of the driven signal model (9.47) is represented by the right coprime MFD

$$v(s) = N_S(s)D_S^{-1}(s)e_{y_c}(s), \tag{9.106}$$

(see (9.73) where $D_S(s)$ has the form $D_S(s) = \det(sI - S^*)I_p$ (see (7.13)). It is assumed that the zeros of $\det N_c(s) = \det(\Xi N(s))$ (see (9.3)) and $\det D_S(s)$ do not have a common zero.

Also given are the $p \times p$ parameterizing polynomial matrix $\tilde{D}(s)$ (state feedback having the properties (9.83) and (9.84)), $\tilde{D}_S(s)$ (controlled signal model having the properties (9.81) and (9.82)), and $\tilde{\bar{D}}(s)$ (reduced-order state observer having the properties (7.107) and (7.108)).

Determine a solution $Q(s)$ with

$$\delta_{ci}[Q(s)] = \delta_{ci}[D(s)], \quad i = 1, 2, \ldots, p, \tag{9.107}$$

$$\Gamma_c[Q(s)] = \Gamma_c[D(s)], \tag{9.108}$$

and $P(s) = P = const$ of the Diophantine equation

$$P(s)N_S(s)N_c(s) + Q(s)D_S(s) = \tilde{D}_S(s)\tilde{D}(s), \tag{9.109}$$

and find solutions $Y(s)$ and $X(s)$ of the Bezout identity

$$Y(s)N(s) + X(s)D(s) = I_p. \tag{9.110}$$

Then, with

$$\bar{V}(s) = \Pi\left\{Q(s)Y(s)\bar{D}^{-1}(s)\tilde{\bar{D}}(s)\right\}, \tag{9.111}$$

and with the prime right-to-left conversion

$$\bar{V}(s)\tilde{\bar{D}}^{-1}(s) = \hat{\Delta}^{-1}(s)V(s) \tag{9.112}$$

the numerator and the denominator matrices of the output feedback structure of this observer-based tracking controller have the forms

$$N_C(s) = D_S(s)[\hat{\Delta}(s)Q(s)Y(s) - V(s)\bar{D}(s)] + \hat{\Delta}(s)PN_S(s)\Xi, \tag{9.113}$$

and

$$D_C(s) = D_S(s)[\hat{\Delta}(s)Q(s)X(s) + V(s)\bar{N}(s)]. \tag{9.114}$$

The error dynamics of the overall observer are characterized by the polynomial matrix

$$\Delta(s) = \hat{\Delta}(s)\tilde{D}_S(s). \tag{9.115}$$

The numerator matrix $N_u(s)$ in the observer structure of this controller as shown in Figure 9.6 can be obtained from

$$N_u(s) = D_C(s) - \Delta(s). \tag{9.116}$$

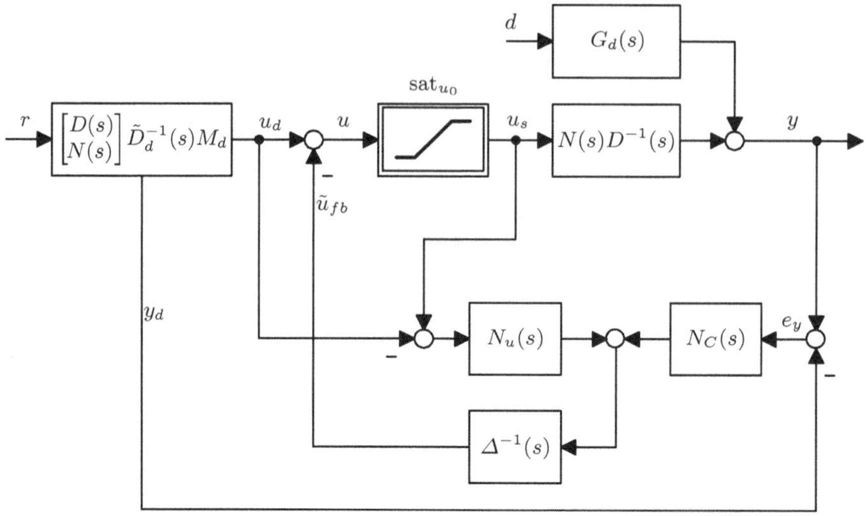

Figure 9.6. Tracking control with disturbance rejection and windup prevention using observer-based state feedback in the frequency domain

Example 9.1. Exact output tracking for a three-tank system
Consider the three-tank system of Example 1.5 with the controlled output $y_c = [y_c^1 \;\; y_c^2]^T = [x_1 \;\; x_3]^T$, which is the measured output, *i.e.*, $y_c = y$ and $N_c(s) = N(s)$ with the selection matrix $\Xi = I_2$. A linearization of the system at the operating point yields the transfer behaviour

$$y_c(s) = G_c(s)u(s) = N(s)D^{-1}(s)u(s) = \bar{D}^{-1}(s)\bar{N}(s)u(s), \qquad (9.117)$$

with

$$N(s) = \begin{bmatrix} 1 & 0 \\ 0 & s+1 \end{bmatrix}, \qquad (9.118)$$

$$D(s) = \begin{bmatrix} s+0.5 & -0.25 \\ -s-1 & s^2+2s+0.75 \end{bmatrix}, \qquad (9.119)$$

and

$$\bar{N}(s) = \begin{bmatrix} s+1 & 0 \\ -1 & 1 \end{bmatrix}, \qquad (9.120)$$

$$\bar{D}(s) = \begin{bmatrix} s^2+1.5s+0.5 & -0.25 \\ -2s-1.5 & s+1 \end{bmatrix} \qquad (9.121)$$

(see Example 1.5, where the right MFD has been normalized to obtain $\det \Gamma_c[D(s)] = 1$).

The model-based feedforward controller can be designed to obtain a decoupled reference transfer behaviour (see Section 9.2). Thus, the controlled output y_c tracks the reference r asymptotically, *i.e.*, $\lim_{t\to\infty} y_c(t) = r(t)$. However, it may be required that y_c exactly tracks a reference trajectory y_{cd}, *i.e.*, $y_c(t) = y_{cd}(t)\ \forall t \in [0, T]$, $T < \infty$. In what follows it is shown that this problem can also be solved on the basis of the decoupling approach presented in Section 9.2.

In order to obtain an exact output tracking the input–output behaviour of the system has to satisfy

$$y_c(s) = G_c(s)u(s) = y_{cd}(s). \tag{9.122}$$

Thus, the feedforward control achieving exact tracking is

$$u(s) = G_c^{-1}(s)y_{cd}(s). \tag{9.123}$$

When using this approach it is, however, not obvious how to implement the feedforward controller. An alternative approach is to use a model-based feedforward controller. By applying the input u_d generated by this controller also to the plant and provided that no disturbances are present the reference transfer behaviour

$$y_c(s) = G_r(s)r(s) \tag{9.124}$$

results. Thus, exact output tracking is achieved if

$$y_c(s) = G_r(s)r(s) = y_{cd}(s) \tag{9.125}$$

is satisfied. Consequently, the reference inputs r have to satisfy

$$r(s) = G_r^{-1}(s)y_{cd}(s). \tag{9.126}$$

Obviously, the model-based feedforward controller has the advantage that the transfer matrix $G_r(s)$ can be assigned such that its inverse in (9.126) is easy to compute. This is, for example, the case if $G_r(s)$ is diagonal, *i.e.*, if the reference behaviour of the model-based feedforward controller is decoupled. To determine the corresponding decoupling controller the results of Theorem 6.1 are used to compute the relative degrees $\delta_1 = 1$ and $\delta_2 = 1$ of the outputs y_c^1 and y_c^2. The three-tank system is minimum phase with respect to the controlled output y_c since

$$\det N_c(s) = s + 1 = 0 \tag{9.127}$$

has one zero at $s = -1$ (see (9.118)). Thus, all assumptions are satisfied for the frequency-domain design of a decoupling controller. By assigning the reference transfer matrix

$$G_r(s) = \begin{bmatrix} \frac{1}{s} & 0 \\ 0 & \frac{1}{s} \end{bmatrix} \tag{9.128}$$

to the closed-loop system the reference inputs for exact tracking are

$$r(s) = G_r^{-1}(s)y_{cd}(s) = sy_{cd}(s), \tag{9.129}$$

i.e., $r = \dot{y}_{cd}$. This reference transfer behaviour is achieved by choosing

$$\Lambda_d(s) = \begin{bmatrix} s & 0 \\ 0 & s \end{bmatrix}, \tag{9.130}$$

such that the decoupling controller can be computed using the results of Theorem 6.3. The matrix M_d of the decoupling controller reads

$$M_d = \Gamma_c[D(s)]\Gamma_c^{-1}[\Lambda_d(s)N_c(s)] = \begin{bmatrix} 1 & 0 \\ -1 & 1 \end{bmatrix}, \tag{9.131}$$

(see (9.34)) and the parameterizing polynomial matrix $\tilde{D}_d(s)$ is given by

$$\tilde{D}_d(s) = M_d\Lambda_d(s)N_c(s) = \begin{bmatrix} s & 0 \\ -s & s^2+s \end{bmatrix} \tag{9.132}$$

(see (9.33)). Then, the frequency-domain representation of the model-based feedforward tracking controller is

$$\begin{bmatrix} u_d(s) \\ y_{cd}(s) \end{bmatrix} = \begin{bmatrix} D(s) \\ N_c(s) \end{bmatrix} \tilde{D}_d^{-1}(s)M_d r(s) \tag{9.133}$$

(see (9.30)).

A reference trajectory $y_c^* = [y_{cd}^1 \ \ y_{cd}^2]^T$ can, for example, be planned to realize a finite transition time between the start points

$$y_{cd}^1(0) = 0, \quad y_{cd}^2(0) = 0 \tag{9.134}$$

and the end points

$$y_{cd}^1(T) = y_{1E}, \quad y_{cd}^2(T) = y_{2E}, \quad 0 < T < \infty. \tag{9.135}$$

Since (9.129) implies that y_{cd} has to be differentiable one can consider

$$y_{cd}^i(t) = a_{i0} + a_{i1}t, \quad t \in [0, T] \tag{9.136}$$

in the time domain, where a_{i0} and a_{i1} are constant parameters.

The coefficients in (9.136) are computed by taking the boundary conditions (9.134) and (9.135) into account that leads to the reference trajectories

$$y_{cd}^i(t) = \frac{y_{iE}}{T}t, \quad i = 1, 2, \quad t \in [0, T]. \tag{9.137}$$

Then, the corresponding reference inputs r_i of the model-based feedforward controller that ensure $y_c(t) = y_{cd}(t)$, $T \in [0, T]$, in the disturbance-free case are

$$r_i(t) = \dot{y}^i_{cd}(t) = \tfrac{y_{iE}}{T}, \quad i = 1, 2, \quad t \in [0, T]. \tag{9.138}$$

If $r_i(t) = 0$, $i = 1, 2$, for all $t > T$, then the outputs satisfy $y^i_c(t) = y_{iE}$ for all $t > T$. Thus, (9.138) shows that by applying the model-matching principle combined with a decoupling controller for the reference model, the exact output tracking problem for the controlled system becomes a trivial task.

It is assumed that two disturbances act at the input of the system, namely $d_1(t) = d_{S1}1(t - t_{d1})$ and $d_2(t) = d_{S2}1(t - t_{d2}) + d_{Si2}\sin(3t)1(t - t_{di2})$ that need to be suppressed by a feedback controller, which is designed along the internal model principle. The transfer behaviour (9.106) of the disturbance signal model is characterized by the polynomial matrices

$$N_S(s) = \begin{bmatrix} 1 & 0 \\ s & 0 \\ s^2 & 0 \\ 0 & 1 \\ 0 & s \\ 0 & s^2 \end{bmatrix} \quad \text{and} \quad D_S(s) = \begin{bmatrix} s^3 + 9s & 0 \\ 0 & s^3 + 9s \end{bmatrix}. \tag{9.139}$$

The disturbance rejecting state feedback is parameterized by the polynomial matrix

$$\tilde{D}(s) = \begin{bmatrix} s + 5 & 0 \\ -s - 5 & s^2 + 6s + 5 \end{bmatrix}. \tag{9.140}$$

Since the order of the system is $n = 3$ and the number of outputs is $m = 2$ a minimal-order state observer of the order $n_O = 1$ is required. Given the polynomial matrix

$$\bar{D}_2(s) = \begin{bmatrix} s + 1.5 & 0 \\ -2 & 1 \end{bmatrix} \tag{9.141}$$

a minimal-order state observer with a pole at $s = -5$ is parameterized by

$$\tilde{\bar{D}}(s) = \begin{bmatrix} s + 5 & 0 \\ -2 & 1 \end{bmatrix}. \tag{9.142}$$

The controlled signal model is parameterized by the polynomial matrix

$$\tilde{D}_S(s) = \begin{bmatrix} (s + 5)^3 & 0 \\ 0 & (s + 5)^3 \end{bmatrix}. \tag{9.143}$$

Applying Theorem 9.1 one obtains the polynomial matrices $N_C(s)$ and $D_C(s)$

characterizing the output feedback structure of the feedback compensator along the following lines. First, compute the solutions

$$P = \begin{bmatrix} 625 & 320 & 141 & 0 & 0 & 0 \\ -625 & -320 & -141 & 625 & 320 & 141 \end{bmatrix}, \tag{9.144}$$

and

$$Q(s) = \begin{bmatrix} s+20 & 0 \\ -s-20 & s^2+21s+20 \end{bmatrix} \tag{9.145}$$

of the Diophantine equation (9.109) and subsequently determine the solutions

$$Y(s) = \begin{bmatrix} 1 & 0 \\ 4s+2 & 0 \end{bmatrix} \quad \text{and} \quad X(s) = \begin{bmatrix} 0 & 0 \\ -4 & 0 \end{bmatrix} \tag{9.146}$$

of the Bezout identity (9.110). With these results compute

$$\bar{V}(s) = \begin{bmatrix} 1 & 0 \\ 4s^2+100s+399 & 1 \end{bmatrix}, \tag{9.147}$$

(see (9.111)) and carry out the right-to-left conversion (9.112) which leads to

$$V(s) = \begin{bmatrix} 1 & 0 \\ 4s+80 & 1 \end{bmatrix}, \tag{9.148}$$

and

$$\hat{\Delta}(s) = \begin{bmatrix} s+5 & 0 \\ -1 & 1 \end{bmatrix}. \tag{9.149}$$

The polynomial matrix $\Delta(s)$ (see (9.115)) then has the form

$$\Delta(s) = \hat{\Delta}(s)\tilde{D}_S(s) = \begin{bmatrix} (s+5)^4 & 0 \\ -(s+5)^3 & (s+5)^3 \end{bmatrix}. \tag{9.150}$$

With these results one obtains the polynomial matrices

$$N_C(s) = \begin{bmatrix} 23.5s^4+240.5s^3+1236.5s^2+3120.5s+3125 & 0.25s^3+2.25s \\ -38.5s^3-282s^2-986.5s-1250 & 19s^3+141s^2+491s+625 \end{bmatrix}, \tag{9.151}$$

and

$$D_C(s) = \begin{bmatrix} s^4+s^3+9s^2+9s & 0 \\ -s^3-9s & s^3+9s \end{bmatrix} \tag{9.152}$$

of the observer-based feedback compensator for the rejection of constant and sinusoidal disturbances.

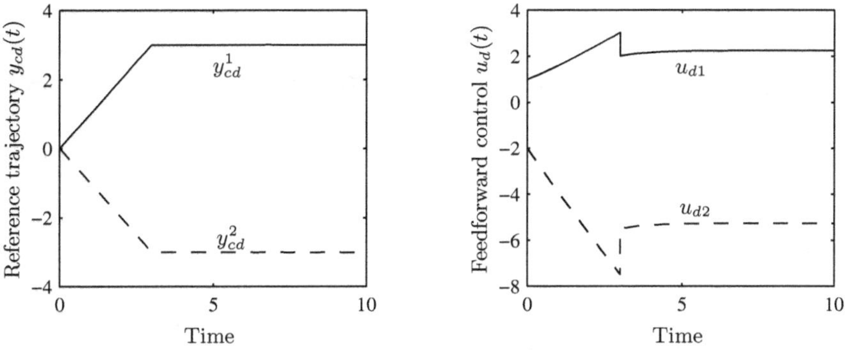

(a) Reference trajectory y_{cd} for the controlled output y_c

(b) Corresponding feedforward control signal u_d

Figure 9.7. Reference trajectories and feedforward control

With these results the polynomial matrix $N_u(s) = D_C(s) - \Delta(s)$ finally becomes

$$N_u(s) = \begin{bmatrix} -19s^3 - 141s^2 - 491s - 625 & 0 \\ 15s^2 + 66s + 125 & -15s^2 - 66s - 125 \end{bmatrix}. \qquad (9.153)$$

First, assume that the reference trajectories result from setting $y^1_{cd}(0) = 0$, $y^1_{cd}(T) = 3$ $y^2_{cd}(0) = 0$, $y^2_{cd}(T) = -3$, and $T = 3$, so that in view of (9.137) $r_1 = 1$ and $r_2 = -1$ are applied during the time interval $0 < T \leq 3$ and $r_1 = 0$ and $r_2 = 0$ for $3 < T < \infty$. Figure 9.7 shows the resulting reference trajectories and the corresponding feedforward control.

The influence of initial errors is investigated by assuming that the reference trajectories start at $y^1_{cd}(0) = 0$ and $y^2_{cd}(0) = 0$, wheras the initial conditions of the outputs of the system are $y^1_c(0) = 0.3$ and $y^2_c(0) = -0.3$. Figure 9.8 shows the reference transients of the closed loop system and the corresponding feedback control. After the transients caused by the initial errors have settled the feedback control $\tilde{u}_{fb}$ vanishes, so that only the feedforward controller is active.

To demonstrate the disturbance-rejecting behaviour of the closed-loop system assume there are disturbances with amplitudes $d_{S1} = 1$, $d_{S2} = -1$ and $d_{Si2} = 1$ and that they attack at different times, namely $t_{d1} = 0$, $t_{d2} = 4$ and $t_{di2} = 8$. Figure 9.9 shows the resulting transients of the closed-loop system. Figure 9.9b demonstrates that the internal model contained in the feedback controller exactly produces an input to the plant that coincides with the negative values of the disturbances required for an asymptotic rejection of the disturbances.

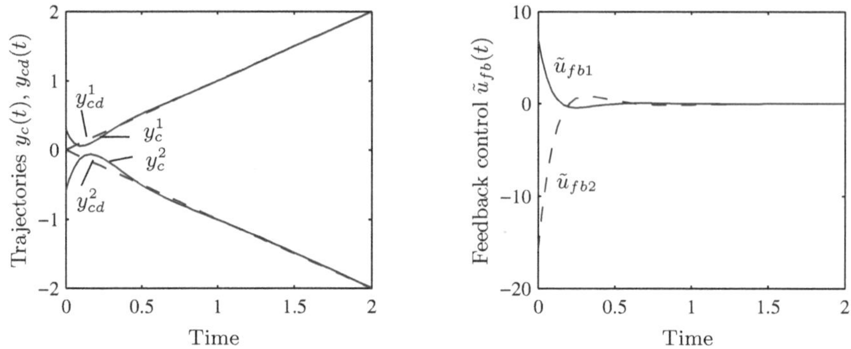

(a) Trajectories y_c and y_{cd}

(b) Corresponding feedback control signal $\tilde{u}_{fb}$

Figure 9.8. Tracking behaviour in the presence of initial errors $y_c^1(0) - y_{cd}^1(0) = 0.3$ and $y_c^2(0) - y_{cd}^2(0) = -0.3$

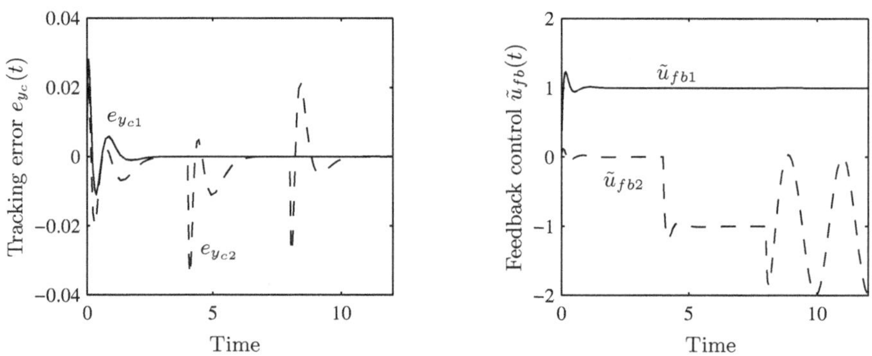

(a) Tracking error e_{y_c}

(b) Corresponding feedback control signal $\tilde{u}_{fb}$

Figure 9.9. Tracking error e_{y_c} in the presence of input disturbances $d_1(t) = 1(t)$ and $d_2(t) = -1(t-4) + \sin(3t)1(t-8)$

10

Observer-based Compensators with Disturbance Rejection for Discrete-time Systems

Nearly all applications of control systems are now based on computer control. Nevertheless, the main portion of this book is devoted to continuous-time systems. In view of the design of observer-based compensators, this is no major problem, because the design algorithms for continuous-time and discrete-time systems are very similar, especially when using the frequency-domain approach.

An introduction to sampled-data control is not intended here. It is assumed that the reader is familiar with the control theory of sampled-data systems as contained in standard references like, *e.g.*, [1, 6, 19]. In the following sections on discrete-time systems the presentation starts with the discrete-time state-space models and their corresponding transfer behaviour in the z-domain. These models describe the behaviour of a continuous-time system at the sampling instants. How to derive such models for continuous-time systems can be found in the above references. The discrete-time descriptions are very similar to their continuous-time counterparts. Therefore, the parameterizations and the design procedures of observer-based compensators are more or less identical for continuous-time and discrete-time systems. And this is true for both the time and the frequency domains, so that nearly all results presented so far also apply to sampled-data systems.

Thus, the following presentations of observer-based compensators in the time domain and in the frequency domain can be kept short. In Section 10.1 the design of observer-based compensators with signal models for the rejection of persistently acting disturbances is described in the time domain. These results contain the standard state feedback control in the presence of measurable states and the observer-based state feedback compensator as special cases. Section 10.2 contains the frequency-domain equivalent of the results in Section 10.1.

10.1 Discrete-time Control in the Time Domain

The compensator discussed in this section is the most general one considered in this book, namely the observer-based compensator incorporating signal models for disturbance rejection. In this formulation, the observer-based compensator without signal model and also the static state feedback controller without observer are contained as special cases.

Considered are linear, time-invariant, discrete-time systems with the state equations

$$x(k+1) = Ax(k) + Bu(k) + B_d d(k), \tag{10.1}$$

$$y(k) = Cx(k) + D_d d(k), \tag{10.2}$$

where $x \in \mathbb{R}^n$ is the state, $u \in \mathbb{R}^p$ is the input, $y \in \mathbb{R}^m$ is the measurement and $d \in \mathbb{R}^\rho$ is an unmeasurable disturbance. The notion $x(k)$ is a shorthand notation of $x(kT)$, where T is the constant time interval between two sampling instants. The time interval T is also called the sampling time.

In view of designing reduced-order observers of the order $n_O = n - \kappa$, $0 \le \kappa \le m$, the $m \times 1$ output vector y of these systems is arranged according to

$$y(k) = \begin{bmatrix} y_1(k) \\ y_2(k) \end{bmatrix} = \begin{bmatrix} C_1 \\ C_2 \end{bmatrix} x(k) + \begin{bmatrix} D_{d1} \\ D_{d2} \end{bmatrix} d(k). \tag{10.3}$$

Here, $y_2 \in \mathbb{R}^\kappa$ with $0 \le \kappa \le m$ contains the measurements directly used in the construction of the estimate $\hat{x}$ and $y_1 \in \mathbb{R}^{m-\kappa}$ contains the remaining $m - \kappa$ measurements (see also Chapter 3). On the basis of the discrete-time model (10.1) and (10.2) the design of the state feedback control and the observer is completely analogous to the continuous-time case.

To be able to apply the design of disturbance rejecting controllers as presented in Chapter 7 the controlled output $y_c \in \mathbb{R}^p$ is assumed to be measurable. Therefore, a $p \times m$ selection matrix Ξ (see (2.3)) exists, so that

$$y_c(k) = \Xi y(k) = \Xi Cx(k) + \Xi D_d d(k) = C_c x(k) + D_{cd} d(k). \tag{10.4}$$

It is further assumed that the disturbances d can be modelled in a signal process

$$v^*(k+1) = S^* v^*(k), \tag{10.5}$$

$$d(k) = Hv^*(k), \tag{10.6}$$

with $v^* \in \mathbb{R}^q$, $v^*(0) = v_0^*$ unknown and (H, S^*) observable with the characteristic polynomial

$$\det(zI - S^*) = z^q + \psi_{q-1} z^{q-1} + \ldots + \psi_1 z + \psi_0. \tag{10.7}$$

The characteristic polynomial (10.7) can be obtained from the continuous-time disturbance models by the transformation

$$\tilde{z}_i = e^{\tilde{s}_i T} \tag{10.8}$$

applied to the eigenvalues $\tilde{s}_i$ of the continuous-time disturbance model (see, *e.g.*, [9]). An application of (10.8) to the signals listed in Table 7.1 yields the following collection of characteristic polynomials.

Table 10.1. Typical types of signals and corresponding characteristic polynomials

Signal form	$\det(zI - S^*)$
step	$z - 1$
ramp	$(z-1)^2$
parabola	$(z-1)^3$
sine	$z^2 - 2\cos(\omega_0 T)z + 1$
sine with non-zero mean	$(z-1)(z^2 - 2\cos(\omega_0 T)z + 1)$
exponential increase	$z - e^{\alpha T}, \alpha > 0$

In what follows the internal model principle for the rejection of persistently acting disturbances of Chapter 7 is applied to discrete-time systems. This approach accomplishes disturbance rejection with the aid of a driven signal model consisting of p models of the assumed signal process according to

$$v(k+1) = Sv(k) + B_\varepsilon y_c(k) - B_\sigma Mr(k), \tag{10.9}$$

with $v \in \mathbb{R}^{pq}$. This model is driven by the controlled outputs y_c and by the signal Mr. It is controllable by y_c if the matrices S and B_ε in (10.9) are defined as

$$S = \begin{bmatrix} S^* & & & \\ & S^* & & \\ & & \ddots & \\ & & & S^* \end{bmatrix} \quad \text{and} \quad B_\varepsilon = \begin{bmatrix} b_\varepsilon & & & \\ & b_\varepsilon & & \\ & & \ddots & \\ & & & b_\varepsilon \end{bmatrix}, \tag{10.10}$$

where S^* and b_ε have the forms

$$S^* = \begin{bmatrix} 0 & 1 & \cdots & 0 \\ \vdots & \vdots & \ddots & \vdots \\ 0 & 0 & \cdots & 1 \\ -\psi_0 & -\psi_1 & \cdots & -\psi_{q-1} \end{bmatrix} \quad \text{and} \quad b_\varepsilon = \begin{bmatrix} 0 \\ \vdots \\ 0 \\ 1 \end{bmatrix} \tag{10.11}$$

(see also (7.13) and (7.15)).

An asymptotic rejection of the modelled disturbances in y_c results if the system (10.1) and the appending signal model (10.9) are stabilized by a state feedback control

$$u(k) = -\tilde{u}_{aug}(k) + Mr(k) = -\begin{bmatrix} K_x & K_v \end{bmatrix} \begin{bmatrix} x(k) \\ v(k) \end{bmatrix} + Mr(k). \tag{10.12}$$

As in Chapter 7 the controlled signal model is treated as an observer, so that the reference transfer behaviour of the closed-loop system with internal signal model for the rejection of persistently acting disturbances is the same as the reference transfer behaviour of a constant state feedback loop, where

$$u(k) = -Kx(k) + Mr(k) \tag{10.13}$$

is applied to the system (10.1) with $r \in \mathbb{R}^p$ being the reference input. Provided that the system (C_c, A, B) has no invariant zero at $z = 1$

$$M = \left[C_c(I - A + BK)^{-1}B\right]^{-1} \tag{10.14}$$

assures a vanishing steady-state error $y_c(\infty) - r_0$ for stationary constant reference signals $r(\infty) = r_0 = const.$

The design procedure is summarized in the following theorem.

Theorem 10.1 (Disturbance rejection using the internal model principle). *Consider the system (10.1) and (10.4)*

$$x(k+1) = Ax(k) + Bu(k) + B_d d(k), \tag{10.15}$$

$$y_c(k) = C_c x(k) + D_{cd} d(k), \tag{10.16}$$

and assume that the disturbance d can be modelled by the signal process (10.5) and (10.6), namely

$$v^*(k+1) = S^* v^*(k), \tag{10.17}$$

$$d(k) = Hv^*(k). \tag{10.18}$$

Given further a driven model

$$v(k+1) = Sv(k) + B_\varepsilon y_c(k) - B_\sigma Mr(k) \tag{10.19}$$

of the disturbance signal process. Let the following assumptions be satisfied.

(a1) The pairs (A, B) and (S, B_ε) are controllable.

(a2) The condition rank $\begin{bmatrix} \lambda_i I - A & B \\ C_c & 0 \end{bmatrix} = n + p$ *holds for all eigenvalues λ_i, $i = 1, 2, \ldots, q$ of S^*,* i.e., *no eigenvalue of the signal model (10.17) is a transmission zero of the transfer behaviour of the system (10.15) and (10.16) between u and y_c.*

Compute a feedback gain K such that $A - BK$ is Schur and find the unique solution Σ of

$$\Sigma(A - BK) - S\Sigma = B_\varepsilon C_c, \tag{10.20}$$

to obtain

$$B_\sigma = -\Sigma B. \tag{10.21}$$

Now, determine a gain K_v*, which always exists, such that* $S - B_\sigma K_v$ *is Schur, and finally compute*

$$K_x = K - K_v \Sigma. \tag{10.22}$$

Then, the feedback

$$u(k) = -\begin{bmatrix} K_x & K_v \end{bmatrix} \begin{bmatrix} x(k) \\ v(k) \end{bmatrix} + Mr(k) \tag{10.23}$$

of the states x *of the system (10.15) and the states* v *of the signal model (10.19) with* M *as defined in (10.14) stabilizes the system (10.15) and (10.16) and (10.19) and assures an asymptotic rejection of the disturbance modelled by (10.17) and (10.18). The reference transfer matrix of the closed-loop system is given by*

$$G_r(z) = C_c(zI - A + BK)^{-1} BM, \tag{10.24}$$

i.e., *the reference transfer behaviour is not affected by the signal model for disturbance rejection.*

The proof of this theorem follows exactly the arguments presented in Section 7.1.

So far it has been assumed that all states of the system can be measured. If this is not the case an estimate $\hat{x}$ of the state x can be obtained by the discrete-time version of an observer of the order $n_O = n - \kappa$, $0 \leq \kappa \leq m$, as described for continuous-time systems in Chapter 3. This observer has the state equations

$$\hat{\zeta}(k+1) = T(A - L_1 C_1)\Theta\hat{\zeta}(k) + \begin{bmatrix} TL_1 & T(A - L_1 C_1)\Psi_2 \end{bmatrix} \begin{bmatrix} y_1(k) \\ y_2(k) \end{bmatrix} + TBu(k), \tag{10.25}$$

and it yields an estimate $\hat{\zeta}$ for the linear combination $\zeta = Tx$, so that with $y_2 = C_2 x$ the state estimate has the form

$$\hat{x}(k) = \begin{bmatrix} C_2 \\ T \end{bmatrix}^{-1} \begin{bmatrix} y_2(k) \\ \hat{\zeta}(k) \end{bmatrix} = \Psi_2 y_2(k) + \Theta\hat{\zeta}(k), \tag{10.26}$$

(see (3.5)). As in the case of the continuous-time observer, the relation

$$C_2 L_1 = 0 \tag{10.27}$$

can be assumed.

If the state x is replaced by the estimate (10.26) the control signal (10.12) becomes $u = -\hat{\tilde{u}}_{aug} + Mr$ with

$$\hat{\tilde{u}}_{aug}(k) = K_x \Theta\hat{\zeta}(k) + K_x \Psi_2 y_2(k) + K_v v(k), \tag{10.28}$$

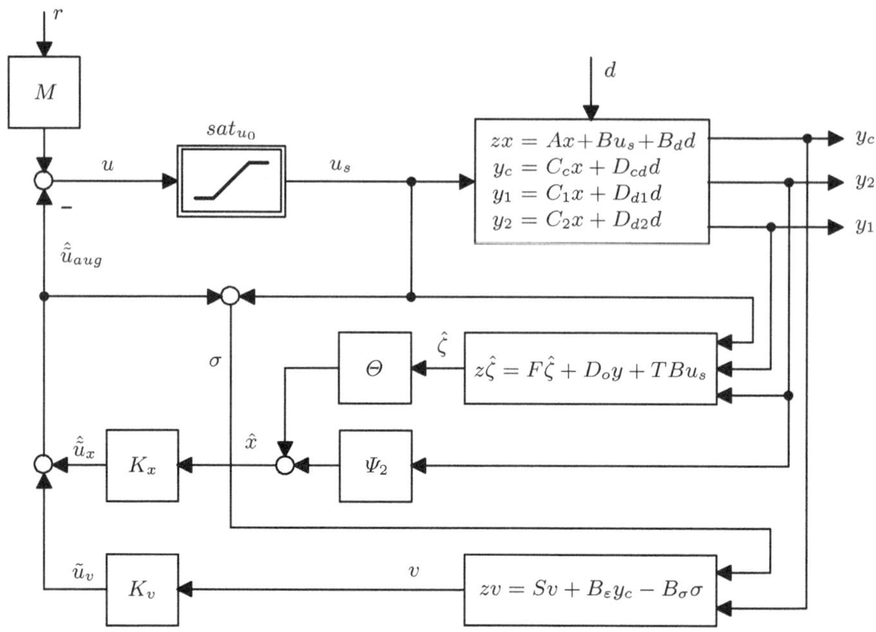

Figure 10.1. Block diagram of the discrete-time closed-loop system with signal model for disturbance rejection and measures for the prevention of controller windup

so that Equations (10.9), (10.25) and (10.28) constitute the state equations of the observer-based compensator with signal model for disturbance rejection. Figure 10.1 shows the block diagram of the closed-loop system, where the abbreviations

$$F = T(A - L_1C_1)\Theta, \tag{10.29}$$

and

$$D_o = [TL_1 \quad T(A - L_1C_1)\Psi_2] \tag{10.30}$$

have been used. It has also been assumed that the compensator is realized in the observer structure, *i.e.*, a model $u_s = \mathrm{sat}_{u_0}(u)$ (see (4.108)) of the actuator saturation is added to the compensator and the saturated output u_s of this model is fed into the state observer (*via* TBu_s) and into the signal model (*via* $-B_\sigma(\hat{\tilde{u}}_{aug} + u_s)$). This prevents controller windup as described in Chapter 4 and Chapter 7.

In an observer-based controller (see Chapter 4), the order n_C of the compensator coincides with the order of the state observer, *i.e.*, $n_C = n_O$. Here, the signal model is also part of the compensator, so that the order of the observer-based compensator with signal model for disturbance rejection is

$$n_C = n_O + pq. \tag{10.31}$$

The dynamics of the controlled signal model and of the state observer define the dynamics of the so-called *overall observer*, because the matrix B_σ has been designed, so that the controlled signal model obtains the property of an observer, *i.e.*, the reference behaviour of the closed-loop system shown in Figure 10.1 is the same as if only the constant state feedback (10.13) had been applied to the system (10.15) and (10.16). This overall observer also has the order n_C.

10.2 Discrete-time Control in the Frequency Domain

In the frequency domain the system (10.1) and (10.2) is described by

$$y(z) = G(z)u(z) + G_d(z)d(z), \tag{10.32}$$

with

$$G(z) = C(zI - A)^{-1}B, \tag{10.33}$$

and

$$G_d(z) = C(zI - A)^{-1}B_d + D_d. \tag{10.34}$$

The transfer matrix $G(z)$ of the system can be represented both in a right and a left MFD according to

$$G(z) = N(z)D^{-1}(z) = \bar{D}^{-1}(z)\bar{N}(z), \tag{10.35}$$

where $(N(z), D(z))$ is a right coprime pair, such that $D(z)$ is column reduced (see Chapter 2 with s replaced by z) and $(\bar{N}(z), \bar{D}(z))$ is a left coprime pair, such that $\bar{D}(z)$ and

$$\bar{D}_\kappa(z) = \Pi\left\{\bar{D}(z)\begin{bmatrix} I_{m-\kappa} & 0 \\ 0 & z^{-1}I_\kappa \end{bmatrix}\right\} \tag{10.36}$$

are row reduced (see also Section 3.3).

The transfer behaviour between u and the controlled output y_c is characterized by

$$y_c(z) = G_c(z)u(z) = N_c(z)D^{-1}(z)u(z) = \Xi N(z)D^{-1}(z)u(z) \tag{10.37}$$

(see also (10.4)).

The signal model (10.9) is driven by the controlled variable y_c and the transfer behaviour between y_c and its states v is

$$v(z) = G_v(z)y_c(z). \tag{10.38}$$

Due to the companion form (10.11) of the signal model the left and right coprime MFDs of the transfer matrix $G_v(z)$ are

$$G_v(z) = (zI - S)^{-1}B_\varepsilon = N_S(z)D_S^{-1}(z), \tag{10.39}$$

with the $pq \times p$ polynomial matrix

$$N_S(z) = \begin{bmatrix} 1 & & \\ z & & 0 \\ \vdots & & \\ z^{q-1} & & \\ & \ddots & \\ & & 1 \\ & & z \\ 0 & & \vdots \\ & & z^{q-1} \end{bmatrix}, \tag{10.40}$$

and the $p \times p$ polynomial matrix

$$D_S(z) = \text{diag}[\det(zI - S^*)]. \tag{10.41}$$

To specify the reference transfer matrix

$$G_r(z) = C_c(zI - A + BK)^{-1}BM = N_c(z)\tilde{D}^{-1}(z)M \tag{10.42}$$

of the closed-loop system (see (10.24)) in the frequency domain, the polynomial matrix $\tilde{D}(z)$ and the matrix M have to be determined. The parameterizing polynomial matrix $\tilde{D}(z)$ of the state feedback in the frequency domain and the corresponding feedback matrix K in the time domain are related by

$$\tilde{D}(z)D^{-1}(z) = I + K(zI - A)^{-1}B, \tag{10.43}$$

and this relation specifies the structure of $\tilde{D}(z)$ in the following way.

Theorem 10.2 (Parameterizing polynomial matrix of the state feedback). *The $p \times p$ polynomial matrix $\tilde{D}(z)$ characterizing the dynamics of the state feedback loop in the frequency domain has the properties*

$$\delta_{ci}[\tilde{D}(z)] = \delta_{ci}[D(z)], \quad i = 1, 2, \ldots, p, \tag{10.44}$$

and

$$\Gamma_c[\tilde{D}(z)] = \Gamma_c[D(z)], \tag{10.45}$$

and it has exactly the same number of free parameters as the state feedback gain K, namely pn.

The proof of this theorem is analogous to the proof of Theorem 2.1.

Thus, a stabilizing state feedback controller is parameterized in the frequency domain by the polynomial matrix $\tilde{D}(z)$, with $\det \tilde{D}(z)$ being a Schur polynomial. In view of Theorem 10.2 the results on parametric compensator design in Chapter 5 can also be applied in the discrete-time case. By the constant matrix

$$M = \tilde{D}(1)N_c^{-1}(1) \tag{10.46}$$

a reference behaviour with vanishing steady-state error $y_c(\infty) - r_0$ is assured for stationary constant reference inputs $r(\infty) = r_0$ provided that $z = 1$ is not a zero of $\det N_c(z)$.

The overall observer consists of the reduced-order state observer (10.25) and of the driven signal model (10.19). In the time domain the reduced-order observer of the order $n_O = n - \kappa$, $0 \leq \kappa \leq m$, is parameterized by the free parameters in L_1 and Ψ_2. In the frequency domain the $m \times m$ polynomial matrix $\tilde{\bar{D}}(z)$ plays the same role. The connecting relation between both parameterizations is

$$\bar{D}^{-1}(z)\tilde{\bar{D}}(z) = \begin{bmatrix} C_1 \\ C_2 \end{bmatrix} (zI - A)^{-1}[L_1 \quad \Psi_2] + \begin{bmatrix} I_{m-\kappa} & 0 \\ 0 & 0_\kappa \end{bmatrix} \tag{10.47}$$

(see Section 3.3). This relation specifies the structure of the polynomial matrix $\tilde{\bar{D}}(z)$ in the following way.

Theorem 10.3 (Parameterizing polynomial matrix of the observer). *The $m \times m$ polynomial matrix $\tilde{\bar{D}}(z)$ characterizing the dynamics of an observer of the order $n_O = n - \kappa$, $0 \leq \kappa \leq m$, in the frequency domain has the properties*

$$\delta_{rj}[\tilde{\bar{D}}(z)] = \delta_{rj}[\bar{D}_\kappa(z)], \quad j = 1, 2, \ldots, m, \tag{10.48}$$

and

$$\Gamma_r[\tilde{\bar{D}}(z)] = \Gamma_r[\bar{D}_\kappa(z)], \tag{10.49}$$

and it contains exactly $m(n-\kappa)$ free parameters. The polynomial matrix $\bar{D}_\kappa(z)$ in (10.48) and (10.49) is row reduced and it has the form defined in (10.36).

The proof of this theorem follows the same lines as the proof of Theorem 3.1.

Since $\tilde{\bar{D}}(z)$ has the same structure as its continuous-time counterpart the parametric approach of Chapter 5 can also be used to parameterize stabilizing polynomial matrices $\tilde{\bar{D}}(z)$, where $\det \tilde{\bar{D}}(z)$ is a Schur polynomial. The second part of the overall observer, namely the driven signal model is parameterized by the feedback gain K_v in the time domain.

The $p \times p$ polynomial matrix $\tilde{D}_S(z)$ characterizing the dynamics of the controlled signal model in the frequency domain has the properties

$$\delta_{ci}[\tilde{D}_S(z)] = \delta_{ci}[D_S(z)] = q, \quad i = 1, 2, \ldots, p, \tag{10.50}$$

and

$$\Gamma_c[\tilde{D}_S(z)] = \Gamma_c[D_S(z)] = I. \tag{10.51}$$

This frequency-domain parameterization of K_v directly follows from Theorem 10.2 and from (7.73) and (7.74) in the continuous-time case.

An observer-based controller with signal model for disturbance rejection can now be designed along the lines of the following theorem, which is the discrete-time version of Theorem 7.2.

Theorem 10.4 (Design of observer-based compensators with signal models for disturbance rejection). *Given is the system (10.15) and (10.16) described by its transfer behaviour $y(z) = G(z)u(z)$ and $y_c(z) = G_c(z)u(z)$, where the $m \times p$ transfer matrix $G(z)$ is represented in a right coprime and a left coprime MFD*

$$G(z) = N(z)D^{-1}(z) = \bar{D}^{-1}(z)\bar{N}(z), \tag{10.52}$$

and the $p \times p$ transfer matrix $G_c(z)$ in a right MFD

$$G_c(z) = N_c(z)D^{-1}(z) = \Xi N(z)D^{-1}(z). \tag{10.53}$$

In (10.52) and (10.53) the polynomial matrix $D(z)$ is column reduced and $\bar{D}(z)$ is such that

$$\bar{D}_\kappa(z) = \Pi\left\{\bar{D}(z)\begin{bmatrix} I_{m-\kappa} & 0 \\ 0 & z^{-1}I_\kappa \end{bmatrix}\right\} \tag{10.54}$$

is row reduced, where $n_O = n - \kappa$ with $0 \leq \kappa \leq m$ is the order of the state observer for the system. The transfer behaviour between $y_c \in \mathbb{R}^p$ and $v \in \mathbb{R}^{pq}$ of the driven signal model (see (10.9)) is represented by the right coprime MFD

$$v(z) = N_S(z)D_S^{-1}(z)y_c(z), \tag{10.55}$$

(see (10.39)) where $D_S(z)$ has the form $D_S(z) = \det(zI - S^*)I_p$. It is assumed that $\det N_c(z) = \det(\Xi N(z))$ and $\det D_S(z)$ do not have a common zero.

Given are also the $p \times p$ parameterizing polynomial matrices $\tilde{D}(z)$ (state feedback having the properties (10.44) and(10.45)), $\tilde{D}_S(z)$ (controlled signal model having the properties (10.50) and (10.51)), and $\tilde{\bar{D}}(z)$ (reduced-order state observer having the properties (10.48) and (10.49)).

Determine solutions $Q(z)$ with

$$\delta_{ci}[Q(z)] = \delta_{ci}[D(z)], \quad i = 1, 2, \ldots, p, \tag{10.56}$$

$$\Gamma_c[Q(z)] = \Gamma_c[D(z)], \tag{10.57}$$

and $P(z) = P = const$ of the Diophantine equation

$$P(z)N_S(z)N_c(z) + Q(z)D_S(z) = \tilde{D}_S(z)\tilde{D}(z), \tag{10.58}$$

and find solutions $Y(z)$ and $X(z)$ of the Bezout identity

$$Y(z)N(z) + X(z)D(z) = I_p. \tag{10.59}$$

Then, with

$$\bar{V}(z) = \Pi \left\{ Q(z)Y(z)\bar{D}^{-1}(z)\tilde{\bar{D}}(z) \right\}, \tag{10.60}$$

and with the prime right-to-left conversion

$$\bar{V}(z)\tilde{\bar{D}}^{-1}(z) = \hat{\Delta}^{-1}(z)V(z) \tag{10.61}$$

the numerator and the denominator matrices of the output feedback structure of this observer-based controller have the forms

$$N_C(z) = D_S(z)[\hat{\Delta}(z)Q(z)Y(z) - V(z)\bar{D}(z)] + \hat{\Delta}(z)PN_S(z)\Xi, \tag{10.62}$$

and

$$D_C(z) = D_S(z)[\hat{\Delta}(z)Q(z)X(z) + V(z)\bar{N}(z)]. \tag{10.63}$$

The error dynamics of the overall observer are characterized by the polynomial matrix

$$\Delta(z) = \hat{\Delta}(z)\tilde{D}_S(z). \tag{10.64}$$

The numerator matrix $N_u(z)$ in the observer structure of this controller as shown in Figure 10.2 can be obtained from

$$N_u(z) = D_C(z) - \Delta(z). \tag{10.65}$$

This compensator asymptotically rejects all modelled disturbances in the controlled output y_c and it assures the reference behaviour

$$y_c(z) = N_c(z)\tilde{D}^{-1}(z)Mr(z). \tag{10.66}$$

When implementing the compensator in the observer structure the prevention of controller windup of Section 4.5 can be directly applied. This leads to the structure of the closed-loop system of Figure 10.1 as shown in Figure 10.2.

Example 10.1. Parameterization of an observer-based compensator with sinusoidal signal model for disturbance rejection in the frequency domain

Considered is a system of the order three with two measured outputs y. The right coprime MFD of its transfer matrix (10.52) is

$$G(z) = N(z)D^{-1}(z) = \begin{bmatrix} 1 & 0 \\ 0 & 1 \end{bmatrix} \begin{bmatrix} z^2 - z + 0.25 & -1 \\ 0 & z-1 \end{bmatrix}^{-1}, \tag{10.67}$$

and the left coprime MFD is

$$G(z) = \bar{D}^{-1}(z)\bar{N}(z) = \begin{bmatrix} z^2 - z + 0.25 & -1 \\ 0 & z-1 \end{bmatrix}^{-1} \begin{bmatrix} 1 & 0 \\ 0 & 1 \end{bmatrix}. \tag{10.68}$$

The controlled output is $y_c = y$, so that $\Xi = I$ (see (10.4)). It is assumed that there exist two input disturbances (*i.e.*, $G_d(z) = G(z)$) of the form

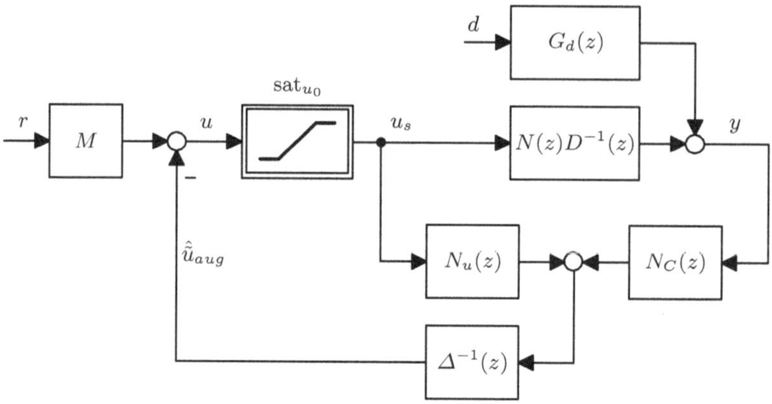

Figure 10.2. Frequency-domain representation of the observer-based compensator in Figure 10.1

$d_i(t) = d_{0i}\sin(3t + \varphi_i)$, $i = 1, 2$. The sampling time is assumed to be $T = \arccos(0.96)/3$, so that the characteristic polynomial (10.7) of the disturbance process is $\det(zI - S^*) = z^2 - 1.92z + 1$. The frequency-domain description (10.40) and (10.41) of the signal model is

$$N_S(z) = \begin{bmatrix} 1 & 0 \\ z & 0 \\ 0 & 1 \\ 0 & z \end{bmatrix} \quad \text{and } D_S(z) = \begin{bmatrix} z^2 - 1.92z + 1 & 0 \\ 0 & z^2 - 1.92z + 1 \end{bmatrix}. \tag{10.69}$$

The state feedback is intended to place the eigenvalues of the discrete-time system at $\tilde{z}_1 = 0.1$, $\tilde{z}_2 = 0.3$ and $\tilde{z}_3 = 0.4$ that is, *e.g.*, achieved by the parameterizing polynomial matrix

$$\tilde{D}(z) = \begin{bmatrix} z^2 - 0.7z + 0.12 & 0 \\ 0 & z - 0.1 \end{bmatrix}. \tag{10.70}$$

The minimal-order state observer is characterized by $\kappa = m = 2$, so that it has the order $n_O = n - \kappa = 1$. Since

$$\bar{D}_2(z) = \Pi\left\{\bar{D}(z)\begin{bmatrix} z^{-1} & 0 \\ 0 & z^{-1} \end{bmatrix}\right\} = \begin{bmatrix} z - 1 & 0 \\ 0 & 1 \end{bmatrix} \tag{10.71}$$

(see (10.36)) is row reduced, a parameterizing polynomial matrix of the observer that places its eigenvalue at $z = 0.3$ is, *e.g.*,

$$\tilde{\bar{D}}(z) = \begin{bmatrix} z - 0.3 & 1 \\ 0 & 1 \end{bmatrix}. \tag{10.72}$$

It is further assumed that the four eigenvalues of the controlled signal model are located at $z = 0.5$. This is assured by

$$\tilde{D}_S(z) = \begin{bmatrix} z^2 - z + 0.25 & 0 \\ 0 & z^2 - z + 0.25 \end{bmatrix}. \tag{10.73}$$

The observer-based controller with signal model for the rejection of the sinusoidal disturbances can now be designed along the lines of Theorem 10.4. The solutions P and $Q(z)$ of (10.58) are

$$P = \begin{bmatrix} -0.4624 & 0.430408 & 0 & 0 \\ 0 & 0 & -0.845 & 0.9244 \end{bmatrix}, \tag{10.74}$$

$$Q(z) = \begin{bmatrix} 0.4924 + 0.22z + z^2 & 0 \\ 0 & 0.82 + z \end{bmatrix}. \tag{10.75}$$

Now, the solutions

$$Y(z) = \begin{bmatrix} 1 & 0 \\ 0 & 1 \end{bmatrix} \text{ and } X(z) = \begin{bmatrix} 0 & 0 \\ 0 & 0 \end{bmatrix} \tag{10.76}$$

of the Bezout identity (10.59) can be used to compute

$$\bar{V}(z) = \Pi\left\{Q(z)Y(z)\bar{D}^{-1}(z)\bar{\tilde{D}}(z)\right\} = \begin{bmatrix} z+0.92 & 1 \\ 0 & 1 \end{bmatrix}, \tag{10.77}$$

and the right-to-left conversion (10.61) then yields

$$\hat{\Delta}(z) = \begin{bmatrix} z - 0.3 & 0 \\ 0 & 1 \end{bmatrix} \quad \text{and} \quad V(z) = \begin{bmatrix} z + 0.92 & -1.22 \\ 0 & 1 \end{bmatrix}. \tag{10.78}$$

With these results the polynomial matrices of the left coprime MFD of the compensator are

$$\begin{aligned} N_C(z) &= D_S(z)[\hat{\Delta}(z)Q(z)Y(z) - V(z)\bar{D}(z)] + \hat{\Delta}(z)PN_S(z)\Xi \\ &= \begin{bmatrix} (1.0964z - 0.956)(z - 0.5)^2 & (2.22z - 0.3)(z^2 - 1.92z + 1) \\ 0 & 1.82z^2 - 2.57z + 0.975 \end{bmatrix}, \end{aligned} \tag{10.79}$$

and

$$\begin{aligned} D_C(z) &= D_S(z)[\hat{\Delta}(z)Q(z)X(z) + V(z)\bar{N}(z)] \\ &= \begin{bmatrix} (z + 0.92)(z^2 - 1.92z + 1) & -1.22(z^2 - 1.92z + 1) \\ 0 & z^2 - 1.92z + 1 \end{bmatrix}. \end{aligned} \tag{10.80}$$

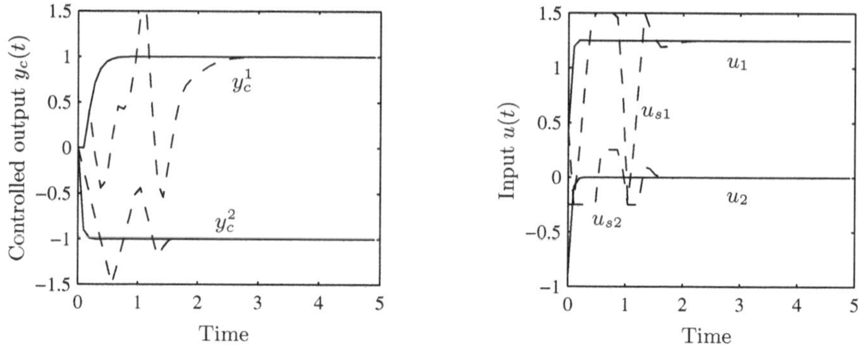

(a) Reference step response of the closed-loop system (*solid lines*: without saturation, *dashed lines*: with input saturation)

(b) Input signal (*solid lines*: without saturation, *dashed lines*: with input saturation)

Figure 10.3. Reference behaviour of the closed-loop system for reference steps $r_1 = 1$ and $r_2 = -1$ without prevention of controller windup

The polynomial matrix (10.64) characterizing the error dynamics of the overall observer has the form

$$\Delta(z) = \hat{\Delta}(z)\tilde{D}_S(z) = \begin{bmatrix} (z-0.3)(z-0.5)^2 & 0 \\ 0 & (z-0.5)^2 \end{bmatrix}. \tag{10.81}$$

With

$$M = \begin{bmatrix} 0.42 & 0 \\ 0 & 0.9 \end{bmatrix} \tag{10.82}$$

(see (10.46)) a reference behaviour (10.66) is assured as if only the constant state feedback, parameterized by $\tilde{D}(z)$, had been applied to the system.

With this compensator the reference step responses of the closed-loop system reach their final values for $t < 1$ without overshoot. The solid lines in Figure 10.3a show the components y_c^1 and y_c^2 of the controlled output y_c caused by reference inputs $r_1(k) = 1$ and $r_2(k) = -1$ and the solid lines in Figure 10.3b the components u_1 and u_2 of the corresponding input signal.

If, however, an input saturation with amplitudes $u_{01} = 1.5$ and $u_{02} = 0.25$ (see (4.108)) is introduced as in Figure 4.5 for continuous-time systems, the reference transients of the closed-loop system are impaired by controller windup, because all five eigenvalues of the compensator lie on or in the vicinity of the unit circle in the complex plane. The dashed lines in Figure 10.3a show the components y_c^1 and y_c^2 of the controlled output y_c and the dashed lines in Figure 10.3b the components u_{s1} and u_{s2} of the constrained input signal to the linear system.

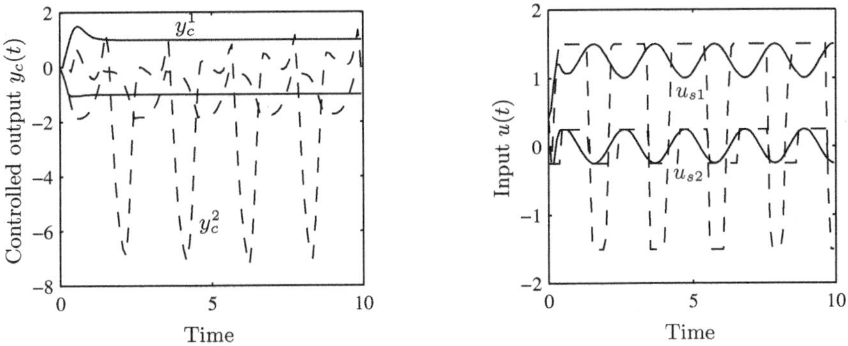

(a) Reference step response in the presence of disturbances (*solid lines*: with prevention of controller windup, *dashed lines*: without prevention of controller windup)

(b) Input signal (*solid lines*: with prevention of controller windup, *dashed lines*: without prevention of controller windup)

Figure 10.4. Reactions of the closed-loop system to joint reference and disturbance inputs

Given the above results the polynomial matrix

$$N_u(z) = \begin{bmatrix} 0.3z^2 - 1.3164z + 0.995 & -1.22z^2 + 2.3424z - 1.22 \\ 0 & -0.92z + 0.75 \end{bmatrix} \tag{10.83}$$

(see (10.65)) can be computed and with this $N_u(z)$ the compensator can be realized in the observer structure of Figure 10.2, where a model of the input saturation is inserted at the output of the controller. This prevents the controller windup completely. As in the linear unconstrained case, the reference step responses (not shown in Figure 10.3) now reach their final values without overshoot. Therefore, whenever there is the danger of input saturation, it is advisable to realize the compensator in the observer structure.

Figure 10.4 shows the reaction of the non-linear closed-loop system with the above input saturation amplitudes to a joint reference and disturbance input, *i.e.*, in addition to the above reference steps, sinusoidal disturbances with $d_{01} = 0.25$, $d_{02} = -0.25$ and $\varphi_1 = \varphi_2 = 0$ are acting. The dashed lines in Figure 10.4a show the controlled outputs, when the compensator is realized in the structure of Figure 4.5, *i.e.*, when controller windup occurs. The corresponding constrained input signals u_{s1} and u_{s2} to the linear system are depicted in dashed lines in Figure 10.4b. If the observer structure of Figure 4.6 is used the transients shown in full lines result. They are not impaired by controller windup and therefore, they show approximately the behaviour as without input constraints. The sinusoidal disturbances are asymptotically compensated in the controlled output y_c of the system.

11

Optimal Control and Estimation for Discrete-time Systems

In this chapter the LQG problem is solved for discrete-time systems in the time and in the frequency domains (see, *e.g.*, [41]). The solution of the discrete-time linear quadratic regulator (LQR) problem is very similar to its continuous-time counterpart. The discrete-time AREs for solving the LQR problem are rather different when compared to the algebraic Riccati equations of the continuous-time case. In the frequency domain, however, the design equations for solving the LQR problem in continuous time and discrete time nearly coincide.

A major difference between continuous-time and discrete-time systems exists in view of the optimal estimator (Kalman filter), where an *a priori* or one-step prediction state estimate $\hat{x}(k)$ and an *a posteriori* or innovated state estimate $\hat{x}^+(k)$ can be obtained. The estimate $\hat{x}(k)$ is computed from the states of the estimator at time k and the undisturbed measurements at time $k-1$, whereas the *a posteriori* estimate $\hat{x}^+(k)$ additionally uses the disturbed measurements at time k. By an optimal injection of the disturbed measurements at time k, the variance of the estimation error can be reduced in comparison to that resulting with the estimate $\hat{x}(k)$.

In Section 11.1 the time-domain solution of the LQR problem is briefly introduced. This is the basis for the derivation of the rational matrix equation characterizing the optimal state feedback in the frequency domain in Section 11.2. This matrix equation can be directly obtained from the algebraic Riccati equation of the time-domain solution by using the connecting relations between the time- and the frequency-domain representations of discrete-time state feedback control. The polynomial matrix $\tilde{D}(z)$ characterizing the linear quadratic regulator results from spectral factorization of this rational matrix equation.

The time-domain version of the stationary Kalman filtering problem is presented in Section 11.3. When using the *a posteriori* estimate an alternative representation of the filter in the time and in the frequency domain is advantageous. In order to formulate the connections between the time- and the frequency-domain representations of such filters a non-minimal time-domain realization is needed. This is discussed in Section 11.4.1. The design equation of the (reduced-order or full-order) optimal estimator in the frequency domain is derived in Section 11.4.2 from the Riccati equation by using the connecting relations between the filter representations in the time and in the frequency domains. The polynomial matrix characterizing the Kalman filter can be obtained from this design equation by discrete-time spectral factorization. If the *a priori* state estimate $\hat{x}(k)$ is of interest the optimal filter can be parameterized by a polynomial matrix $\tilde{\bar{D}}(z)$ that has the same structural properties as the matrix $\tilde{\bar{D}}(s)$ in continuous time. If, however, an *a posteriori* state estimate $\hat{x}^+(k)$ is of interest, the parameterizing matrix $\tilde{\bar{D}}^+(z)$ related to the alternative filter representation is used.

In Section 11.5 a frequency-domain design of observer-based compensators is presented when using the *a posteriori* estimate. The corresponding controller consists of an observer-based compensator, where the polynomial matrix $\tilde{D}(z)$ characterizes the dynamics of the state feedback loop and the polynomial matrix $\tilde{\bar{D}}^+(z)$ is designed to obtain an *a posteriori* estimate $\hat{x}^+(k)$. When using the *a priori* estimate $\hat{x}(k)$ for state feedback, the compensator design goes along the lines presented in Chapter 10.

11.1 The Linear Quadratic Regulator in the Time Domain

Considered are linear, time-invariant, discrete-time systems with the state equations

$$x(k+1) = Ax(k) + Bu(k), \tag{11.1}$$

$$y(k) = Cx(k), \tag{11.2}$$

where $x \in \mathbb{R}^n$ is the state, $u \in \mathbb{R}^p$ is the input, and $y \in \mathbb{R}^m$ is the output. It is assumed that the pair (A, B) is controllable and that the pair (C, A) is observable.

Given a constant positive-semidefinite matrix $Q = Q^T$ and a constant positive-definite matrix $R = R^T$ the discrete-time LQR problem consists of determining a stabilizing control u that minimizes the performance index

$$V = \sum_{k=0}^{\infty} \left(y^T(k)Qy(k) + u^T(k)Ru(k) \right), \tag{11.3}$$

when starting from an initial condition $x(0) = x_0$. The first term y^TQy in (11.3) assures that the minimization process leads to a faster decay of the outputs. The second term u^TRu is required to yield a solution with finite control effort. If a weighting of the controlled output $y_c = \Xi y$ is desired, this can be achieved by choosing the matrix Q in (11.3) as $Q = \Xi^T Q_c \Xi$.

It is, *e.g.*, shown in [22] that the solution to this problem is a constant state feedback

$$u(k) = -Kx(k). \tag{11.4}$$

If $Q = Q_0^T Q_0$ is such that $(Q_0 C, A)$ has no unobservable eigenvalues on the unit circle in the complex plane the stabilizing feedback gain in (11.4) is given by

$$K = \tilde{R}^{-1} B^T P A, \tag{11.5}$$

where $\tilde{R}$ is the abbreviation

$$\tilde{R} = R + B^T P B, \tag{11.6}$$

and $P = P^T$ is the unique positive-semidefinite solution of the *discrete-time algebraic Riccati equation* (DARE)

$$P = A^T \{P - PB\left(R + B^T PB\right)^{-1} B^T P\} A + C^T Q C \tag{11.7}$$

(see [4, 22, 41]).

11.2 The Linear Quadratic Regulator in the Frequency Domain

The LQR problem can also be solved in the frequency domain on the basis of the transfer behaviour

$$y(z) = C(zI - A)^{-1} Bu(z) = G(z)u(z) \tag{11.8}$$

of (11.1) and (11.2). In (11.8) the transfer matrix $G(z)$ is represented by the right coprime MFD

$$G(z) = N(z)D^{-1}(z), \tag{11.9}$$

where $D(z)$ is column reduced.

The state feedback (11.4) is characterized in the frequency domain by the denominator matrix $\tilde{D}(z)$ and this polynomial matrix has the properties (10.44) and (10.45). The rational matrix equation characterizing the optimal solution $\tilde{D}(z)$ of the LQR problem in the frequency domain can be obtained from the DARE (11.7) by using the connecting relation (10.43) in a similar way as in Chapter 8. This matrix equation leads to the spectral factorization of a $p \times p$ rational matrix $H(z)$ in the complex variable $z = \rho e^{j\delta}$, which is symmetric and positive, *i.e.*,

$$H^T(z^{-1}) = H(z), \tag{11.10}$$

and

$$H(e^{j\delta}) > 0 \text{ for real } \delta, \quad -\pi < \delta \leq \pi. \tag{11.11}$$

The Condition (11.10) implies that the zeros of $\det H(z)$ are located symmetrically with respect to the unit circle, *i.e.*, if z_0 is a zero of $\det H(z)$ inside the unit circle then there is also a zero at z_0^{-1} outside the unit circle. Furthermore, (11.11) shows that no zero of $\det H(z)$ lies on the unit circle.

The problem of *spectral factorization* consists of determining a $p \times p$ polynomial matrix $\tilde{\tilde{D}}(z)$ that satisfies

$$H(z) = \tilde{\tilde{D}}^T(z^{-1})\tilde{\tilde{D}}(z), \tag{11.12}$$

with

$$\det \tilde{\tilde{D}}(z) \neq 0 \text{ for all } |z| \geq 1, \tag{11.13}$$

i.e., the determinant $\det \tilde{\tilde{D}}(z)$ is a Schur polynomial and $\det \Gamma_c[\tilde{\tilde{D}}(z)] \neq 0$. Then, $\tilde{\tilde{D}}(z)$ is a *spectral factor* of $H(z)$ containing the zeros of $H(z)$ that lie inside the unit circle. It has been shown in [68] that the spectral factor $\tilde{\tilde{D}}(z)$ exists and that it is unique up to a multiplicative orthogonal matrix U, *i.e.*, $U^T U = I$.

The next theorem shows how the optimal polynomial matrix $\tilde{D}(z)$ can be obtained in the z-domain.

Theorem 11.1 (Frequency-domain solution of the LQR problem). *Consider the transfer behaviour of the system (11.1) and (11.2) in a right coprime MFD*

$$y(z) = C(zI - A)^{-1}Bu(z) = N(z)D^{-1}(z)u(z), \tag{11.14}$$

with $D(z)$ column reduced. Given are the $m \times m$ weighting matrix Q and the $p \times p$ weighting matrix R in the performance index (11.3), where $Q = Q_0^T Q_0$ is positive-semidefinite and $R = R^T$ is positive-definite. The linear quadratic regulator problem is solvable iff any greatest common right divisor of $D(z)$ and $Q_0 N(z)$ has no zeros on the unit circle.

Then, the unique polynomial matrix $\tilde{D}(z)$ characterizing the optimal and stabilizing state feedback in the frequency domain exists and it can be obtained by determining the spectral factor $\tilde{\tilde{D}}(z)$ in

$$D^T(z^{-1})RD(z) + N^T(z^{-1})QN(z) = \tilde{\tilde{D}}^T(z^{-1})\tilde{\tilde{D}}(z), \tag{11.15}$$

and subsequent computation of

$$\tilde{D}(z) = \Gamma_c[D(z)]\Gamma_c^{-1}[\tilde{\tilde{D}}(z)]\tilde{\tilde{D}}(z). \tag{11.16}$$

Proof. The rational equation (11.15) can be obtained from the Riccati equation (11.7). Pre-multiplication of (11.5) by $\tilde{R}$ leads to

$$\tilde{R}K = B^T PA, \tag{11.17}$$

so that the DARE (11.7) can also be expressed as

$$P = A^T PA - K^T \tilde{R} K + C^T QC \tag{11.18}$$

(see (11.6)). It is easy to show that

$$P - A^T PA = (z^{-1}I - A^T)P(zI - A) + (z^{-1}I - A^T)PA + A^T P(zI - A). \tag{11.19}$$

Therefore, (11.18) can be written as

$$(z^{-1}I - A^T)P(zI - A) + (z^{-1}I - A^T)PA + A^T P(zI - A) + K^T \tilde{R} K = C^T QC. \tag{11.20}$$

Pre-multiplying this result by $B^T(z^{-1}I - A^T)^{-1}$ and postmultiplying it by $(zI - A)^{-1}B$ leads to

$$\begin{aligned} &B^T PB + \tilde{R}K(zI - A)^{-1}B + B^T(z^{-1}I - A^T)^{-1}K^T\tilde{R} \\ &\quad + B^T(z^{-1}I - A^T)^{-1}K^T\tilde{R}K(zI - A)^{-1}B \\ &\quad = B^T(z^{-1}I - A^T)^{-1}C^T QC(zI - A)^{-1}B, \end{aligned} \tag{11.21}$$

where (11.17) has been used. This can be reordered as

$$\begin{aligned} &[I + B^T(z^{-1}I - A^T)^{-1}K^T]\tilde{R}[I + K(zI - A)^{-1}B] \\ &\quad = R + B^T(z^{-1}I - A^T)^{-1}C^T QC(zI - A)^{-1}B, \end{aligned} \tag{11.22}$$

in view of (11.6) when adding R on both sides of (11.21). Using the connecting relation

$$\tilde{D}(z)D^{-1}(z) = I + K(zI - A)^{-1}B \tag{11.23}$$

(see (10.43)) and (11.14) this takes the form

$$\begin{aligned} &(D^T(z^{-1}))^{-1}\tilde{D}^T(z^{-1})\tilde{R}\tilde{D}(z)D^{-1}(z) \\ &\qquad = R + (D^T(z^{-1}))^{-1}N^T(z^{-1})QN(z)D^{-1}(z). \end{aligned} \tag{11.24}$$

By pre-multiplying this result by $D^T(z^{-1})$ and postmultiplying it by $D(z)$ one obtains the rational matrix equation

$$H(z) = \tilde{D}^T(z^{-1})\tilde{R}\tilde{D}(z) = D^T(z^{-1})RD(z) + N^T(z^{-1})QN(z). \tag{11.25}$$

As R is positive-definite and P is positive-semidefinite $\tilde{R} = R + B^T PB$ is also positive-definite and symmetric. And as there exists no greatest common right divisor of $D(z)$ and $Q_0N(z)$ with zeros on the unit circle, which is the frequency-domain characterization of the fact, that the pair (Q_0C, A) has no unobservable eigenvalues on the unit circle, there exists a stabilizing solution P of the DARE (11.7) (see [22]) and the rational matrix $H(z)$ has the properties (11.10) and (11.11). Conversely, if the pair (Q_0C, A) has no unobservable eigenvalues on the unit circle $Q_0N(z)$ and $D(z)$ do not have a greatest common right divisor that has a zero on the unit circle. In order to prove this assume that such a right divisor exists. This would imply that the corresponding pole is not controllable, because there are no unobservable eigenvalues on the unit circle. This, however, would contradict the fact that there exist a stabilizing solution of the DARE (11.7) and an optimal spectral factor $\tilde{D}(z)$ of (11.15) because the pre-requisites for the spectral factorization would not be satisfied. Thus, the time-domain conditions also imply the corresponding frequency-domain conditions for the solution of the LQR problem. Therefore, the existence of an optimal time-domain solution also guarantees that the corresponding frequency-domain solution exists.

Because $\tilde{R} = \tilde{R}^T$ is a real-valued positive-definite matrix there exists a unique factorization $\tilde{R} = SS^T = S^2$, so that the square root $\tilde{R}^{1/2}$ of $\tilde{R}$ is $\tilde{R}^{1/2} = S$ (see, *e.g.*, [37]). Therefore, a spectral factor $\tilde{R}^{1/2}\tilde{D}(z)$ of the rational matrix $H(z)$ also exists in (11.25), since the DARE (11.7) has a stabilizing solution. Because the spectral factorization is unique up to an orthogonal matrix U the spectral factor $\tilde{\tilde{D}}(z)$ in (11.15) satisfies

$$U\tilde{\tilde{D}}(z) = \tilde{R}^{1/2}\tilde{D}(z), \tag{11.26}$$

or

$$\tilde{D}(z) = \tilde{R}^{-1/2}U\tilde{\tilde{D}}(z). \tag{11.27}$$

In view of

$$\Gamma_c[\tilde{D}(z)] = \Gamma_c[D(z)] \tag{11.28}$$

(see (10.45)) the orthogonal matrix U in (11.27) has the form

$$U = \tilde{R}^{1/2}\Gamma_c[D(z)]\Gamma_c^{-1}[\tilde{\tilde{D}}(z)], \tag{11.29}$$

as $\tilde{\tilde{D}}(z)$ is column reduced. Inserting (11.29) in (11.27) leads to (11.16). □

Remark 11.1. In the time-domain solution it was required that the pair (Q_0C, A) has no unobservable eigenvalues on the unit circle. This corresponds to the frequency-domain condition that the polynomial matrices $D(z)$ and $Q_0N(z)$ do not have a greatest common right divisor with a zero on the unit circle. This, however, only assures that the eigenvalues of the closed-loop system are located inside the unit circle. All eigenvalues can only be influenced by the optimal control if the pair (Q_0C, A) is observable or, equivalently, if the greatest common right divisor of $D(z)$ and $Q_0N(z)$ is a unimodular matrix (see Remark 1.5).

Remark 11.2. The left-hand side of (11.15) leads to the spectral factorization of a rational matrix $H(z)$ since $D^T(z^{-1})$ and $N^T(z^{-1})$ are rational matrices. The algorithms for spectral factorization, however, are based on a representation of $H(z)$ as a polynomial matrix. This is obtained by multiplying all elements of $D^T(z^{-1})$ and $N^T(z^{-1})$ by z^{ν} if $z^{-\nu}$ is the highest negative power of z appearing in $D^T(z^{-1})$. This can lead to additional zeros at $z = 0$ in $\tilde{\tilde{D}}(z)$

resulting from the discrete-time spectral factorization, *i.e.*, $\det \tilde{\tilde{D}}(z) = z^k p(z)$, $k > 0$, where $p(z)$ is a Schur polynomial of degree n. To obtain the correct solution these additional zeros at $z = 0$ have to be removed from $\tilde{\tilde{D}}(z)$, so that for the resulting $\tilde{\tilde{D}}_{red}(z)$ both conditions $\det \tilde{\tilde{D}}_{red}(z) = p(z)$ and $\tilde{\tilde{D}}^T_{red}(z^{-1})\tilde{\tilde{D}}_{red}(z) = H(z)$ are satisfied (see [32]). In the remainder of this chapter, it is always assumed that the additional zeros at $z = 0$ have already been removed from the factorization result.

Example 11.1. Frequency-domain solution of the LQR problem
Given is a discrete-time system of the order 2 with one input u and two outputs y. Its transfer behaviour $y(z) = G(z)u(z)$ is characterized by

$$G(z) = N(z)D^{-1}(z) = \begin{bmatrix} z \\ 1 \end{bmatrix} \frac{1}{z^2 - z + 0.25}. \tag{11.30}$$

Assume that the weighting matrices Q and R in the performance index (11.3) have the forms

$$Q = \begin{bmatrix} 6.281875/3 & 0 \\ 0 & 6.281875/3 \end{bmatrix} \quad \text{and} \quad R = 1. \tag{11.31}$$

Then, the left-hand side of the design equation (11.15) takes the form

$$\begin{aligned} &D^T(z^{-1})RD(z) + N^T(z^{-1})QN(z) \\ &= \frac{0.25}{3}\left[3z^2 - 15z + 75.005 - \frac{15}{z} + \frac{3}{z^2}\right]. \end{aligned} \tag{11.32}$$

Pre-requisite for the existence of a stable spectral factor of (11.32) is that the pair $(D(z), Q_0N(z))$ does not have a greatest common right divisor with a zero on the unit circle. Multiplying the rational function in the square brackets of (11.32) by z^2 leads to the polynomial $3z^4 - 15z^3 + 70.005z^2 - 15z + 3$. This polynomial has no zero on the unit circle, so that a stable spectral factor of (11.32) exists. The discrete-time spectral factorization of (11.32) yields the Schur polynomial

$$\tilde{\tilde{D}}(z) = (6z^2 - 1.2z + 0.25)/\sqrt{6}, \tag{11.33}$$

with zeros at $\tilde{\tilde{z}}_{1,2} = -0.1 \pm \mathrm{j}\sqrt{0.095/3}$.

In the single-input case the highest column-degree-coefficient matrices are the leading coefficients of the polynomials. Here, they have the forms

$$\Gamma_c[\tilde{\tilde{D}}(z)] = \sqrt{6} \quad \text{and} \quad \Gamma_c[D(z)] = 1, \tag{11.34}$$

so that the Relation (11.16) finally yields the polynomial

$$\tilde{D}(z) = z^2 - 0.2z + 0.125/3, \tag{11.35}$$

which parameterizes the optimal state feedback control in the frequency domain. The eigenvalues of the open-loop system at $z_{1,2} = 0.5$ have been shifted towards $\tilde{z}_{1,2} = 0.1 \pm \mathrm{j}0.178$, *i.e.*, they are located closer to the origin of the z-plane.

11.3 The Stationary Kalman Filter in the Time Domain

The discrete-time systems considered are linear and time invariant and they are described by the difference equation

$$x(k+1) = Ax(k) + Bu(k) + Gw(k), \tag{11.36}$$

where $x \in \mathbb{R}^n$ is the state, $u \in \mathbb{R}^p$ is the input and $w \in \mathbb{R}^q$ is a stochastic disturbance. It is assumed that the pairs (A, B) and (A, G) are controllable. To handle the full-order filter and the reduced-order filter in one scheme it is assumed that only part of the measured output $y \in \mathbb{R}^m$ is corrupted by noise, so that one can write

$$y_1(k) = C_1 x(k) + v_1(k), \tag{11.37}$$
$$y_2(k) = C_2 x(k), \tag{11.38}$$

where $y_2 \in \mathbb{R}^\kappa$, $0 \le \kappa \le m$, is the measurement not corrupted by noise (ideal measurement), $y_1 \in \mathbb{R}^{m-\kappa}$ is the disturbed measurement and $v_1 \in \mathbb{R}^{m-\kappa}$ is the measurement noise. It is assumed that the system, *i.e.*, the pair (C, A) is observable, where the $m \times n$ matrix C is defined as

$$C = \begin{bmatrix} C_1 \\ C_2 \end{bmatrix}. \tag{11.39}$$

In view of a well-defined order of the filter the system is assumed not to have eigenvalues at $z = 0$ (see [60]). For $\kappa = 0$ all measurements are corrupted by noise. This case is characterized by $C_1 = C$ and a vanishing C_2. For $\kappa = m$ all measurements are free of noise, characterized by $C_2 = C$ and a vanishing C_1. The stochastic inputs $\{w(k)\}$ and $\{v_1(k)\}$ are independent, zero-mean, stationary Gaussian white noise with

$$\mathrm{E}\{w(k)w^T(\ell)\} = \bar{Q}\delta_{k\ell}, \tag{11.40}$$

$$\mathrm{E}\{v_1(k)v_1^T(\ell)\} = \bar{R}_1\delta_{k\ell}, \tag{11.41}$$

where

$$\delta_{k\ell} = \begin{cases} 1, & k = \ell, \\ 0, & \text{otherwise}. \end{cases} \tag{11.42}$$

It is assumed that the covariance matrices $\bar{Q}$ and $\bar{R}_1$ are real and symmetric and that $\bar{Q}$ is positive-semidefinite and $\bar{R}_1$ is positive-definite. It is further assumed that the initial state $x(0) = x_0$ is not correlated with the disturbances, *i.e.*, $\mathrm{E}\{x_0 w^T(k)\} = 0$ and $\mathrm{E}\{x_0 v_1^T(k)\} = 0$ for all $k \ge 0$.

The *stationary discrete-time Kalman filter* (see, *e.g.*, [3]) yields an estimate $\hat{x}$ for the state of (11.36) with minimal stationary variance $\mathrm{E}\{(x-\hat{x})^T(x-\hat{x})\}$ for the estimation error $x - \hat{x}$. If parts of the measurements are free of noise a (stationary) reduced-order Kalman filter can be used to obtain a state estimate with minimal stationary variance.

In order to derive this filter consider $n - \kappa$ linear combinations

$$\zeta(k) = Tx(k) \tag{11.43}$$

of the states of the system. Together with the κ ideal measurements y_2 these linear combinations can be used to represent the state x of the system as

$$x(k) = \begin{bmatrix} C_2 \\ T \end{bmatrix}^{-1} \begin{bmatrix} y_2(k) \\ \zeta(k) \end{bmatrix} = \Psi_2 y_2(k) + \Theta\zeta(k). \tag{11.44}$$

In the solution of this filtering problem it is assumed that the covariance matrix

$$\Phi = C_2 G \bar{Q} G^T C_2^T \tag{11.45}$$

of the measurement noise in $y_2(k+1) = C_2 x(k+1)$ is positive-definite. Thus, the covariance matrix

$$\bar{R} = \begin{bmatrix} \bar{R}_1 & 0 \\ 0 & \Phi \end{bmatrix} \tag{11.46}$$

of the measurement noise with respect to $y_1(k)$ and $y_2(k+1)$ is also positive-definite, which is a standing assumption in the design of reduced-order optimal estimators (see, *e.g.*, [3, 34, 50]). The stationary minimal variance estimate $\hat{x}$ results from (11.44) when the estimate $\hat{\zeta}$ for ζ is obtained from a *reduced-order optimal estimator* (*reduced-order Kalman filter*)

$$\hat{\zeta}(k+1) = T(A - L_1^+ C_1)\Theta\hat{\zeta}(k) + \begin{bmatrix} TL_1^+ & T(A - L_1^+ C_1)\Psi_2 \end{bmatrix} \begin{bmatrix} y_1(k) \\ y_2(k) \end{bmatrix} + TBu(k), \tag{11.47}$$

$$\hat{x}(k) = \Theta\hat{\zeta}(k) + \Psi_2 y_2(k) \tag{11.48}$$

(see [34]). The optimal estimate $\hat{\zeta}$ results if the matrices L_1^+ and Ψ_2 are chosen such that

$$L_1^+ = A\bar{P}C_1^T \hat{R}^{-1}, \tag{11.49}$$

and

$$\Psi_2 = \tilde{P}C_2^T X^{-1}. \tag{11.50}$$

In (11.49) and (11.50) the abbreviations

$$\hat{R} = \bar{R}_1 + C_1 \bar{P} C_1^T, \tag{11.51}$$

$$\tilde{P} = A\bar{P}A^T + G\bar{Q}G^T - A\bar{P}C_1^T \hat{R}^{-1} C_1 \bar{P} A^T, \tag{11.52}$$

and

$$X = C_2 \tilde{P} C_2^T \tag{11.53}$$

have been used. By assumption, $\bar{R}_1$ is positive-definite and this implies that $\hat{R}$ in (11.51) is also positive-definite if $\bar{P}$ is positive-semidefinite. It can further be shown that if Φ in (11.45) is positive-definite and $\bar{P}$ is positive-semidefinite then X is also positive-definite (see Problem 1.2 in [3]). Consequently, the inverse matrices in (11.49) and (11.50) exist. The stationary value $\bar{P}$ of the error covariance

$$\bar{P}(k) = \mathrm{E}\left\{(x(k) - \hat{x}(k))(x(k) - \hat{x}(k))^T\right\} \tag{11.54}$$

can be obtained as the positive-semidefinite solution of the DARE

$$\bar{P} = A\bar{P}A^T + G\bar{Q}G^T - A\bar{P}C_1^T\hat{R}^{-1}C_1\bar{P}A^T - \tilde{P}C_2^TX^{-1}C_2\tilde{P} \tag{11.55}$$

(see [34, 50]).

The matrices L_1^+ and Ψ_2 completely characterize the optimal estimator (11.47) and (11.48). The matrices T and Θ that are needed for an explicit parameterization can be obtained as a solution of

$$T\Psi_2 = 0, \tag{11.56}$$

with $\operatorname{rank} T = n - \kappa$ and subsequent computation of

$$\Theta = \begin{bmatrix} C_2 \\ T \end{bmatrix}^{-1} \begin{bmatrix} 0 \\ I_{n-\kappa} \end{bmatrix} \tag{11.57}$$

(see (3.21) and (3.22)). Instead of the *a priori* estimate $\hat{x}$ defined in (11.48) an *a posteriori* estimate

$$\hat{x}^+(k) = \hat{x}(k) + \Lambda(y_1(k) - C_1\hat{x}(k)) \tag{11.58}$$

can be obtained (see [3]) that also uses the actual measurements at $t = kT$.

Remark 11.3. Also for an observer-based compensator on the basis of the *a posteriori* estimate $\hat{x}^+$ (see Section 11.5) the separation principle is satisfied. This is due to the fact that $x - \hat{x} \to 0$ for $k \to \infty$ and consequently $y_1 - C_1\hat{x} \to 0$ for $k \to \infty$ independently of u.

The *a posteriori* estimate leads to a smaller variance of the estimation error as compared to $\hat{x}$ if the trace of

$$\bar{P}^+(k) = \mathrm{E}\left\{(x(k) - \hat{x}^+(k))(x(k) - \hat{x}^+(k))^T\right\} \tag{11.59}$$

is minimized. Using (11.58) with (11.37) and $\bar{P}(k)$ as defined in (11.54) one obtains the stationary covariance matrix

$$\bar{P}^+ = (I - \Lambda C_1)\bar{P}(I - \Lambda C_1)^T + \Lambda\bar{R}_1\Lambda^T, \tag{11.60}$$

where $\bar{P}^+ = \bar{P}^+(\infty)$ and $\bar{P} = \bar{P}(\infty)$. Expanding the terms in (11.60) and observing (11.51) the matrix $\bar{P}^+$ can be expressed as

$$\bar{P}^+ = \bar{P} - \Lambda C_1\bar{P} - \bar{P}C_1^T\Lambda^T + \Lambda\hat{R}\Lambda^T. \tag{11.61}$$

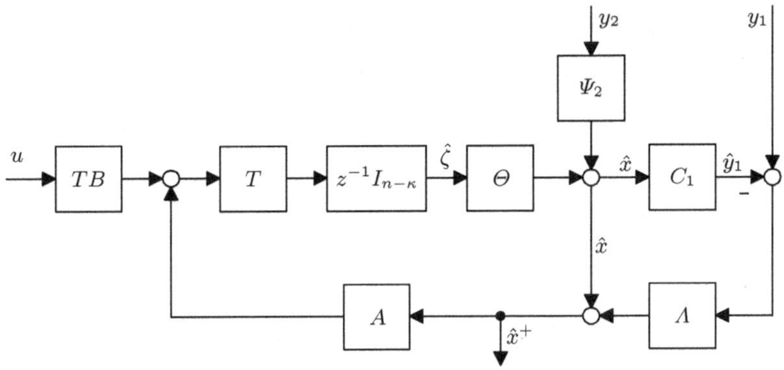

Figure 11.1. Reduced-order Kalman filter with *a priori* and *a posteriori* estimates

By adding $\bar{P}C_1^T\hat{R}^{-1}C_1\bar{P} - \bar{P}C_1^T\hat{R}^{-1}C_1\bar{P}$ this can be written as

$$\bar{P}^+ = \bar{P} - \bar{P}C_1^T\hat{R}^{-1}C_1\bar{P} + (\Lambda\hat{R} - \bar{P}C_1^T)\hat{R}^{-1}(\hat{R}\Lambda^T - C_1\bar{P}). \qquad (11.62)$$

Since $\hat{R}$ is positive-definite the term $(\Lambda\hat{R} - \bar{P}C_1^T)\hat{R}^{-1}(\hat{R}\Lambda^T - C_1\bar{P})$ in (11.62) represents a positive-definite matrix. In view of the well-known fact that the trace of a positive-semidefinite matrix is non-negative the form of (11.62) shows that the trace of $\bar{P}^+$ becomes minimal if

$$\Lambda\hat{R} - \bar{P}C_1^T = 0, \qquad (11.63)$$

or equivalently, if Λ is chosen as

$$\Lambda = \bar{P}C_1^T\hat{R}^{-1}. \qquad (11.64)$$

The variance (11.59) is related with the covariance $\bar{P}$ of the *a priori* estimate by

$$\bar{P}^+ = \bar{P} - \Lambda\hat{R}\Lambda^T, \qquad (11.65)$$

which follows from (11.62) when observing that $\Lambda\hat{R} = \bar{P}C_1^T$. This shows that the *a posteriori* estimate $\hat{x}^+$ usually yields a smaller error variance than the *a priori* estimate $\hat{x}$ since $\hat{R}$ is positive-definite.

By observing that the optimal gain L_1^+ in (11.49) can be expressed as

$$L_1^+ = A\Lambda, \qquad (11.66)$$

in view of (11.64), the *a posteriori* estimate $\hat{x}^+$ can be obtained from a reduced-order estimator (11.47) and (11.48), which is realized in the structure shown in Figure 11.1, where the shift operator is characterized by z^{-1}.

By the same arguments as in the continuous-time case (see Chapter 8) it can be shown that

$$C_2\bar{P} = 0. \tag{11.67}$$

This result, which is needed for the derivation of the frequency-domain equations of the filter, can also be obtained by pre-multiplying (11.55) by C_2 and using (11.52) and (11.53). Pre-multiplying (11.50) by C_2 and observing (11.53) it becomes obvious that

$$C_2\Psi_2 = I, \tag{11.68}$$

which is in agreement with (3.20) when designing reduced-order observers according to Chapter 3. However, as $C_2\bar{P} = 0$ does not imply that $C_2A\bar{P}$ also vanishes an inspection of (11.49) shows that the optimal filter is characterized by $C_2L_1^+ \neq 0$. Therefore, the reduced-order optimal estimator (11.47) and (11.48) is not exactly the same as the reduced-order observer in Chapter 3, where $C_2L_1 = 0$ has been used. However, the observer of Chapter 3 can be applied by modifying the observer gain. This is due to the fact that L_1^+ does not directly appear in (11.47) but only TL_1^+. Then, with $T\Theta = I_{n-\kappa}$ (see (3.6)) and in view of

$$TL_1^+ = T\Theta TL_1^+, \tag{11.69}$$

one can use the observer gain

$$L_1 = \Theta TL_1^+ \tag{11.70}$$

in (11.47) instead of L_1^+. This gain satisfies

$$C_2L_1 = C_2\Theta TL_1^+ = 0 \tag{11.71}$$

in the light of $C_2\Theta = 0$ (see (3.6)).

When observing (11.49) and (11.50) the DARE (11.55) can also be written in the form

$$\bar{P} = A\bar{P}A^T - \begin{bmatrix} L_1^+ & \Psi_2 \end{bmatrix} \begin{bmatrix} \hat{R} & 0 \\ 0 & X \end{bmatrix} \begin{bmatrix} (L_1^+)^T \\ \Psi_2^T \end{bmatrix} + G\bar{Q}G^T. \tag{11.72}$$

This form of the DARE will be used in the derivation of the frequency-domain solution of the Kalman filter.

Remark 11.4. A drawback of the DARE (11.72) is that it cannot be directly solved using standard software. By replacing the undisturbed measurements $y_2(k)$ by $y_2(k+1)$, a regular filtering problem with a positive-definite covariance of the output measurement noise is obtained, if $\bar{R}$ in (11.46) is positive-definite. The resulting DARE can be solved by standard software. This alternative formulation of the reduced-order filtering problem was originally introduced by Bryson and Johansen in [7].

The above optimal estimation scheme contains the stationary full-order Kalman filter of the order n (resulting when all measurements are corrupted by noise, *i.e.*, the case $\kappa = 0$) and the completely reduced-order optimal estimator of the order $n - m$ (resulting when all measurements are free of noise, *i.e.*, the case $\kappa = m$) as special cases.

11.4 The Stationary Kalman Filter in the Frequency Domain

11.4.1 Parameterization of the Stationary Kalman Filter for an *a posteriori* Estimate in the Frequency Domain

The derivation of the Kalman filter with *a posteriori* estimate in the time domain shows that the frequency-domain parameterization of this filter can be based on the results in Chapter 10. However, it is shown in the following that a frequency-domain parameterization of the Kalman filter with $C_2L_1^+ \neq 0$ has the advantage that an observer-based compensator with *a posteriori* state estimate can be designed directly in the frequency domain. In order to derive this frequency-domain parameterization one must consider the non-minimal representations of the optimal estimator in the time and in the frequency domains. Figure 11.2 shows a non-minimal realization of the reduced-order discrete-time estimator in the time domain. Following the same arguments for discrete-time systems as in Section 3.3 the block diagram of Figure 11.2 can be redrawn as shown in Figure 11.3. Using the same reasoning as in Chapter 3 it can also be shown that the dynamics of the system in Figure 11.3 are described by

$$\hat{\zeta}(k+1) = T(A-L_1^+C_1)\Theta\hat{\zeta}(k) + \begin{bmatrix} TL_1^+ & T(A-L_1^+C_1)\Psi_2 \end{bmatrix} \begin{bmatrix} y_1(k) \\ y_2(k) \end{bmatrix} + TBu(k). \tag{11.73}$$

Since $L_1^+ = A\Lambda$ (see (11.66)) this can also be written as

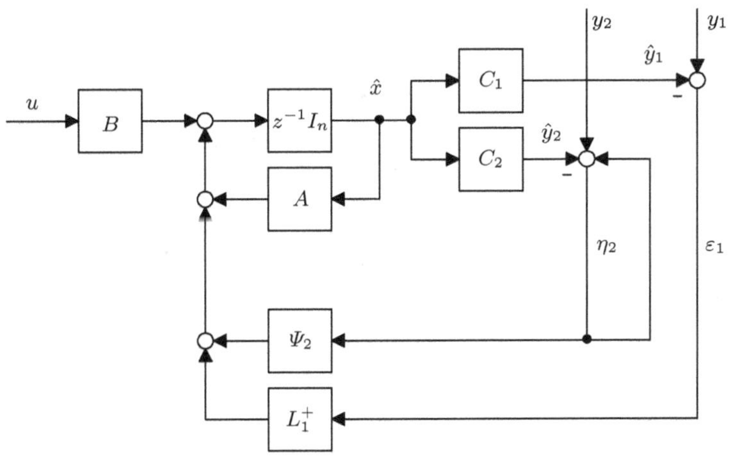

Figure 11.2. Non-minimal realization of the reduced-order discrete-time observer in the time domain

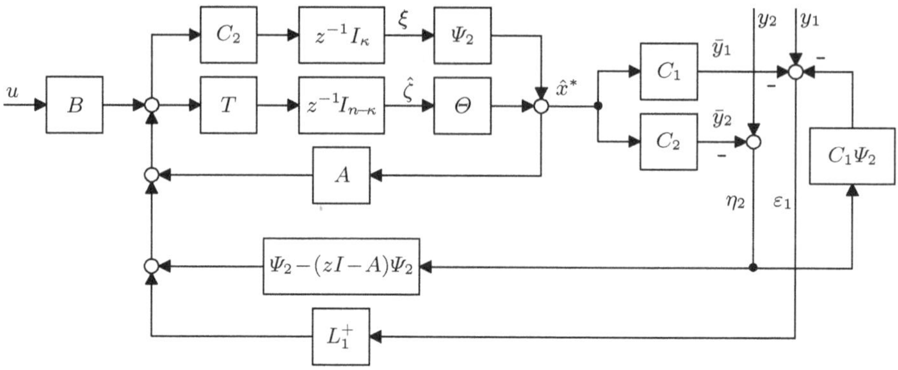

Figure 11.3. Modified version of the block diagram in Figure 11.2

$$\hat{\zeta}(k+1) = TA(I - \Lambda C_1)\Theta\hat{\zeta}(k) + TA\,[\Lambda \quad (I - \Lambda C_1)\Psi_2] \begin{bmatrix} y_1(k) \\ y_2(k) \end{bmatrix} + TBu(k), \tag{11.74}$$

which is a modified representation of the state equation (11.47) of the reduced-order estimator.

Analogous to Chapter 3 it can be verified that the output equation

$$\hat{x}(k) = \Psi_2 y_2(k) + \Theta\hat{\zeta}(k) \tag{11.75}$$

follows from Figure 11.2. Moreover, in view of $\hat{y}_2 = C_2\hat{x}$ (see Figure 11.2) the estimate $\hat{y}_2$ results as

$$\hat{y}_2(k) = C_2(\Psi_2 y_2(k) + \Theta\hat{\zeta}(k)) = y_2(k), \tag{11.76}$$

because of $C_2\Psi_2 = I_\kappa$ and $C_2\Theta = 0$. Thus, (11.73) and (11.76) show that Figure 11.2 is indeed a realization of the reduced-order observer (11.47) and (11.48).

Remark 11.5. The positive unity feedback of η_2 onto itself may be interpreted as an order reduction of the entire system consisting of the plant and the observer of Figure 11.2. This can be shown by inspection of the input to the shift element $z^{-1}I_\kappa$ in Figure 11.3. Observing that $\bar{y}_1 = C_1(\Theta\hat{\zeta} + \Psi_2\xi)$ and $\bar{y}_2 = \xi$ one obtains

$$\begin{aligned} z\xi(z) = C_2 \Big\{ & Bu(z) + A(\Psi_2\xi(z) + \Theta\hat{\zeta}(z)) \\ & + L_1^+ \Big[y_1(z) - C_1\Theta\hat{\zeta}(z) - C_1\Psi_2\xi(z) - C_1\Psi_2 y_2(z) + C_1\Psi_2\xi(z) \Big] \\ & + (\Psi_2 - z\Psi_2 + A\Psi_2)[y_2(z) - \xi(z)] \Big\}. \end{aligned} \tag{11.77}$$

Because of $L_1^+ = A\Lambda$, $C_2\Psi_2 = I_\kappa$ and $C_2\Theta = 0$ (see (11.66) and (3.6)) this takes the form

$$\xi(k) = y_2(k) - y_2(k+1) + C_2 A\Lambda y_1(k) + C_2 A(I - \Lambda C_1)\Psi_2 y_2(k) \quad (11.78)$$
$$+ C_2 A(I - \Lambda C_1)\Theta\hat{\zeta}(k) + C_2 Bu(k)$$

in the time domain. Using $y_1(k) = C_1 x(k)$, $y_2(k) = C_2 x(k)$, $y_2(k+1) = C_2 x(k+1) = C_2 Ax(k) + C_2 Bu(k)$ and $I - \Psi_2 C_2 = \Theta T$ Equation (11.78) can be written as

$$\xi(k) = C_2[I_n - A(I - \lambda_1 C_1)\Theta T]x(k) + C_2 A(I - \lambda_1 C_1)\Theta\hat{\zeta}(k), \quad (11.79)$$

i.e., the output ξ of the κ shift elements can be expressed as a linear combination of the states x of the system and the states $\hat{\zeta}$ of the observer. As the outputs ξ of the shift elements are not independent variables they are consequently no states of the system consisting of the plant (11.36)–(11.38) and the observer shown in Figure 11.2. The order of this system is therefore given by $\dim(x) + \dim(\hat{\zeta}) = n + (n - \kappa)$.

To obtain the non-minimal realization of the reduced-order estimator in the frequency domain, the transfer behaviour of the system (11.36)–(11.38)

$$y_1(z) = G_1(z)u(z) + G_{w1}(z)w(z) + v_1(z), \quad (11.80)$$
$$y_2(z) = G_2(z)u(z) + G_{w2}(z)w(z) \quad (11.81)$$

is considered. The transfer matrix between the input u and the measurement y is represented by the left coprime MFD

$$G(z) = \begin{bmatrix} G_1(z) \\ G_2(z) \end{bmatrix} = \begin{bmatrix} C_1 \\ C_2 \end{bmatrix} (zI - A)^{-1}B = \bar{D}^{-1}(z)\bar{N}(z), \quad (11.82)$$

and the transfer matrix between the disturbance input w and the measurement y by the left coprime MFD

$$G_w(z) = \begin{bmatrix} G_{w1}(z) \\ G_{w2}(z) \end{bmatrix} = \begin{bmatrix} C_1 \\ C_2 \end{bmatrix} (zI - A)^{-1}G = \bar{D}^{-1}(z)\bar{N}_w(z), \quad (11.83)$$

where both $\bar{D}(z)$ in (11.82) and (11.83) and $\bar{D}_\kappa(z)$ according to (10.54) are row reduced. In view of a well-defined order of the optimal filter for the *a posteriori* estimate it is also assumed that $\det \bar{D}(0) \neq 0$ (see [60]). Considered is also the left coprime MFD of the transfer matrix

$$C(zI - A)^{-1} = \bar{D}^{-1}(z)\bar{N}_x(z). \quad (11.84)$$

A frequency-domain representation of the observer in Figure 11.2 can be obtained by representing the model of the system by the MFD (11.84). In the resulting block diagram of Figure 11.4 the positive feedback of η_2 onto itself

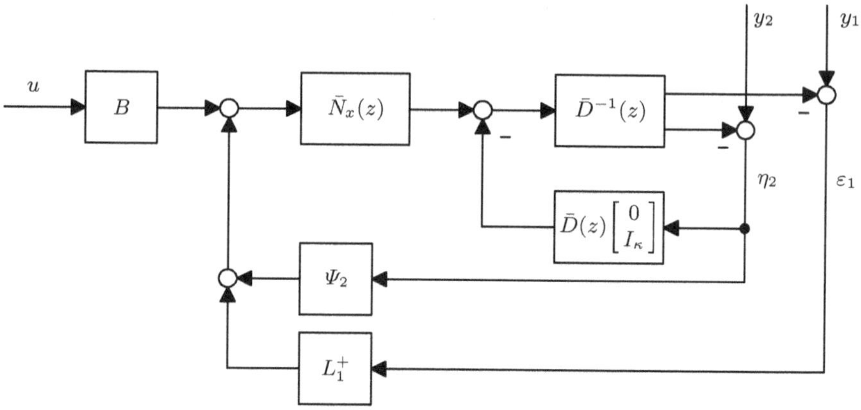

Figure 11.4. Frequency-domain representation of the block diagram in Figure 11.2

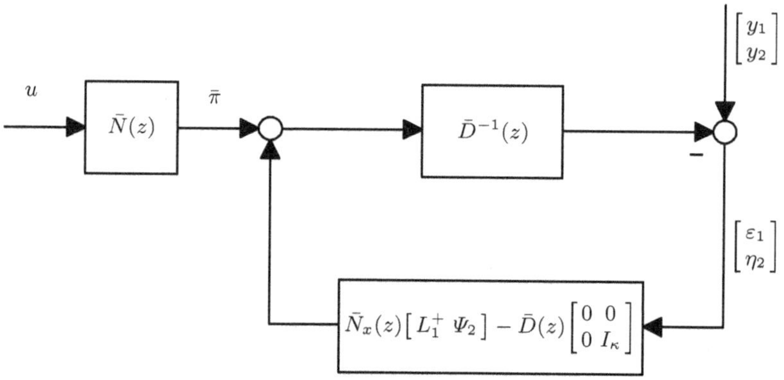

Figure 11.5. Frequency-domain representation of the reduced-order observer

appearing in Figure 11.1 is relocated to the input of the block with transfer behaviour $\bar{D}^{-1}(z)$.

The input–output behaviour of the system in Figure 11.4 can also be represented with the aid of the MFD (11.82). The result is shown in Figure 11.5 and this system has the same transfer behaviour as the non-minimal representation of the reduced-order observer in Figure 11.2.

The transfer behaviour of the reduced-order observer in Figure 11.2 between the input signals y and u and the output signals ε_1 and η_2 is considered first. To this end, insert

$$\xi(z) = y_2(z) - zy_2(z) + C_2A\Lambda y_1(z) + C_2A(I - \Lambda C_1)\Psi_2 y_2(z) + C_2A(I - \Lambda C_1)\Theta\hat{\zeta}(z) + C_2Bu(z) \tag{11.85}$$

(see (11.78)) and

$$\hat{\zeta}(z) = (zI-F)^{-1}TA\begin{bmatrix}\Lambda & (I - \Lambda C_1)\Psi_2\end{bmatrix}\begin{bmatrix}y_1(z)\\ y_2(z)\end{bmatrix} + (zI-F)^{-1}TBu(z) \tag{11.86}$$

(see (11.74)) in $\eta_2(z) = y_2(z) - C_2(\Psi_2\xi(z) + \Theta\hat{\zeta}(z))$ (see Figure 11.3). Here, and in what follows, the abbreviation

$$F = TA(I - \Lambda C_1)\Theta \tag{11.87}$$

is used. By inserting the above-defined quantities $\eta_2(z)$, $\xi(z)$ and $\hat{\zeta}(z)$ in $\varepsilon_1(z) = y_1(z) - C_1\Psi_2\eta_2(z) - C_1(\Psi_2\xi(z) + \Theta\hat{\zeta}(z))$ (see again Figure 11.3) one obtains the transfer behaviour

$$\begin{bmatrix}\varepsilon_1(z)\\ \eta_2(z)\end{bmatrix} = \begin{bmatrix}I_{m-\kappa} - C_1\Theta(zI - F)^{-1}TA\Lambda & C_1\left\{-I - \Theta(zI - F)^{-1}TA(I - \Lambda C_1)\right\}\Psi_2\\ -C_2A\Lambda - C_2A(I - \Lambda C_1)\Theta(zI - F)^{-1}TA\Lambda & C_2\left\{zI - A(I - \Lambda C_1) - A(I - \Lambda C_1)\Theta(zI - F)^{-1}TA(I - \Lambda C_1)\right\}\Psi_2\end{bmatrix}\begin{bmatrix}y_1(z)\\ y_2(z)\end{bmatrix} - \begin{bmatrix}C_1\Theta(zI - F)^{-1}TB\\ C_2\left\{I + A(I - \Lambda C_1)\Theta(zI - F)^{-1}T\right\}B\end{bmatrix}u(z). \tag{11.88}$$

In the system shown in Figure 11.4 this transfer behaviour is characterized by

$$\begin{bmatrix}\varepsilon_1(z)\\ \eta_2(z)\end{bmatrix} = \left\{\bar{N}_x(z)\begin{bmatrix}L_1^+ & \Psi_2\end{bmatrix} + \bar{D}(z)\begin{bmatrix}I_{m-\kappa} & 0\\ 0 & 0_\kappa\end{bmatrix}\right\}^{-1}\left\{\bar{D}(z)\begin{bmatrix}y_1(z)\\ y_2(z)\end{bmatrix} - \bar{N}(z)u(z)\right\}. \tag{11.89}$$

The denominator matrix in (11.89) shows that the $m \times m$ polynomial matrix

$$\tilde{\bar{D}}^+(z) = \bar{N}_x(z)\begin{bmatrix}L_1^+ & \Psi_2\end{bmatrix} + \bar{D}(z)\begin{bmatrix}I_{m-\kappa} & 0\\ 0 & 0_\kappa\end{bmatrix} \tag{11.90}$$

parameterizes the dynamics of the reduced-order observer and a comparison of (11.88) and (11.89) shows that

$$\det\tilde{\bar{D}}^+(z) = \det(zI - F) = \det(zI - TA(I - \Lambda C_1)\Theta). \tag{11.91}$$

If a parameterizing polynomial matrix $\tilde{\bar{D}}^+(z)$ is given and a state-space representation (C, A, B) of the system is defined, then the matrices L_1^+ and Ψ_2 that parameterize the time-domain representation of the observer can be computed by solving the Diophantine equation related to (11.90) and *vice versa* (see (3.91)).

Pre-multiplying (11.90) by $\bar{D}^{-1}(z)$ and observing (11.84) one obtains the connecting relation

$$\bar{D}^{-1}(z)\bar{\tilde{D}}^{+}(z) = \begin{bmatrix} C_1 \\ C_2 \end{bmatrix} (zI - A)^{-1}[L_1^+ \quad \Psi_2] + \begin{bmatrix} I_{m-\kappa} & 0 \\ 0 & 0_\kappa \end{bmatrix} \tag{11.92}$$

between the time- and the frequency-domain parameterizations of a reduced-order observer. A full-order observer (*i.e.*, $n_O = n$) is characterized by $C_1 = C$, $L_1^+ = L$, while the matrices C_2 and Ψ_2 vanish.

In the time domain the reduced-order observer is parameterized by the $m(n-\kappa)$ degrees of freedom in the two matrices L_1^+ and Ψ_2, whereas the freely assignable coefficients of the $m \times m$ polynomial matrix $\bar{\tilde{D}}^{+}(z)$ parameterize the observer in the frequency domain (see also Chapter 3).

The freely assignable coefficients of the parameterizing polynomial matrix $\bar{\tilde{D}}^{+}(z)$ can be identified as follows. Different from the parameterization of the state feedback in the frequency domain (see Section 11.2) the connecting relation (11.92) cannot be directly used to determine the row-degree structure of $\bar{\tilde{D}}^{+}(z)$ since additionally

$$C_2[L_1^+ \quad \Psi_2] = [C_2 L_1^+ \quad I_\kappa] \tag{11.93}$$

must be satisfied (see (11.68)). To determine the structure of the parameterizing matrix $\bar{\tilde{D}}^{+}(z)$ that takes (11.93) into account consider the transfer matrix

$$\bar{\Phi}(z) = \begin{bmatrix} I_{m-\kappa} & 0 \\ -C_2 L_1^+ & zI_\kappa \end{bmatrix} \bar{D}^{-1}(z)\bar{\tilde{D}}^{+}(z) - I_m \tag{11.94}$$

$$= \begin{bmatrix} I_{m-\kappa} & 0 \\ -C_2 L_1^+ & zI_\kappa \end{bmatrix} \begin{bmatrix} C_1 \\ C_2 \end{bmatrix} (zI - A)^{-1}[L_1^+ \quad \Psi_2] - \begin{bmatrix} 0_{m-\kappa} & 0 \\ C_2 L_1^+ & I_\kappa \end{bmatrix},$$

when inserting (11.92). Using the expansion

$$(zI - A)^{-1} = Iz^{-1} + Az^{-2} + \ldots \tag{11.95}$$

it is straightforward to show that

$$\begin{aligned} zC_2(zI - A)^{-1}[L_1^+ \quad \Psi_2] &= C_2[L_1^+ \quad \Psi_2] + C_2 A z^{-1}[L_1^+ \quad \Psi_2] + \ldots \\ &= [C_2 L_1^+ \quad I_\kappa] + C_2 A (zI - A)^{-1}[L_1^+ \quad \Psi_2] \end{aligned} \tag{11.96}$$

in view of $C_2\Psi_2 = I_\kappa$ (see (11.68)). Since also $C_1(zI - A)^{-1}[L_1^+ C_1 \quad \Psi_2]$ is strictly proper, the transfer matrix

$$\bar{\Phi}(z) = \left(\bar{D}(z) \begin{bmatrix} I_{m-\kappa} & 0 \\ z^{-1} C_2 L_1^+ & z^{-1} I_\kappa \end{bmatrix} \right)^{-1} \bar{\tilde{D}}^{+}(z) - I \tag{11.97}$$

is strictly proper, which implies

$$\Pi\left\{\left(\bar{D}(z)\begin{bmatrix} I_{m-\kappa} & 0 \\ z^{-1}C_2L_1^+ & z^{-1}I_\kappa \end{bmatrix}\right)^{-1}\bar{\tilde{D}}^+(z)\right\} = I. \tag{11.98}$$

Consequently, the rational matrix $(\bar{D}_\kappa^+(z))^{-1}\bar{\tilde{D}}^+(z) - I$ is also strictly proper, where $\bar{D}_\kappa^+(z)$ is defined by

$$\bar{D}_\kappa^+(z) = \bar{D}_\kappa(z)\begin{bmatrix} I_{m-\kappa} & 0 \\ C_2L_1^+ & I_\kappa \end{bmatrix}, \tag{11.99}$$

with

$$\bar{D}_\kappa(z) = \Pi\left\{\bar{D}(z)\begin{bmatrix} I_{m-\kappa} & 0 \\ 0 & z^{-1}I_\kappa \end{bmatrix}\right\}. \tag{11.100}$$

This follows from

$$\begin{aligned}
&\Pi\left\{\left(\bar{D}(z)\begin{bmatrix} I_{m-\kappa} & 0 \\ z^{-1}C_2L_1^+ & z^{-1}I_\kappa \end{bmatrix}\right)^{-1}\bar{\tilde{D}}^+(z)\right\} \\
&\quad = \lim_{z\to\infty}\left\{\left(\bar{D}(z)\begin{bmatrix} I_{m-\kappa} & 0 \\ z^{-1}C_2L_1^+ & z^{-1}I_\kappa \end{bmatrix}\right)^{-1}\bar{\tilde{D}}^+(z)\right\} \\
&\quad = \lim_{z\to\infty}\left\{\left(\Pi\left\{\bar{D}(z)\begin{bmatrix} I_{m-\kappa} & 0 \\ 0 & z^{-1}I_\kappa \end{bmatrix}\begin{bmatrix} I_{m-\kappa} & 0 \\ C_2L_1^+ & I_\kappa \end{bmatrix}\right\}\right.\right. \\
&\qquad\quad \left.\left. +SP\left\{\bar{D}(z)\begin{bmatrix} I_{m-\kappa} & 0 \\ 0 & z^{-1}I_\kappa \end{bmatrix}\begin{bmatrix} I_{m-\kappa} & 0 \\ C_2L_1^+ & I_\kappa \end{bmatrix}\right\}\right)^{-1}\bar{\tilde{D}}^+(z)\right\},
\end{aligned} \tag{11.101}$$

and with

$$\lim_{z\to\infty} SP\left\{\bar{D}(z)\begin{bmatrix} I_{m-\kappa} & 0 \\ 0 & z^{-1}I_\kappa \end{bmatrix}\right\} = 0 \tag{11.102}$$

this yields

$$\begin{aligned}
&\lim_{z\to\infty}\left\{\left(D(z)\begin{bmatrix} I_{m-\kappa} & 0 \\ 0 & z^{-1}I_\kappa \end{bmatrix}\begin{bmatrix} I_{m-\kappa} & 0 \\ C_2L_1^+ & I_\kappa \end{bmatrix}\right)^{-1}\tilde{D}^+(z)\right\} \\
&\quad = \lim_{z\to\infty}\left\{\left(\Pi\left\{\bar{D}(z)\begin{bmatrix} I_{m-\kappa} & 0 \\ 0 & z^{-1}I_\kappa \end{bmatrix}\begin{bmatrix} I_{m-\kappa} & 0 \\ C_2L_1^+ & I_\kappa \end{bmatrix}\right\}\right)^{-1}\bar{\tilde{D}}^+(z)\right\} \\
&\quad = \lim_{z\to\infty}\left\{\left(\Pi\left\{\bar{D}(z)\begin{bmatrix} I_{m-\kappa} & 0 \\ 0 & z^{-1}I_\kappa \end{bmatrix}\right\}\begin{bmatrix} I_{m-\kappa} & 0 \\ C_2L_1^+ & I_\kappa \end{bmatrix}\right)^{-1}\bar{\tilde{D}}^+(z)\right\} \\
&\quad = \Pi\left\{(\bar{D}_\kappa^+(z))^{-1}\bar{\tilde{D}}^+(z)\right\}
\end{aligned} \tag{11.103}$$

in view of (11.99). Therefore, the polynomial matrix $\bar{D}_\kappa^+(z)$ satisfies

$$\Pi\left\{\left(\bar{D}_\kappa^+(z)\right)^{-1}\tilde{\bar{D}}^+(z)\right\}=I \tag{11.104}$$

in view of (11.98), (11.101) and (11.103). Equation (11.104) can be used to determine the row-degree structure of $\bar{D}_\kappa^+(z)$ if $\bar{D}_\kappa^+(z)$ is row reduced and this is the case as $\bar{D}_\kappa(z)$ is row reduced (see (11.99)). Not every $\bar{D}(z)$ gives rise to a row-reduced $\bar{D}_\kappa(z)$. But it is shown in Section A.1 that given a row-reduced $\bar{D}(z)$ one can always determine a unimodular matrix $U_L(z)$, so that both $\bar{D}'(z)=U_L(z)\bar{D}(z)$ and

$$\bar{D}_\kappa'(z)=\Pi\left\{\bar{D}'(z)\begin{bmatrix} I_{m-\kappa} & 0\\ 0 & z^{-1}I_\kappa\end{bmatrix}\right\},\quad 0<\kappa\le m \tag{11.105}$$

are row reduced with

$$\deg\det\bar{D}_\kappa'(z)=n-\kappa \tag{11.106}$$

(see also Remark 3.2).

The resulting left coprime MFD $G(z)=(\bar{D}'(z))^{-1}\bar{N}'(z)$ with $\bar{N}'(z)=U_L(z)\bar{N}(z)$, which is equivalent to (11.82), can then be used to design an observer. Without loss of generality it is therefore assumed that both $\bar{D}(z)$ and $\bar{D}_\kappa(z)$ are row reduced.

In the time domain, the optimal reduced-order filter can be parameterized by the optimal matrices L_1^+ and Ψ_2 with $C_2L_1^+\neq 0$. The next theorem shows that the $m\times m$ polynomial matrix $\tilde{\bar{D}}^+(z)$ plays the corresponding role in the frequency domain.

Theorem 11.2 (Parameterizing polynomial matrix of a reduced-order Kalman filter with $C_2L_1^+\neq 0$). *The $m\times m$ polynomial matrix $\tilde{\bar{D}}^+(z)$ characterizing the dynamics of a reduced-order Kalman filter of the order $n_O=n-\kappa$, $0\le\kappa\le m$, with $C_2L_1^+\neq 0$, in the frequency domain has the properties*

$$\delta_{rj}[\tilde{\bar{D}}^+(z)]=\delta_{rj}[\bar{D}_\kappa^+(z)],\quad j=1,2,\dots,m, \tag{11.107}$$

and

$$\Gamma_r[\tilde{\bar{D}}^+(z)]=\Gamma_r[\bar{D}_\kappa^+(z)]. \tag{11.108}$$

Defining the polynomial matrix

$$\bar{D}_\kappa(z)=\Pi\left\{\bar{D}(z)\begin{bmatrix} I_{m-\kappa} & 0\\ 0 & z^{-1}I_\kappa\end{bmatrix}\right\} \tag{11.109}$$

which is assumed to be row reduced, the polynomial matrix $\bar{D}_\kappa^+(z)$ in (11.107) and (11.108) has the form

$$\bar{D}_\kappa^+(z)=\bar{D}_\kappa(z)\begin{bmatrix} I_{m-\kappa} & 0\\ C_2L_1^+ & I_\kappa\end{bmatrix}. \tag{11.110}$$

The polynomial matrix $\tilde{\bar{D}}(z)$ parameterizing the reduced-order observer with $C_2L_1=0$ (see Section 11.3) is related to $\tilde{\bar{D}}^+(z)$ by

$$\tilde{\bar{D}}(z) = \tilde{\bar{D}}^{+}(z)\begin{bmatrix} I_{m-\kappa} & 0 \\ -C_2L_1^{+} & I_\kappa \end{bmatrix}, \tag{11.111}$$

and satisfies

$$\delta_{rj}[\tilde{\bar{D}}(z)] = \delta_{rj}[\bar{D}_\kappa(z)], \quad j = 1, 2, \ldots, m, \tag{11.112}$$

and

$$\Gamma_r[\tilde{\bar{D}}(z)] = \Gamma_r[\bar{D}_\kappa(z)], \tag{11.113}$$

with $\bar{D}_\kappa(z)$ *as defined in (11.109).*

Proof. The proof of the first part of Theorem 11.2 can be carried out analogously to Theorem 3.1. The assumption that $\bar{D}_\kappa(z)$ is row reduced implies that $\bar{D}_\kappa^{+}(z)$ is also row reduced since the matrix multiplying $\bar{D}_\kappa(z)$ in (11.110) is non-singular. In Section 11.3 it is demonstrated that an optimal state estimate $\hat{x}$ can be obtained by using a reduced-order observer as defined in Chapter 3, *i.e.*, by an observer, where the structural relation $C_2L_1 = 0$ is satisfied. To obtain the same results in the frequency domain consider

$$\bar{D}^{-1}(z)\tilde{\bar{D}}^{+}(z)\begin{bmatrix} I_{m-\kappa} & 0 \\ -C_2L_1^{+} & I_\kappa \end{bmatrix} \tag{11.114}$$
$$= \begin{bmatrix} C_1 \\ C_2 \end{bmatrix}(zI - A)^{-1}[L_1^{+} - \Psi_2C_2L_1^{+} \quad \Psi_2] + \begin{bmatrix} I_{m-\kappa} & 0 \\ 0 & 0_\kappa \end{bmatrix},$$

which results from postmultiplying the connecting relation (11.92) by the non-singular matrix $\begin{bmatrix} I_{m-\kappa} & 0 \\ -C_2L_1^{+} & I_\kappa \end{bmatrix}$. By introducing the polynomial matrix

$$\tilde{\bar{D}}(z) = \tilde{\bar{D}}^{+}(z)\begin{bmatrix} I_{m-\kappa} & 0 \\ -C_2L_1^{+} & I_\kappa \end{bmatrix} \tag{11.115}$$

this becomes

$$\bar{D}^{-1}(z)\tilde{\bar{D}}(z) = \begin{bmatrix} C_1 \\ C_2 \end{bmatrix}(zI - A)^{-1}[\Theta T L_1^{+} \quad \Psi_2] + \begin{bmatrix} I_{m-\kappa} & 0 \\ 0 & 0_\kappa \end{bmatrix} \tag{11.116}$$

in the light of $\Theta T = I - \Psi_2C_2$ (see (3.7)). With (11.70) this yields

$$\bar{D}^{-1}(z)\tilde{\bar{D}}(z) = \begin{bmatrix} C_1 \\ C_2 \end{bmatrix}(zI - A)^{-1}[L_1 \quad \Psi_2] + \begin{bmatrix} I_{m-\kappa} & 0 \\ 0 & 0_\kappa \end{bmatrix}, \tag{11.117}$$

where L_1 satisfies $C_2L_1 = 0$ (see (11.71)). Therefore, a reduced-order Kalman filter can also be parameterized by $\tilde{\bar{D}}(z)$, which has the usual properties (11.112) and (11.113). □

11.4.2 Frequency-domain Design of the Stationary Kalman Filter

The optimal estimator is characterized in the frequency domain by the $m \times m$ polynomial matrix $\bar{\bar{D}}^{+}(z)$. Therefore, given the DARE (11.72), the frequency-domain descriptions (11.80)–(11.83) of the system and the connecting relation (11.92), a rational matrix equation must be derived that characterizes the Kalman filtering problem in the frequency domain. This matrix equation leads to the spectral factorization of an $m \times m$ rational matrix $\bar{H}(z)$ in the complex variable $z = \rho e^{j\delta}$, which is symmetric and positive (see (11.10) and (11.11)). This spectral factorization is dual to the spectral factorization described in Section 11.2 and it yields an $m \times m$ polynomial matrix $\bar{\bar{D}}(z)$ that satisfies

$$\bar{H}(z) = \bar{\bar{D}}(z)\bar{\bar{D}}^{T}(z^{-1}), \tag{11.118}$$

with

$$\det \bar{\bar{D}}(z) \neq 0 \text{ for all } |z| \geq 1, \tag{11.119}$$

i.e., the determinant $\det \bar{\bar{D}}(z)$ is a Schur polynomial and $\det \Gamma_r[\bar{\bar{D}}(z)] \neq 0$. Then, $\bar{\bar{D}}(z)$ is a *spectral factor* of $\bar{H}(z)$ and it contains the zeros of $\bar{H}(z)$ that are located inside the unit circle. The spectral factor $\bar{\bar{D}}(z)$ exists and it is unique up to a multiplicative orthogonal matrix $\bar{U}$, *i.e.*, $\bar{U}\bar{U}^T = I$ (see [68]). The dual factorization can be carried out by applying an algorithm for solving (11.12) and (11.13) to the transpose of $\bar{H}(z)$ and taking the transpose of the result.

In a first step a rational matrix equation is solved for the optimal polynomial matrix $\bar{\bar{D}}(z)$ that characterizes the optimal estimator. Given this solution the optimal polynomial matrix $\bar{\bar{D}}^{+}(z)$ can be obtained. This is summarized in the following theorem.

Theorem 11.3 (Design of the reduced-order Kalman filter in the frequency domain). *Given is a system (11.80) and (11.81) driven by white noise with $m - \kappa$, $0 \leq \kappa \leq m$, measurements y_1 also corrupted by white noise. The transfer matrix between the input noise and the measurements is represented in a left coprime MFD $G_w(z) = \bar{D}^{-1}(z)\bar{N}_w(z)$ (see (11.83)). Given are also the symmetric covariance matrices $\bar{Q} = \bar{Q}_0\bar{Q}_0^T$ of the input noise and $\bar{R}_1 = \bar{R}_{10}\bar{R}_{10}^T$ of the measurement noise (see (11.40) and (11.41)), where the $p \times p$ matrix $\bar{Q}$ is positive-semidefinite and the $(m-\kappa) \times (m-\kappa)$ matrix $\bar{R}_1$ is positive-definite. Further, it is assumed that the $\kappa \times \kappa$ constant matrix*

$$\Phi = \Pi\{zG_{w2}(z)\}\bar{Q}\left(\Pi\{zG_{w2}(z)\}\right)^T \tag{11.120}$$

is positive-definite (see (11.83)) and that any greatest common left divisor of

$$\bar{D}(z)\begin{bmatrix} \bar{R}_{10} & 0 \\ 0 & 0 \end{bmatrix} \quad \text{and} \quad \bar{N}_w(z)\bar{Q}_0 \tag{11.121}$$

has no zeros on the unit circle.

If these conditions are satisfied, then the spectral factor $\bar{\bar{D}}(z)$ *in*

$$\bar{D}(z)\begin{bmatrix}\bar{R}_1 & 0\\ 0 & 0\end{bmatrix}\bar{D}^T(z^{-1})+\bar{N}_w(z)\bar{Q}\bar{N}_w^T(z^{-1})=\bar{\bar{D}}(z)\bar{\bar{D}}^T(z^{-1}) \tag{11.122}$$

exists. If the DARE (11.72) has a stabilizing solution $\bar{P}$ *that is positive-semidefinite then*

$$\tilde{\bar{\bar{D}}}(z)=\bar{\bar{D}}(z)\Gamma_r^{-1}[\bar{\bar{D}}(z)]\Gamma_r[\bar{D}_\kappa(z)] \tag{11.123}$$

characterizes the Kalman filter, where $C_2L_1 = 0$ *holds. In order to obtain a frequency-domain parameterization of the observer with* $L_1^+ = A\Lambda$ *and* $C_2L_1^+ \neq 0$*, the polynomial matrix* $\tilde{\bar{\bar{D}}}^+(z)$ *is required. It is related with the polynomial matrix* $\tilde{\bar{\bar{D}}}(z)$ *by*

$$\tilde{\bar{\bar{D}}}^+(z)=\tilde{\bar{\bar{D}}}(z)\begin{bmatrix}I_{m-\kappa} & 0\\ C_2L_1^+ & I_\kappa\end{bmatrix} \tag{11.124}$$

(see (11.115)). The unknown quantity $C_2L_1^+$ *can be obtained from the spectral factorization result* $\bar{\bar{D}}(z)$ *(see (11.122)) by*

$$\begin{bmatrix}I\\ C_2L_1^+\end{bmatrix}=\{\Gamma_r[\bar{D}_\kappa(z)]\}^{-1}\Gamma_r[\bar{\bar{D}}(z)]\Gamma_r^T[\bar{\bar{D}}(z)]\{\Gamma_r^T[\bar{D}_\kappa(z)]\}^{-1}\begin{bmatrix}\hat{R}^{-1}\\ 0\end{bmatrix}. \tag{11.125}$$

Proof. Using the expansion

$$(zI-A)^{-1}=Iz^{-1}+Az^{-2}+\dots \tag{11.126}$$

it is straightforward to show that

$$zG_{w2}(z)=zC_2(zI-A)^{-1}G=C_2G+C_2A(zI-A)^{-1}G, \tag{11.127}$$

so that the polynomial part of $zG_{w2}(z)$ is given by $\Pi\{zG_{w2}(z)\}=C_2G$, which verifies that (11.120) is equivalent to (11.45).

The rational matrix equation (11.122) can be obtained from the Riccati equation (11.72). It is easy to show that

$$\bar{P}-A\bar{P}A^T=(zI-A)\bar{P}(z^{-1}I-A^T)+A\bar{P}(z^{-1}I-A^T)+(zI-A)\bar{P}A^T. \tag{11.128}$$

Therefore, the DARE (11.72) can be written as

$$\begin{aligned}(zI-A)\bar{P}(z^{-1}I-A^T)+A\bar{P}(z^{-1}I-A^T)+(zI-A)\bar{P}A^T&\\ +L_1^+\hat{R}(L_1^+)^T+\Psi_2X\Psi_2^T=G\bar{Q}G^T.&\end{aligned} \tag{11.129}$$

Pre-multiplying this result by $\begin{bmatrix}C_1\\ C_2\end{bmatrix}(zI-A)^{-1}$ and postmultiplying it by $(z^{-1}I-A^T)^{-1}\begin{bmatrix}C_1^T & C_2^T\end{bmatrix}$ one obtains

$$\begin{bmatrix} C_1 \\ C_2 \end{bmatrix} \bar{P} \begin{bmatrix} C_1^T & C_2^T \end{bmatrix} + \begin{bmatrix} C_1 \\ C_2 \end{bmatrix} (zI - A)^{-1} A\bar{P} \begin{bmatrix} C_1^T & C_2^T \end{bmatrix}$$
$$+ \begin{bmatrix} C_1 \\ C_2 \end{bmatrix} \bar{P} A^T (z^{-1}I - A^T)^{-1} \begin{bmatrix} C_1^T & C_2^T \end{bmatrix} \tag{11.130}$$
$$+ \begin{bmatrix} C_1 \\ C_2 \end{bmatrix} (zI - A)^{-1} \left[L_1^+ \hat{R} (L_1^+)^T + \Psi_2 X \Psi_2^T \right] (z^{-1}I - A^T)^{-1} \begin{bmatrix} C_1^T & C_2^T \end{bmatrix}$$
$$= \begin{bmatrix} C_1 \\ C_2 \end{bmatrix} (zI - A)^{-1} G \bar{Q} G^T (z^{-1}I - A^T)^{-1} \begin{bmatrix} C_1^T & C_2^T \end{bmatrix}.$$

Observing (11.49), (11.51) and (11.67) Equation (11.130) can be written as

$$\left\{ \begin{bmatrix} C_1 \\ C_2 \end{bmatrix} (zI - A)^{-1} \begin{bmatrix} L_1^+ & \Psi_2 \end{bmatrix} + \begin{bmatrix} I_{m-\kappa} & 0 \\ 0 & 0 \end{bmatrix} \right\} \begin{bmatrix} \hat{R} & 0 \\ 0 & X \end{bmatrix}$$
$$\cdot \left\{ \begin{bmatrix} (L_1^+)^T \\ \Psi_2^T \end{bmatrix} (z^{-1}I - A^T)^{-1} \begin{bmatrix} C_1^T & C_2^T \end{bmatrix} + \begin{bmatrix} I_{m-\kappa} & 0 \\ 0 & 0 \end{bmatrix} \right\} \tag{11.131}$$
$$= \begin{bmatrix} \bar{R}_1 & 0 \\ 0 & 0 \end{bmatrix} + \begin{bmatrix} C_1 \\ C_2 \end{bmatrix} (zI - A)^{-1} G \bar{Q} G^T (z^{-1}I - A^T)^{-1} \begin{bmatrix} C_1^T & C_2^T \end{bmatrix}.$$

Now introducing (11.83) and (11.92) this takes the form

$$\bar{D}^{-1}(z) \tilde{\bar{D}}^+(z) \begin{bmatrix} \hat{R} & 0 \\ 0 & X \end{bmatrix} \{\tilde{\bar{D}}^+(z^{-1})\}^T \{\bar{D}^T(z^{-1})\}^{-1} \tag{11.132}$$
$$= \begin{bmatrix} \bar{R}_1 & 0 \\ 0 & 0 \end{bmatrix} + \bar{D}^{-1}(z) \bar{N}_w(z) \bar{Q} \bar{N}_w^T(z^{-1}) \{\bar{D}^T(z^{-1})\}^{-1}.$$

When pre-multiplying this result by $\bar{D}(z)$ and postmultiplying it by $\bar{D}^T(z^{-1})$ one obtains the rational matrix equation

$$\tilde{\bar{D}}^+(z) \tilde{\bar{R}} \{\tilde{\bar{D}}^+(z^{-1})\}^T = \bar{D}(z) \begin{bmatrix} \bar{R}_1 & 0 \\ 0 & 0 \end{bmatrix} \bar{D}^T(z^{-1}) + \bar{N}_w(z) \bar{Q} \bar{N}_w^T(z^{-1}), \tag{11.133}$$

(see (11.122)) where the abbreviation

$$\tilde{\bar{R}} = \begin{bmatrix} \hat{R} & 0 \\ 0 & X \end{bmatrix} \tag{11.134}$$

has been used. To obtain the parameterizing polynomial matrix $\tilde{\bar{D}}^+(z)$ from the spectral factor $\tilde{\bar{D}}(z)$ (see (11.122)) one needs to know its row-degree structure. Unfortunately, this property depends on the optimal time-domain solution (see Theorem 11.2). Since the row-degree structure of the polynomial matrix $\tilde{\bar{D}}(z)$ characterizing the filter with $C_2 L_1 = 0$ in the frequency domain is known (see Theorem 11.2) the polynomial matrix $\tilde{\bar{D}}(z)$ can be obtained by spectral factorization. Therefore, the polynomial matrix $\tilde{\bar{D}}(z)$ is computed first and subsequently $\tilde{\bar{D}}^{+}(z)$ has to be obtained without recourse to the time-domain solution. To this end, (11.133) is reformulated in terms of $\tilde{\bar{D}}(z)$. Observing (11.124) and (11.134) in (11.133) one obtains

$$\tilde{\bar{D}}(z)\hat{\bar{R}}^{1/2}\hat{\bar{R}}^{1/2}\tilde{\bar{D}}^{T}(z^{-1}) = \bar{D}(z)\begin{bmatrix}\bar{R}_1 & 0\\ 0 & 0\end{bmatrix}\bar{D}^{T}(z^{-1}) + \bar{N}_w(z)\bar{Q}\bar{N}_w^{T}(z^{-1}), \tag{11.135}$$

where the abbreviation

$$\hat{\bar{R}} = \begin{bmatrix}\hat{R} & \hat{R}L_1^{+T}C_2^T\\ C_2L_1^{+}\hat{R} & X + C_2L_1^{+}\hat{R}L_1^{+T}C_2^T\end{bmatrix} \tag{11.136}$$

has been used. With $\tilde{\bar{R}}$ in (11.134) this can be written in the form

$$\hat{\bar{R}} = \tilde{\bar{R}} + \begin{bmatrix}0 & \hat{R}L_1^{+T}C_2^T\\ C_2L_1^{+}\hat{R} & C_2L_1^{+}\hat{R}L_1^{+T}C_2^T\end{bmatrix}, \tag{11.137}$$

which shows that $\hat{\bar{R}}$ is positive-definite, because $\tilde{\bar{R}}$ in (11.134) is positive-definite if $\bar{R}_1$ and Φ are positive-definite (see (11.120) and Section 11.3) and the other matrix in (11.137) is positive-semidefinite, which can be shown using the results in [40]. Consequently, the square root $\hat{\bar{R}}^{1/2}$ in (11.135) exists. The spectral factor $\bar{\bar{D}}(z)$ (see (11.122)) with $\det \bar{\bar{D}}(z)$ a Schur polynomial can be obtained by spectral factorization from the right-hand side of (11.135). It exists because the greatest common left divisor of the pair (11.121) has no zero on the unit circle and it is unique up to an orthogonal matrix $\bar{U}$ (see [68]).

Since there exists a stabilizing solution of the DARE (11.72) this implies that the spectral factor

$$\bar{\bar{D}}'(z) = \tilde{\bar{D}}(z)\hat{\bar{R}}^{1/2} \tag{11.138}$$

in (11.135) exists. This spectral factor is unique up to an orthogonal matrix $\bar{U}$, *i.e.*,

$$\bar{\bar{D}}(z)\bar{U} = \tilde{\bar{D}}(z)\hat{\bar{R}}^{1/2}, \tag{11.139}$$

or

$$\tilde{\bar{D}}(z) = \bar{\bar{D}}(z)\bar{U}\hat{\bar{R}}^{-1/2}. \tag{11.140}$$

As $\Gamma_r[\tilde{\bar{D}}(z)] = \Gamma_r[\bar{D}_\kappa(z)]$ (see (11.112)) the orthogonal matrix $\bar{U}$ is given by

$$\bar{U} = \Gamma_r^{-1}[\bar{\bar{D}}(z)]\Gamma_r[\bar{D}_\kappa(z)]\hat{\bar{R}}^{1/2}. \tag{11.141}$$

Then, by inserting (11.141) in (11.140) one obtains (11.123).

Given the optimal polynomial matrix $\tilde{\bar{D}}(z)$, the parameterizing polynomial matrix $\tilde{\bar{D}}^{+}(z)$ of the filter yielding the *a posteriori* state estimate can only be computed if $C_2L_1^{+}$ is known (see (11.110)). Though this is a time-domain quantity, it can be computed from the optimal $\tilde{\bar{D}}(z)$. The matrix $C_2L_1^{+}$ can be obtained in the following way. Since by assumption a solution of the DARE (11.72) exists one has

$$\bar{\bar{D}}(z)\bar{\bar{D}}^T(z^{-1}) = \tilde{\bar{\bar{D}}}^+(z)\begin{bmatrix}\hat{R} & 0\\ 0 & X\end{bmatrix}\{\tilde{\bar{\bar{D}}}^+(z^{-1})\}^T \tag{11.142}$$

(see (11.122), (11.133) and (11.134)). Now, introducing

$$\tilde{\bar{\bar{D}}}^+(z) = \tilde{\bar{\bar{D}}}(z)\begin{bmatrix}I_{m-\kappa} & 0\\ C_2L_1^+ & I_\kappa\end{bmatrix} \tag{11.143}$$

(see (11.115)) in (11.142) leads to

$$\bar{\bar{D}}(z)\bar{\bar{D}}^T(z^{-1}) = \tilde{\bar{\bar{D}}}(z)\begin{bmatrix}I_{m-\kappa} & 0\\ C_2L_1^+ & I_\kappa\end{bmatrix}\begin{bmatrix}\hat{R} & 0\\ 0 & X\end{bmatrix}\begin{bmatrix}I_{m-\kappa} & (L_1^+)^TC_2^T\\ 0 & I_\kappa\end{bmatrix}\tilde{\bar{\bar{D}}}^T(z^{-1}), \tag{11.144}$$

and when substituting (11.123) this finally yields

$$\begin{aligned}\bar{\bar{D}}(z)\bar{\bar{D}}^T(z^{-1}) = \bar{\bar{D}}(z)\{\Gamma_r[\bar{\bar{D}}(z)]\}^{-1}\Gamma_r[\bar{D}_\kappa(z)]\begin{bmatrix}I_{m-\kappa} & 0\\ C_2L_1^+ & I_\kappa\end{bmatrix}\begin{bmatrix}\hat{R} & 0\\ 0 & X\end{bmatrix}\\ \cdot\begin{bmatrix}I_{m-\kappa} & (L_1^+)^TC_2^T\\ 0 & I_\kappa\end{bmatrix}\Gamma_r^T[\bar{D}_\kappa(z)]\{\Gamma_r^T[\bar{\bar{D}}(z)]\}^{-1}\bar{\bar{D}}^T(z^{-1}).\end{aligned} \tag{11.145}$$

This shows that the matrix on the right-hand side of (11.145), which is bracketed by $\bar{\bar{D}}(z)$ and $\bar{\bar{D}}^T(z^{-1})$, is the identity matrix. Therefore, one obtains

$$\begin{aligned}&\{\Gamma_r[\bar{D}_\kappa(z)]\}^{-1}\Gamma_r[\bar{\bar{D}}(z)]\Gamma_r^T[\bar{\bar{D}}(z)]\{\Gamma_r^T[\bar{D}_\kappa(z)]\}^{-1}\\ &\qquad = \begin{bmatrix}\hat{R} & \hat{R}(L_1^+)^TC_2^T\\ C_2L_1^+\hat{R} & C_2L_1^+\hat{R}(L_1^+)^TC_2^T + X\end{bmatrix}\end{aligned} \tag{11.146}$$

when pre-multiplying this matrix by $\{\Gamma_r[\bar{D}_\kappa(z)]\}^{-1}\Gamma_r[\bar{\bar{D}}(z)]$ and postmultiplying it by $\Gamma_r^T[\bar{\bar{D}}(z)]\{\Gamma_r^T[\bar{D}_\kappa(z)]\}^{-1}$. The quantities on the left-hand side of (11.146) are known and therefore, the positive-definite $(m-\kappa)\times(m-\kappa)$ matrix $\hat{R}$ (see (11.51)) on the left upper corner of this matrix can be used to compute

$$\begin{aligned}\begin{bmatrix}I\\ C_2L_1^+\end{bmatrix} &= \begin{bmatrix}\hat{R} & \hat{R}(L_1^+)^TC_2^T\\ C_2L_1^+\hat{R} & C_2L_1^+\hat{R}(L_1^+)^TC_2^T + X\end{bmatrix}\begin{bmatrix}\hat{R}^{-1}\\ 0\end{bmatrix}\\ &= \{\Gamma_r[\bar{D}_\kappa(z)]\}^{-1}\Gamma_r[\bar{\bar{D}}(z)]\Gamma_r^T[\bar{\bar{D}}(z)]\{\Gamma_r^T[\bar{D}_\kappa(z)]\}^{-1}\begin{bmatrix}\hat{R}^{-1}\\ 0\end{bmatrix}.\end{aligned} \tag{11.147}$$

Using this result and $\tilde{\bar{\bar{D}}}(z)$ one obtains the optimal polynomial matrix $\tilde{\bar{\bar{D}}}^+(z)$, which parameterizes the (reduced-order) Kalman filter for an *a posteriori* state estimate by means of (11.124). Then, $\tilde{\bar{D}}^+(z)$ also satisfies the Condition (11.108). To see this, insert (11.123) in (11.124), which gives

$$\tilde{\bar{D}}^+(z) = \bar{\bar{D}}(z)\Gamma_r^{-1}[\bar{\bar{D}}(z)]\Gamma_r[\bar{D}_\kappa(z)] \begin{bmatrix} I_{m-\kappa} & 0 \\ C_2 L_1^+ & I_\kappa \end{bmatrix}. \tag{11.148}$$

Since (11.110) implies

$$\Gamma_r[\bar{D}_\kappa^+(z)] = \Gamma_r[\bar{D}_\kappa(z)] \begin{bmatrix} I_{m-\kappa} & 0 \\ C_2 L_1^+ & I_\kappa \end{bmatrix} \tag{11.149}$$

the result

$$\tilde{\bar{D}}^+(z) = \bar{\bar{D}}(z)\Gamma_r^{-1}[\bar{\bar{D}}(z)]\Gamma_r[\bar{D}_\kappa^+(z)] \tag{11.150}$$

holds, which implies (11.108). Since the spectral factor in (11.138) is unique up to a right multiplication with an orthogonal matrix the row degrees of $\tilde{\bar{D}}(z)$ in (11.140) satisfy (11.112). The polynomial matrices $\bar{D}_\kappa^+(z)$ and $\bar{D}_\kappa(z)$ have the same row degrees (see (11.110)) since $\bar{D}_\kappa^+(z)$ results from $\bar{D}_\kappa(z)$ by right multiplication with a non-singular matrix. Consequently, $\tilde{\bar{D}}^+(z)$ in (11.150) has the property (11.107) as $\tilde{\bar{D}}^+(z)$ and $\tilde{\bar{D}}(z)$ are related *via* (11.124). □

As it is usually not known beforehand if the DARE (11.72) has a stabilizing solution $\bar{P}$, one must check whether $\tilde{\bar{D}}^+(z)$ in (11.124) is optimal by determining the corresponding time-domain solution from the frequency-domain results. Given the quantities $\hat{R}$ and $C_2 L_1^+$, X can also be obtained, which is obvious by inspection of the lower right-hand element in the matrix on the right-hand side of (11.146). The observer gains L_1^+ and Ψ_2 are attainable by solving the Diophantine equation

$$\bar{N}_x(z)\bar{Y}(z) + \bar{D}(z)\bar{X}(z) = \tilde{\bar{D}}^+(z) \tag{11.151}$$

for

$$\bar{Y}(z) = [L_1^+ \quad \Psi_2] \text{ and } \bar{X}(z) = \begin{bmatrix} I_{m-\kappa} & 0 \\ 0 & 0_\kappa \end{bmatrix} \tag{11.152}$$

(see (11.90)). Now, assume that (C, A, G) describes a minimal realization of (11.83). Then, by inserting the observer gains L_1^+ and Ψ_2 and the known quantities $\hat{R}$, $\bar{Q}$ and X in (11.72) one obtains the matrix equation

$$\begin{aligned} \bar{P} - A\bar{P}A^T &= -[L_1^+ \quad \Psi_2] \begin{bmatrix} \hat{R} & 0 \\ 0 & X \end{bmatrix} \begin{bmatrix} (L_1^+)^T \\ \Psi_2^T \end{bmatrix} + G\bar{Q}G^T \\ &= -L_1^+\hat{R}(L_1^+)^T - \Psi_2 X \Psi_2^T + G\bar{Q}G^T \end{aligned} \tag{11.153}$$

for $\bar{P}$. Now, observe that

$$\begin{aligned} &(A - L_1^+C_1)\bar{P}(A^T - C_1^T(L_1^+)^T) \\ &= A\bar{P}A^T - A\bar{P}C_1^T(L_1^+)^T - L_1^+C_1\bar{P}A^T + L_1^+C_1\bar{P}C_1^T(L_1^+)^T. \end{aligned} \tag{11.154}$$

Since $L_1^+ \hat{R} = A\bar{P}C_1^T$ (see (11.49)) and $C_1 \bar{P} C_1^T = \hat{R} - \bar{R}_1$ (see (11.51)) this becomes

$$(A - L_1^+ C_1)\bar{P}(A^T - C_1^T (L_1^+)^T) = A\bar{P}A^T - L_1^+ \hat{R}(L_1^+)^T - L_1^+ \bar{R}_1 (L_1^+)^T, \quad (11.155)$$

or

$$A\bar{P}A^T = L_1^+ \hat{R}(L_1^+)^T + (A - L_1^+ C_1)\bar{P}(A^T - C_1^T (L_1^+)^T) + L_1^+ \bar{R}_1 (L_1^+)^T. \quad (11.156)$$

By inserting this result in (11.153) it takes the form

$$\bar{P} - (A - L_1^+ C_1)\bar{P}(A^T - C_1^T (L_1^+)^T) = L_1^+ \bar{R}_1 (L_1^+)^T - \Psi_2 X \Psi_2^T + G\bar{Q}G^T. \quad (11.157)$$

Pre-multiplying (11.157) by ΘT and postmultiplying it by $T^T \Theta^T$ yields

$$\begin{aligned} &\Theta T \bar{P} T^T \Theta^T - \Theta T (A - L_1^+ C_1)\bar{P}(A^T - C_1^T (L_1^+)^T) T^T \Theta^T \\ &\quad = \Theta T L_1^+ \bar{R}_1 (L_1^+)^T T^T \Theta^T + \Theta T G \bar{Q} G^T T^T \Theta^T, \end{aligned} \quad (11.158)$$

when taking $T\Psi_2 = 0$ into account (see (11.56)). Since

$$\Theta T \bar{P} T^T \Theta^T = (I - \Psi_2 C_2)\bar{P}(I - C_2^T \Psi_2^T) = \bar{P} \quad (11.159)$$

in view of $C_2 \bar{P} = 0$ (see (11.67)) Equation (11.158) can be written as

$$\begin{aligned} &\bar{P} - \Theta T (A - L_1^+ C_1)\Theta T \bar{P} T^T \Theta^T (A^T - C_1^T (L_1^+)^T) T^T \Theta^T \\ &\quad = \Theta T L_1^+ \bar{R}_1 (L_1^+)^T T^T \Theta^T + \Theta T G \bar{Q} G^T T^T \Theta^T. \end{aligned} \quad (11.160)$$

Since L_1^+ and Ψ_2 result from the spectral factor $\bar{\tilde{D}}^+(z)$ (see (11.151)) the matrix $F = T(A - L_1^+ C_1)\Theta$ (see also (11.87) and (11.66)) is a Schur matrix. Then, (11.160) can also be expressed as

$$\bar{P} - \Theta F T \bar{P} T^T F^T \Theta^T = \Theta T L_1^+ \bar{R}_1 (L_1^+)^T T^T \Theta^T + \Theta T G \bar{Q} G^T T^T \Theta^T. \quad (11.161)$$

Pre-multiplying this result by T and postmultiplying it by T^T finally yields

$$\bar{P}_{red} - F \bar{P}_{red} F^T = T L_1^+ \bar{R}_1 (L_1^+)^T T^T + T G \bar{Q} G^T T^T, \quad (11.162)$$

with $\bar{P}_{red} = T\bar{P}T^T$ because $T\Theta = I$ (see (11.44)). This is a discrete-time Lyapunov equation for $\bar{P}_{red}$, which has a unique positive-definite solution because F is Schur and the right-hand side of (11.162) is positive-definite (see, *e.g.*, [37]). Observing

$$\begin{bmatrix} C_2 \\ T \end{bmatrix} \bar{P} \begin{bmatrix} C_2^T & T^T \end{bmatrix} = \begin{bmatrix} 0_\kappa & 0 \\ 0 & T\bar{P}T^T \end{bmatrix} = \begin{bmatrix} 0_\kappa & 0 \\ 0 & \bar{P}_{red} \end{bmatrix} \quad (11.163)$$

in view of (11.67) the matrix $\bar{P}$ is related to $\bar{P}_{red}$ by

$$\bar{P} = \begin{bmatrix} \Psi_2 & \Theta \end{bmatrix} \begin{bmatrix} 0 & 0 \\ 0 & \bar{P}_{red} \end{bmatrix} \begin{bmatrix} \Psi_2^T \\ \Theta^T \end{bmatrix} = \Theta \bar{P}_{red} \Theta^T. \quad (11.164)$$

By inserting (11.164) in (11.153) it has to be verified that (11.164) solves (11.153). Since $\bar{P}_{red}$ is always positive-definite, the matrix $\bar{P}$ in (11.164) is also positive-semidefinite. If the matrices L_1^+ and Ψ_2 obtained from (11.49) and (11.50) coincide with the solution of (11.151) the polynomial matrix $\tilde{\bar{D}}^+(z)$ characterizes the optimal solution.

In order to parameterize the *a posteriori* filter in the time domain the matrix $\bar{P}$ in (11.164) can be used to determine the gain Λ from (11.64).

Remark 11.6. It is interesting to note that all examples show that the spectral factor (11.124) yields the optimal solution. Therefore, it seems that the conditions for the existence of the spectral factor in Theorem 11.3 are also sufficient for the optimality of the filter. That this is always the case for full-order Kalman filters can be proven with similar arguments as in Theorem 11.1 using the results in [22].

Remark 11.7. Also in discrete time the rational matrix equations of the reduced-order optimal estimator can be obtained from the equations of the full-order filter by inserting the singular covariance matrix of the measurement noise (see also Remark 8.2). Moreover, in the time domain the Riccati equations and the optimal filter gains are clearly different in the continuous time and in the discrete time. In the frequency domain, however, the discrete-time equations simply result from the continuous-time equations by replacing $[\cdot](s)$ by $[\cdot](z)$ and $[\cdot]^T(-s)$ by $[\cdot]^T(z^{-1})$.

Example 11.2. Parameterization of a reduced-order Kalman filter in the frequency domain

Consider a system of the order two with one input u and two outputs y. There is an input disturbance w and only the first output is corrupted by noise, so that one has the case $\kappa = 1$. The transfer behaviour (11.80) and (11.81) is characterized by

$$y_1(z) = \frac{z}{z^2 - z + 0.25}u(z) + \frac{[2z + 0.25 \quad -z + 0.5]}{z^2 - z + 0.25}w(z) + v_1(z), \quad (11.165)$$

$$y_2(z) = \frac{1}{z^2 - z + 0.25}u(z) + \frac{[-z + 3 \quad -2z + 1]}{z^2 - z + 0.25}w(z). \quad (11.166)$$

Assume that the covariance matrix of w has the form

$$\bar{Q} = \begin{bmatrix} 1 & 0 \\ 0 & 1 \end{bmatrix}, \quad (11.167)$$

and that the covariance of the measurement noise v_1 is $\bar{R}_1 = 100/142$. The input–output behaviour $y(z) = G(z)u(z)$ of the system is characterized by the left MFD

$$G(z) = \bar{D}^{-1}(z)\bar{N}(z) = \begin{bmatrix} 1 & -z \\ z & -z + 0.25 \end{bmatrix}^{-1} \begin{bmatrix} 0 \\ 1 \end{bmatrix}, \quad (11.168)$$

and the transfer behaviour $y(z) = G_w(z)w(z)$ by the left MFD

$$G_w(z) = \bar{D}^{-1}(z)\bar{N}_w(z) = \begin{bmatrix} 1 & -z \\ z & -z+0.25 \end{bmatrix}^{-1} \begin{bmatrix} 1 & 2 \\ 3 & 1 \end{bmatrix}. \tag{11.169}$$

This $\bar{D}(z)$ assures that the polynomial matrix

$$\bar{D}_1(z) = \Pi\left\{\bar{D}(z)\begin{bmatrix} 1 & 0 \\ 0 & z^{-1} \end{bmatrix}\right\} = \begin{bmatrix} 1 & -1 \\ z & -1 \end{bmatrix} \tag{11.170}$$

is row reduced. With $\Pi\{zG_{w2}(z)\} = [-1 \quad -2]$ (see (11.166)) one obtains $\Phi = 5$ (see (11.120)), which is positive. With $\bar{R}_{10} = 10/\sqrt{142}$ and $\bar{Q}_0 = I$ the greatest common left divisor of the pair (11.121), namely

$$\bar{D}(z)\begin{bmatrix} \bar{R}_{10} & 0 \\ 0 & 0 \end{bmatrix} = \frac{10}{\sqrt{142}}\begin{bmatrix} 1 & 0 \\ z & 0 \end{bmatrix} \text{ and } \bar{N}_w(z)\bar{Q}_0 = \begin{bmatrix} 1 & 2 \\ 3 & 1 \end{bmatrix} \tag{11.171}$$

is a unimodular matrix, so that the conditions for the existence of a stable spectral factor are satisfied. The left-hand side of (11.122) has the form

$$\bar{N}_w(z)\bar{Q}\bar{N}_w^T(z^{-1}) + \bar{D}(z)\begin{bmatrix} \bar{R}_1 & 0 \\ 0 & 0 \end{bmatrix}\bar{D}^T(z^{-1}) = \begin{bmatrix} 5+\bar{R}_1 & 5+\dfrac{\bar{R}_1}{z} \\ \bar{R}_1 z + 5 & 10+\bar{R}_1 \end{bmatrix}. \tag{11.172}$$

Spectral factorization of this result yields

$$\bar{\bar{D}}(z) = \begin{bmatrix} \dfrac{0.1125}{\sqrt{0.15975}} & -\sqrt{5.625} \\ \dfrac{z}{\sqrt{0.15975}} & -\dfrac{\sqrt{5.625}}{1.125} \end{bmatrix} \tag{11.173}$$

with $\det \bar{\bar{D}}(z) = \frac{\sqrt{25}}{\sqrt{0.71}}(z-0.1)$. Since the matrix $\bar{D}_\kappa(z)$ has the form (11.170) the matrix

$$\{\Gamma_r[\bar{D}_\kappa(z)]\}^{-1}\Gamma_r[\bar{\bar{D}}(z)]\Gamma_r^T[\bar{\bar{D}}(z)]\{\Gamma_r^T[\bar{D}_\kappa(z)]\}^{-1} \tag{11.174}$$
$$= \begin{bmatrix} \dfrac{1}{0.15975} & \dfrac{50}{9} \\ \dfrac{50}{9} & \dfrac{95}{9} \end{bmatrix}$$

(see (11.146)) can be computed. From (11.125) then follows $C_2L_1^+ = 0.8875$.

If this observer is applied to parameterize an observer-based controller on the basis of the *a priori* estimate $\hat{x}$, its parameterizing polynomial matrix is

$$\tilde{\bar{D}}(z) = \begin{bmatrix} 1 & -1 \\ z+71/90 & -8/9 \end{bmatrix} \tag{11.175}$$

(see (11.123)). If, however, the *a posteriori* estimate $\hat{x}^+$ is used the parameterizing polynomial matrix takes the form

$$\bar{\tilde{D}}^+(z) = \begin{bmatrix} 0.1125 & -1 \\ z & -8/9 \end{bmatrix} \tag{11.176}$$

(see (11.124)).

To verify that this frequency-domain solution characterizes the optimal time-domain filter, the minimal realization

$$A = \begin{bmatrix} 1 & -1 \\ 0.25 & 0 \end{bmatrix}, \quad G = \begin{bmatrix} 1 & 2 \\ 3 & 1 \end{bmatrix}, \quad C = \begin{bmatrix} -1 & 1 \\ -1 & 0 \end{bmatrix} \tag{11.177}$$

of the system (11.169) is considered. With the above $\bar{D}(z)$ the polynomial matrix $\bar{N}_x(z)$ in $C(zI - A)^{-1} = \bar{D}^{-1}(z)\bar{N}_x(z)$ obtains the form $\bar{N}_x(z) = I_2$. The Diophantine equation (11.151) is solved by the pair

$$\bar{Y}(z) = [L_1^+ \quad \Psi_2] = \begin{bmatrix} -0.8875 & -1 \\ 0 & -8/9 \end{bmatrix} \text{ and } \bar{X}(z) = \begin{bmatrix} 1 & 0 \\ 0 & 0 \end{bmatrix}, \tag{11.178}$$

and $T\Psi_2 = 0$ (see (11.56)) is, *e.g.*, satisfied for $T = [-8/9 \quad 1]$. Then, (11.57) leads to $\Theta = [0 \quad 1]^T$, so that one obtains $F = T(A - L_1^+C_1)\Theta = 0.1$. Inserting the above results in (11.162) it can be solved for $\bar{P}_{red} = 50/9$. This finally yields

$$\bar{P} = \Theta\bar{P}_{red}\Theta^T = \begin{bmatrix} 0 & 0 \\ 0 & 50/9 \end{bmatrix}, \tag{11.179}$$

so that (11.49) and (11.50) lead to the same L_1^+ and Ψ_2 as in (11.178). Given the result (11.179) also the gain Λ can be computed using (11.64) and this yields

$$\Lambda = \begin{bmatrix} 0 \\ 0.8875 \end{bmatrix}. \tag{11.180}$$

11.5 Observer-based Compensators with *a posteriori* State Estimate in the Frequency Domain

The transfer behaviour of an observer-based compensator on the basis of an optimal *a priori* estimate $\hat{x}$ can be designed along the lines of Chapter 10. If, however, the observer-based compensator is based on an optimal *a posteriori* state estimate $\hat{x}^+$ the frequency-domain design takes on a modified structure because in order to design the compensator the observer parameterization has to be modified (see Section 11.4.1). Therefore, only the design equations of observer-based compensators with *a posteriori* state estimate are derived in the following.

In the time domain, the system is described by the state equations

$$x(k+1) = Ax(k) + Bu(k) + Gw(k), \tag{11.181}$$
$$y_1(k) = C_1x(k) + v_1(k), \tag{11.182}$$
$$y_2(k) = C_2x(k), \tag{11.183}$$

where $x \in \mathbb{R}^n$ is the state, $u \in \mathbb{R}^p$ is the input, $y_1 \in \mathbb{R}^{m-\kappa}$ is the measured output corrupted by noise and $y_2 \in \mathbb{R}^\kappa$ is the ideal measurement. In order to obtain a well-defined order of the filter it is also assumed that the system does not have an eigenvalue at $z = 0$. The stochastic disturbances are the input noise $w \in \mathbb{R}^q$ and the measurement noise $v_1 \in \mathbb{R}^{m-\kappa}$ (see Section 11.3). In the frequency domain, the system (11.181)–(11.183) is described by its transfer behaviour

$$y(z) = G(z)u(z) + G_w(z)w(z) + \begin{bmatrix} v_1(z) \\ 0 \end{bmatrix}, \tag{11.184}$$

where $G(z)$ is represented in the right and the left coprime MFDs

$$G(z) = N(z)D^{-1}(z) = \bar{D}^{-1}(z)\bar{N}(z), \tag{11.185}$$

with $D(z)$ column reduced and $\bar{D}(z)$, such that $\bar{D}_\kappa(z)$ (see (11.109)) is row reduced.

The state feedback

$$u(k) = -Kx(k) + Mr(k) \tag{11.186}$$

is either designed by pole placement (see Section 10.1) or to assure an optimal control as described in Section 11.1. In the frequency domain the state feedback is parameterized by the polynomial matrix $\tilde{D}(z)$ having the properties (10.44) and (10.45).

The estimator is assumed to be designed as a reduced-order Kalman filter and its state equations are

$$\hat{\zeta}(k+1) = TA(I - \Lambda C_1)\Theta\hat{\zeta}(k) + TA\,[\Lambda \quad (I - \Lambda C_1)\Psi_2]\begin{bmatrix} y_1(k) \\ y_2(k) \end{bmatrix} + TBu(k). \tag{11.187}$$

The optimal *a posteriori* estimate $\hat{x}^+$ is

$$\hat{x}^+(k) = (I - \Lambda C_1)\Theta\hat{\zeta}(k) + [\Lambda \quad (I - \Lambda C_1)\Psi_2]\begin{bmatrix} y_1(k) \\ y_2(k) \end{bmatrix} \tag{11.188}$$

(see (11.48) and (11.58)). In the frequency domain the reduced-order Kalman filter with *a posteriori* estimate is parameterized by the polynomial matrix $\tilde{\bar{D}}^+(z)$ having the properties (11.107) and (11.108).

When replacing x by $\hat{x}^+$ (11.186) can be written as

$$u(k) = -\hat{\tilde{u}}^+(k) + Mr(k), \tag{11.189}$$

with

$$\hat{\tilde{u}}^+(k) = K\hat{x}^+(k) = K(I - \Lambda C_1)\Theta\hat{\zeta}(k) + K[\Lambda \quad (I - \Lambda C_1)\Psi_2]\begin{bmatrix} y_1(k) \\ y_2(k) \end{bmatrix}, \tag{11.190}$$

when using (11.188). The transfer behaviour between the input signals y and u and the output signal $\hat{\tilde{u}}^+$ has the form

$$\begin{aligned}\hat{\tilde{u}}^+(z) = \{&K(I - \Lambda C_1)\Theta(zI - F)^{-1} \\ &\cdot [TA\Lambda \quad TA(I - \Lambda C_1)\Psi_2] + K[\Lambda \quad (I - \Lambda C_1)\Psi_2]\}\begin{bmatrix} y_1(z) \\ y_2(z) \end{bmatrix} \\ &+K(I - \Lambda C_1)\Theta(zI - F)^{-1}TBu(z),\end{aligned} \tag{11.191}$$

where the abbreviation

$$F = TA(I - \Lambda C_1)\Theta \tag{11.192}$$

has been used. By introducing the left MFDs

$$\begin{aligned}K(I - \Lambda C_1)\Theta(zI - F)^{-1}[TA\Lambda \quad TA(I - \Lambda C_1)\Psi_2] + K[\Lambda \quad (I - \Lambda C_1)\Psi_2] \\ = \Delta^{-1}(z)N_C^+(z),\end{aligned} \tag{11.193}$$

and

$$K(I - \Lambda C_1)\Theta(zI - F)^{-1}TB = \Delta^{-1}(z)N_u^+(z) \tag{11.194}$$

the transfer behaviour (11.191) can also be expressed as

$$\hat{\tilde{u}}^+(z) = \Delta^{-1}(z)[N_C^+(z) \quad N_u^+(z)]\begin{bmatrix} y(z) \\ u(z) \end{bmatrix}, \tag{11.195}$$

and it is assumed that the pair $(\Delta(z), [N_C^+(z) \quad N_u^+(z)])$ is left coprime (see also Remark 4.2) and that $\Delta(z)$ is row reduced. Then, the closed-loop system can be represented by the block diagram of Figure 11.6 that defines the observer structure of the compensator in the frequency domain. By an elimination of the feedback $\Delta^{-1}(z)N_u^+(z)u(z)$ and an introduction of

$$D_C^+(z) = N_u^+(z) + \Delta(z), \tag{11.196}$$

one obtains the output feedback structure

$$u(z) = -G_C^+(z)y(z) + G_{Cr}^+(z)r(z) \tag{11.197}$$

of the compensator in Figure 11.7, where

$$N_C^{*+}(z) = \Delta^{-1}(z)N_C^+(z), \tag{11.198}$$

(see (11.193)) and

$$D_C^{*+}(z) = \Delta^{-1}(z)D_C^+(z) \tag{11.199}$$

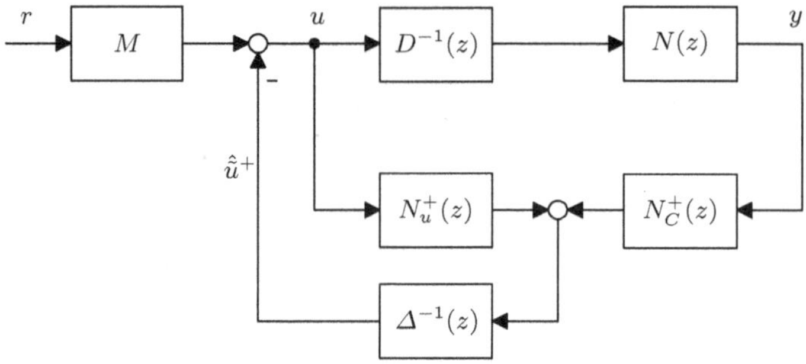

Figure 11.6. Closed-loop system in the frequency domain with compensator in the *observer structure*

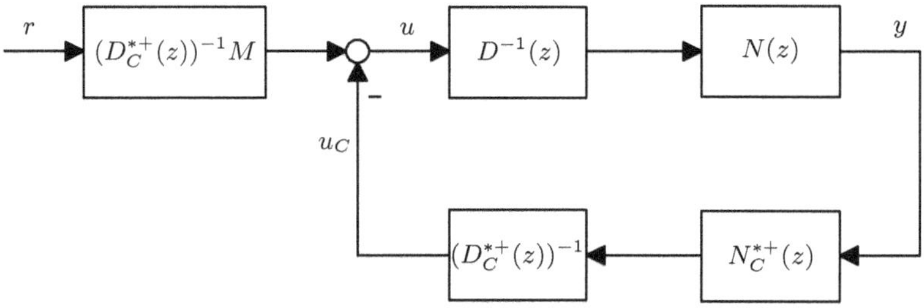

Figure 11.7. Closed-loop system with the compensator in the *output feedback structure* represented as a fraction of transfer matrices

(see (11.194) and (11.196)). By inserting (11.189) and (11.190) in (11.187) and taking the z-transform of the result it can be solved for $\hat{\zeta}(z)$. Inserting this in (11.189) one obtains the transfer matrices

$$\begin{aligned} G_C^+(z) &= K(I - \Lambda C_1)\Theta \left\{zI - T(A - BK)(I - \Lambda C_1)\Theta\right\}^{-1} T(A - BK) \\ &\qquad \cdot [\Lambda \quad (I - \Lambda C_1)\Psi_2] + [K\Lambda \quad K(I - \Lambda C_1)\Psi_2] \\ &= (D_C^{*+}(z))^{-1} N_C^{*+}(z), \end{aligned} \tag{11.200}$$

and

$$\begin{aligned} G_{Cr}^+(z) &= -K(I - \Lambda C_1)\Theta \left\{zI - T(A - BK)(I - \Lambda C_1)\Theta\right\}^{-1} TBM + M \\ &= (D_C^{*+}(z))^{-1} M \end{aligned} \tag{11.201}$$

in (11.197).

The input–output behaviour (11.197) of the compensator can also be represented by fractions of polynomial matrices. By introducing

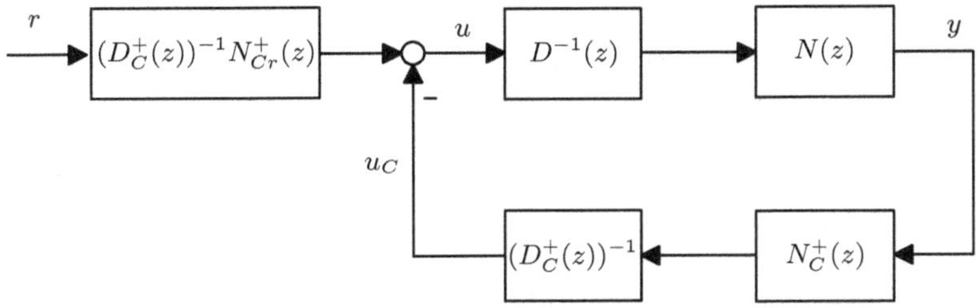

Figure 11.8. Closed-loop system in the frequency domain with compensator in the *output feedback structure*

$$N_{Cr}^+(z) = \Delta(z)M, \tag{11.202}$$

and redrawing Figure 11.7 the output feedback structure in Figure 11.8 is obtained. Here, the controller is described by

$$u(z) = -(D_C^+(z))^{-1}N_C^+(z)y(z) + (D_C^+(z))^{-1}N_{Cr}^+(z)r(z). \tag{11.203}$$

This defines the usual output feedback structure of the compensator.

The factorization (11.200) and the left MFD $G_C^+(z) = (D_C^+(z))^{-1}N_C^+(z)$ of this compensator can be obtained as described in the following theorem.

Theorem 11.4 (Computation of the left MFD of the compensator with *a posteriori* state estimate). *Given are the right and the left coprime MFDs*

$$G(z) = N(z)D^{-1}(z) = \bar{D}^{-1}(z)\bar{N}(z) \tag{11.204}$$

of the transfer matrix in $y(z) = G(z)u(z)$ of the system (11.181)–(11.183), where $D(z)$ is column reduced and $\bar{D}(z)$ is such that $\bar{D}_\kappa(z)$ (see (11.109)) is row reduced. In order to obtain a well-defined filter order it is further assumed that $\det D(0) \neq 0$ *and* $\det \bar{D}(0) \neq 0$. *Also given are the parameterizing polynomial matrices $\tilde{D}(z)$ (state feedback as defined in Theorem 11.1) and $\tilde{\bar{D}}^+(z)$ (reduced-order Kalman filter with $C_2L_1^+ \neq 0$ as introduced in Section 11.4).*

Then, with a solution $Y(z)$ and $X(z)$ of the Bezout identity

$$Y(z)N(z) + X(z)D(z) = I_p, \tag{11.205}$$

and the polynomial matrix

$$\begin{aligned}\bar{V}^+(z) = \Pi\left\{\tilde{D}(z)Y(z)\bar{D}^{-1}(z)\tilde{\bar{D}}^+(z)\right\} \\ + \left(SP\left\{\tilde{D}(z)Y(z)\bar{D}^{-1}(z)\tilde{\bar{D}}^+(z)\right\}\begin{bmatrix} I_{m-\kappa} & 0 \\ 0 & 0_\kappa \end{bmatrix}\right)\Bigg|_{z=0}\end{aligned} \tag{11.206}$$

the transfer matrices $N_C^{*+}(z)$ *and* $D_C^{*+}(z)$ *characterizing the transfer matrix* $G_C^{+}(z) = (D_C^{*+}(z))^{-1}N_C^{*+}(z)$ *(see also Figure 11.7) of the observer-based compensator (11.197) are given by*

$$N_C^{*+}(z) = \Delta^{-1}(z)N_C^{+}(z) = \tilde{D}(z)Y(z) - \bar{V}^{+}(z)(\tilde{\bar{D}}^{+}(z))^{-1}\bar{D}(z), \quad (11.207)$$

and

$$D_C^{*+}(z) = \Delta^{-1}(z)D_C^{+}(z) = \tilde{D}(z)X(z) + \bar{V}^{+}(z)(\tilde{\bar{D}}^{+}(z))^{-1}\bar{N}(z). \quad (11.208)$$

Introducing the prime right-to-left conversion

$$\bar{V}^{+}(z)(\tilde{\bar{D}}^{+}(z))^{-1} = \Delta^{-1}(z)V^{+}(z) \quad (11.209)$$

the polynomial matrices of the left MFD (11.203) of the observer-based compensator (11.197) are given by

$$N_C^{+}(z) = \Delta(z)\tilde{D}(z)Y(z) - V^{+}(z)\bar{D}(z), \quad (11.210)$$

and

$$D_C^{+}(z) = \Delta(z)\tilde{D}(z)X(z) + V^{+}(z)\bar{N}(z). \quad (11.211)$$

Proof. First, some preliminary results are formulated. Inserting $L_1^{+} = A\Lambda$ in (A.51), observing (11.92) and substituting s by z one obtains

$$\begin{aligned}
&SP\left\{\tilde{D}(z)Y(z)C(zI-A)^{-1}[L_1^{+} \quad \Psi_2]\right\} \\
&= SP\left\{\tilde{D}(z)Y(z)\left(C(zI-A)^{-1}[L_1^{+} \quad \Psi_2] + \begin{bmatrix} I_{m-\kappa} & 0 \\ 0 & 0_\kappa \end{bmatrix}\right)\right\} \\
&= SP\left\{\tilde{D}(z)Y(z)\bar{D}^{-1}(z)\tilde{\bar{D}}^{+}(z)\right\} = K(zI-A)^{-1}\left[A\Lambda \quad \Psi_2\right]. \quad (11.212)
\end{aligned}$$

Since it is assumed that the system has no poles at the origin of the z-plane the strictly proper part of $\tilde{D}(z)Y(z)\bar{D}^{-1}(z)\tilde{\bar{D}}^{+}(z)\begin{bmatrix} I_{m-\kappa} & 0 \\ 0 & 0_\kappa \end{bmatrix}$ taken at $z=0$ is

$$\left(SP\left\{\tilde{D}(z)Y(z)\bar{D}^{-1}(z)\tilde{\bar{D}}^{+}(z)\right\}\begin{bmatrix} I_{m-\kappa} & 0 \\ 0 & 0_\kappa \end{bmatrix}\right)\Bigg|_{z=0} = [-K\Lambda \quad 0], \quad (11.213)$$

which is obvious by inspection of (11.212).

By observing Definition 2.1 in Chapter 2 Equation (11.206) can be written as

$$\begin{aligned}
\bar{V}^{+}(z) &= \Pi\left\{\tilde{D}(z)Y(z)\bar{D}^{-1}(z)\tilde{\bar{D}}^{+}(z)\right\} \\
&\quad + \left(SP\left\{\tilde{D}(z)Y(z)\bar{D}^{-1}(z)\tilde{\bar{D}}^{+}(z)\right\}\begin{bmatrix} I_{m-\kappa} & 0 \\ 0 & 0_\kappa \end{bmatrix}\right)\Bigg|_{z=0} \\
&= \tilde{D}(z)Y(z)\bar{D}^{-1}(z)\tilde{\bar{D}}^{+}(z) - SP\left\{\tilde{D}(z)Y(z)\bar{D}^{-1}(z)\tilde{\bar{D}}^{+}(z)\right\} \\
&\quad + \left(SP\left\{\tilde{D}(z)Y(z)\bar{D}^{-1}(z)\tilde{\bar{D}}^{+}(z)\right\}\begin{bmatrix} I_{m-\kappa} & 0 \\ 0 & 0_\kappa \end{bmatrix}\right)\Bigg|_{z=0}. \quad (11.214)
\end{aligned}$$

Now, inserting (11.92) and observing (11.212) and (11.213) one obtains

$$\bar{V}^{+}(z) = \tilde{D}(z)Y(z)\left\{C(zI-A)^{-1}\left[A\Lambda \quad \Psi_2\right] + \begin{bmatrix} I_{m-\kappa} & 0 \\ 0 & 0_{\kappa}\end{bmatrix}\right\} \\ -K(zI-A)^{-1}\left[A\Lambda \quad \Psi_2\right] - [K\Lambda \quad 0], \tag{11.215}$$

or

$$\bar{V}^{+}(z) = \tilde{D}(z)Y(z)\begin{bmatrix} I_{m-\kappa} & 0 \\ 0 & 0_{\kappa}\end{bmatrix} - [K\Lambda \quad 0] \\ + \left(\tilde{D}(z)Y(z)C - K\right)(zI-A)^{-1}\left[A\Lambda \quad \Psi_2\right]. \tag{11.216}$$

After substituting (11.90) in (11.89) a comparison of (11.88) and (11.89) yields

$$(\tilde{\bar{D}}^{+}(z))^{-1}\bar{D}(z) = \left[\begin{matrix} I_{m-\kappa} - C_1\Theta(zI-F)^{-1}TA\Lambda \\ -C_2A\Lambda - C_2A(I-\Lambda C_1)\Theta(zI-F)^{-1}TA\Lambda \end{matrix}\right. \vdots \\ \left.\begin{matrix} -C_1\left\{I + \Theta(zI-F)^{-1}TA(I-\Lambda C_1)\right\}\Psi_2 \\ C_2\left\{zI - A(I-\Lambda C_1) - A(I-\Lambda C_1)\Theta(zI-F)^{-1}TA(I-\Lambda C_1)\right\}\Psi_2 \end{matrix}\right], \tag{11.217}$$

and

$$(\tilde{\bar{D}}^{+}(z))^{-1}\bar{N}(z) = \begin{bmatrix} C_1\Theta(zI-F)^{-1}TB \\ C_2\left\{I + A(I-\Lambda C_1)\Theta(zI-F)^{-1}T\right\}B \end{bmatrix}, \tag{11.218}$$

where the abbreviation (11.192) has been used. This concludes the preliminary results.

Inserting the Relations (11.216) and (11.217) in (11.207) the numerator transfer matrix of the compensator obtains the form

$$N_C^{*+}(z) = \tilde{D}(z)Y(z) - \tilde{D}(z)Y(z)\left[\begin{matrix} I_{m-\kappa} - C_1\Theta(zI-F)^{-1}TA\Lambda \\ 0 \end{matrix}\right. \vdots \\ \left.\begin{matrix} -C_1\left\{I + \Theta(zI-F)^{-1}T\breve{A}\right\}\Psi_2 \\ 0 \end{matrix}\right] \\ + \left[K\Lambda - K\Lambda C_1\Theta(zI-F)^{-1}TA\Lambda \vdots - K\Lambda C_1\left\{I + \Theta(zI-F)^{-1}T\breve{A}\right\}\Psi_2\right] \\ + \left(\tilde{D}(z)Y(z)C - K\right)(zI-A)^{-1} \\ \cdot\left[-A\Lambda\{I - C_1\Theta(zI-F)^{-1}TA\Lambda\} + \Psi_2C_2\{A\Lambda + \breve{A}\Theta(zI-F)^{-1}TA\Lambda\} \vdots \right. \\ \left. A\Lambda C_1\{I + \Theta(zI-F)^{-1}T\breve{A}\}\Psi_2 - \Psi_2C_2\{zI - \breve{A} - \breve{A}\Theta(zI-F)^{-1}T\breve{A}\}\Psi_2\right], \tag{11.219}$$

where additionally the abbreviation

$$\breve{A} = A(I - \Lambda C_1) \tag{11.220}$$

has been used. Because of $\Psi_2 C_2 = I - \Theta T$ (see (3.7)), $T\Psi_2 = 0$ (see (11.56)) and (11.220) this can be rewritten as

$$
\begin{aligned}
N_C^{*+}(z) = \tilde{D}(z)Y(z) - \tilde{D}(z)Y(z) & \begin{bmatrix} I_{m-\kappa} - C_1\Theta(zI-F)^{-1}TA\Lambda & \vdots & -C_1\left\{I + \Theta(zI-F)^{-1}T\breve{A}\right\}\Psi_2 \\ 0 & \vdots & 0 \end{bmatrix} \\
& + \left[K\Lambda - K\Lambda C_1\Theta(zI-F)^{-1}TA\Lambda \vdots - K\Lambda C_1\left\{I + \Theta(zI-F)^{-1}T\breve{A}\right\}\Psi_2\right] \\
& + \left(\tilde{D}(z)Y(z)C - K\right)(zI-A)^{-1} \\
& \cdot \Big[\left(-\Theta + \{A\Theta - \Theta TA\Theta + \Theta TA\Lambda C_1\Theta\}(zI-F)^{-1}\right)TA\Lambda \vdots \\
& \quad \left\{-zI + A - \Theta T\breve{A} + (A\Theta - \Theta TA\Theta + \Theta TA\Lambda C_1\Theta)(zI-F)^{-1}T\breve{A}\right\}\Psi_2\Big].
\end{aligned}
\tag{11.221}
$$

Now, observing $F = TA(I - \Lambda C_1)\Theta$ (see (11.192)),

$$
\Theta = \Theta(zI-F)(zI-F)^{-1} = (z\Theta - \Theta TA\Theta + \Theta TA\Lambda C_1\Theta)(zI-F)^{-1}, \tag{11.222}
$$

and

$$
\begin{aligned}
\Theta T\breve{A} &= \Theta(zI-F)(zI-F)^{-1}T\breve{A} \\
&= (z\Theta - \Theta TA\Theta + \Theta TA\Lambda C_1\Theta)(zI-F)^{-1}T\breve{A},
\end{aligned}
\tag{11.223}
$$

(11.221) can be expressed as

$$
\begin{aligned}
N_C^{*+}(z) = \tilde{D}(z)Y(z) - \tilde{D}(z)Y(z) & \begin{bmatrix} I_{m-\kappa} - C_1\Theta(zI-F)^{-1}TA\Lambda & \vdots & -C_1\left\{I + \Theta(zI-F)^{-1}T\breve{A}\right\}\Psi_2 \\ 0 & \vdots & 0 \end{bmatrix} \\
& + \left[K\Lambda - K\Lambda C_1\Theta(zI-F)^{-1}TA\Lambda \vdots - K\Lambda C_1\left\{I + \Theta(zI-F)^{-1}T\breve{A}\right\}\Psi_2\right] \\
& + \left(\tilde{D}(z)Y(z)C - K\right)(zI-A)^{-1} \\
& \cdot \left[(-zI+A)\Theta(zI-F)^{-1}TA\Lambda \vdots (-zI+A)\left\{I + \Theta(zI-F)^{-1}T\breve{A}\right\}\Psi_2\right].
\end{aligned}
\tag{11.224}
$$

Because of $C = \begin{bmatrix} C_1 \\ C_2 \end{bmatrix}$, $C_2\Theta = 0$ and $C_2\Psi_2 = I$ this is equivalent to

$$
\begin{aligned}
N_C^{*+}(z) = {} & \tilde{D}(z)Y(z) - \tilde{D}(z)Y(z)\begin{bmatrix} I_{m-\kappa} - C_1\Theta(zI-F)^{-1}TA\Lambda & \vdots & -C_1\left\{I+\Theta(zI-F)^{-1}T\breve{A}\right\}\Psi_2 \\ 0 & \vdots & 0 \end{bmatrix} \\
& + \left[K\Lambda - K\Lambda C_1\Theta(zI-F)^{-1}TA\Lambda \,\vdots\, - K\Lambda C_1\left\{I+\Theta(zI-F)^{-1}T\breve{A}\right\}\Psi_2\right] \\
& + \tilde{D}(z)Y(z)\begin{bmatrix} -C_1\Theta(zI-F)^{-1}TA\Lambda & \vdots & -C_1\left\{I+\Theta(zI-F)^{-1}T\breve{A}\right\}\Psi_2 \\ 0 & \vdots & I \end{bmatrix} \\
& + \left[K\Theta(zI-F)^{-1}TA\Lambda \,\vdots\, \left\{K+K\Theta(zI-F)^{-1}T\breve{A}\right\}\Psi_2\right],
\end{aligned}
\tag{11.225}
$$

or

$$
\begin{aligned}
N_C^{*+}(z) = {} & \tilde{D}(z)Y(z) - \tilde{D}(z)Y(z) \\
& + \left[K\Lambda - K\Lambda C_1\Theta(zI-F)^{-1}TA\Lambda \,\vdots\, - K\Lambda C_1\left\{I+\Theta(zI-F)^{-1}T\breve{A}\right\}\Psi_2\right] \\
& + \left[K\Theta(zI-F)^{-1}TA\Lambda \,\vdots\, \left\{K+K\Theta(zI-F)^{-1}T\breve{A}\right\}\Psi_2\right].
\end{aligned}
\tag{11.226}
$$

By an appropriate rearrangement and after resubstituting (11.220) this finally takes the form

$$
\begin{aligned}
N_C^{*+}(z) &= K(I-\Lambda C_1)\Theta(zI-F)^{-1}\left[TA\Lambda \quad TA(I-\Lambda C_1)\Psi_2\right] \\
&\qquad + K[\Lambda \quad (I-\Lambda C_1)\Psi_2] \\
&= \Delta^{-1}(z)N_C^{+}(z)
\end{aligned}
\tag{11.227}
$$

(see (11.193)).

Inserting the Relations (11.216) and (11.218) in (11.208) the denominator transfer matrix of the compensator obtains the form

$$
\begin{aligned}
D_C^{*+}(z) = {} & \tilde{D}(z)X(z) + \tilde{D}(z)Y(z)\begin{bmatrix} C_1\Theta(zI-F)^{-1}TB \\ 0 \end{bmatrix} \\
& \qquad\qquad K\Lambda C_1\Theta(zI \quad F)^{-1}TB \\
& + \left(\tilde{D}(z)Y(z)C - K\right)(zI-A)^{-1} \\
& \quad \cdot \left[A\Lambda C_1\Theta(zI-F)^{-1}TB + \Psi_2C_2\left\{I+\breve{A}\Theta(zI-F)^{-1}T\right\}B\right],
\end{aligned}
\tag{11.228}
$$

where (11.220) has been used. By observing $\Psi_2C_2 = I - \Theta T$ (see (3.7)) the term $\Psi_2C_2\left\{I+\breve{A}\Theta(zI-F)^{-1}T\right\}$ in (11.228) can be written as

$$
\begin{aligned}
\Psi_2C_2\left\{I+\breve{A}\Theta(zI-F)^{-1}T\right\} &= I - \Theta T + (\breve{A}\Theta - \Theta T\breve{A}\Theta)(zI-F)^{-1}T \\
&= I + (-zI + A - A\Lambda C_1)\Theta(zI-F)^{-1}T,
\end{aligned}
\tag{11.229}
$$

where (11.220) and (11.222) have been used. Therefore, (11.228) takes the form

$$D_C^{*+}(z) = \tilde{D}(z)X(z) + \tilde{D}(z)Y(z)\begin{bmatrix} C_1\Theta(zI-F)^{-1}TB \\ 0 \end{bmatrix} - KAC_1\Theta(zI-F)^{-1}TB + \left(\tilde{D}(z)Y(z)C - K\right)(zI-A)^{-1}\left\{I + (-zI+A)\Theta(zI-F)^{-1}T\right\}B. \tag{11.230}$$

Solving (11.205) for $X(z)$ yields

$$X(z) = D^{-1}(z) - Y(z)N(z)D^{-1}(z), \tag{11.231}$$

and with (11.8) and (11.9) one obtains

$$X(z) = D^{-1}(z) - Y(z)C(zI-A)^{-1}B. \tag{11.232}$$

Pre-multiplying (11.232) by $\tilde{D}(z)$ leads to

$$\tilde{D}(z)X(z) = \tilde{D}(z)D^{-1}(z) - \tilde{D}(z)Y(z)C(zI-A)^{-1}B, \tag{11.233}$$

and with (11.23) one obtains

$$\tilde{D}(z)X(z) = I + K(zI-A)^{-1}B - \tilde{D}(z)Y(z)C(zI-A)^{-1}B. \tag{11.234}$$

Inserting this in (11.230) and observing $C_2\Theta = 0$ (11.230) takes the form

$$\begin{aligned} D_C^{*+}(z) &= I + \tilde{D}(z)Y(z)\begin{bmatrix} C_1\Theta(zI-F)^{-1}TB \\ 0 \end{bmatrix} - KAC_1\Theta(zI-F)^{-1}TB \\ &\quad - \left(\tilde{D}(z)Y(z)C - K\right)\Theta(zI-F)^{-1}TB \\ &= I + K(I - AC_1)\Theta(zI-F)^{-1}TB \\ &= \Delta^{-1}(z)D_C^{+}(z) \end{aligned} \tag{11.235}$$

(see (11.194)). □

Remark 11.8. An inspection of (11.190) shows that the use of an *a posteriori* state estimate $\hat{x}^+$ leads to an additional feedthrough KA in the transfer behaviour between y_1 and $\hat{\bar{u}}^+$. Comparing (11.216) and (A.56) it becomes obvious that the polynomial matrix $\bar{V}^+(z)$ can be written as

$$\bar{V}^+(z) = \bar{V}(z) - [KA \quad 0], \tag{11.236}$$

with

$$\bar{V}(z) = \tilde{D}(z)Y(z)\begin{bmatrix} I_{m-\kappa} & 0 \\ 0 & 0_\kappa \end{bmatrix} + \left(\tilde{D}(z)Y(z)C - K\right)(zI-A)^{-1}\left[AA \quad \Psi_2\right], \tag{11.237}$$

i.e., the parameterization of the compensator with *a posteriori* state estimate could be obtained by the procedure described in Chapter 10 by adding a feedthrough according to (4.51) with $F_1 = K\Lambda$. The polynomial matrix $\bar{\tilde{D}}^+(z)$ in (11.213) shows how this feedthrough can be obtained in the frequency domain. The derivation of (11.213) also demonstrates that this is only the case when using the parameterizing matrix $\bar{\tilde{D}}^+(z)$, *i.e.*, when using the observer gain $L_1^+ = A\Lambda$, which leads to $C_2 L_1^+ \neq 0$.

Example 11.3. Design of an observer-based compensator using the a posteriori state estimate

Given is the system of the order $n = 2$ with one input u and two outputs y already considered in Examples 11.1 and 11.2.

The right and left MFDs of this system are

$$N(z)D^{-1}(z) = \begin{bmatrix} z \\ 1 \end{bmatrix} \frac{1}{z^2 - z + 0.25}, \tag{11.238}$$

and

$$\bar{D}^{-1}(z)\bar{N}(z) = \begin{bmatrix} 1 & -z \\ z & -z + 0.25 \end{bmatrix}^{-1} \begin{bmatrix} 0 \\ 1 \end{bmatrix} \tag{11.239}$$

(see (11.30) and (11.168)).

In Example 11.1 an optimal state feedback is designed for this system and it is parameterized by the polynomial

$$\tilde{D}(z) = z^2 - 0.2z + 0.125/3 \tag{11.240}$$

(see (11.35)).

A reduced-order Kalman filter for this system is considered in Example 11.2. The optimal polynomial matrix parameterizing the filter in view of an *a posteriori* state estimate $\hat{x}^+$ has the form

$$\bar{\tilde{D}}^+(z) = \begin{bmatrix} 0.1125 & -1 \\ z & -8/9 \end{bmatrix} \tag{11.241}$$

(see (11.176)). Thus, all parameters are known that are required for the design of an observer-based compensator along the lines of Theorem 11.4.

The pair $Y(z) = [0 \quad 1]$ and $X(z) = 0$ solves (11.205). When inserting this in (11.206) one obtains

$$\bar{V}^+(z) = \begin{bmatrix} 0.8875z & z & 8/90 \end{bmatrix}. \tag{11.242}$$

The right-to-left conversion (11.209) yields

$$V^+(z) = \begin{bmatrix} -z^2 - 0.7z & z - 0.01 \end{bmatrix} \quad \text{and} \quad \Delta(z) = z - 0.1, \tag{11.243}$$

and with this (11.210) and (11.211) lead to

$$N_C^+(z) = \begin{bmatrix} 0.71z & -\frac{595}{3000}z - \frac{5}{3000} \end{bmatrix} \quad \text{and} \quad D_C^+(z) = z - 0.01. \tag{11.244}$$

The polynomial matrix $N_u^+(z) = D_C^+(z) - \Delta(z)$, which characterizes the observer structure (see Figure 11.6) has the form

$$N_u^+(z) = 0.09. \tag{11.245}$$

A

Appendix

A.1 Computing a Row-reduced Polynomial Matrix $\bar{D}_\kappa(s)$

The following theorem shows that, for any row-reduced $m \times m$ denominator matrix $\bar{D}(s)$ in the MFD $G(s) = \bar{D}^{-1}(s)\bar{N}(s)$, there always exists a unimodular matrix $U_L(s)$, such that both $\bar{D}^{'}(s) = U_L(s)\bar{D}(s)$ and

$$\bar{D}^{'}_\kappa(s) = \Pi\left\{\bar{D}^{'}(s)\begin{bmatrix} I_{m-\kappa} & 0 \\ 0 & s^{-1}I_\kappa \end{bmatrix}\right\}, \quad 0 < \kappa \le m \tag{A.1}$$

are row reduced. The transfer matrix $G(s)$ with respect to $\bar{D}'(s)$ can be obtained by

$$G(s) = \bar{D}^{-1}(s)U_L^{-1}(s)U_L(s)\bar{N}(s) = \bar{D}^{'-1}(s)\bar{N}^{'}(s), \tag{A.2}$$

where $\bar{N}'(s) = U_L(s)\bar{N}(s)$. If $G(s)$ is strictly proper (*i.e.*, $G(\infty) = 0$) and $\bar{D}(s)$ is row reduced then $\delta_{ri}[\bar{D}(s)] > \delta_{ri}[\bar{N}(s)]$, $i = 1, 2, \ldots, m$, is satisfied (see [9]). Since $\delta_{ri}[\bar{N}(s)] \ge 0$ always holds this implies $\delta_{ri}[\bar{D}(s)] \ge 1$. This result is needed in the following theorem, which has a constructive proof that can be used to determine the unimodular matrix $U_L(s)$.

Theorem A.1. *For any row-reduced $m \times m$ polynomial matrix $\bar{D}(s)$ with $\delta_{ri}[\bar{D}(s)] \ge 1$, $i = 1, 2, \ldots, m$, and $\deg\det\bar{D}(s) = n$ there exists a unimodular polynomial matrix $U_L(s)$ such that both*

$$\det\Gamma_r[U_L(s)\bar{D}(s)] \ne 0, \tag{A.3}$$

and

$$\det\Gamma_r[\bar{D}^{'}_\kappa(s)] \ne 0. \tag{A.4}$$

Proof. Let γ_{1j}, $j = 1, 2, \ldots, m$, be the elements of the first column of the matrix $\Gamma_r[\bar{D}(s)]$ and denote its jth row by ρ_j. Consider a non-vanishing element γ_{1l_1} of $\Gamma_r[\bar{D}(s)]$, where the row corresponding to γ_{1l_1} has the lowest row degree among all rows with $\gamma_{1j} \neq 0$. Then, all other non-vanishing elements of $\Gamma_r[\bar{D}(s)]$ in the first column can be eliminated by the elementary row operation

$$\rho_j \quad \Rightarrow \quad \tilde{\rho}_j = \rho_j - \frac{\gamma_{1j}}{\gamma_{1l_1}} s^{\delta_{rj}[\bar{D}(s)]-\delta_{rl_1}[\bar{D}(s)]} \rho_{l_1}, \; j \neq l_1, \; \gamma_{1j}, \gamma_{1l_1} \neq 0. \quad \text{(A.5)}$$

When all non-vanishing elements $\gamma_{1j} \neq 0$, $j \neq l_1$, in the first column of $\Gamma_r[\bar{D}(s)]$ have been eliminated using (A.5) the last row is interchanged with the l_1th row if $l_1 \neq m$, *i.e.*,

$$\tilde{\rho}_m \quad \Leftrightarrow \quad \rho_{l_1}. \quad \text{(A.6)}$$

These row operations can be represented by pre-multiplying $\bar{D}(s)$ with a unimodular polynomial matrix $U_1(s)$ (see, *e.g.*, [67]) yielding

$$\Gamma_r[U_1(s)\bar{D}(s)] = \begin{bmatrix} 0 & * \cdots * \\ \vdots & \vdots \quad\; \vdots \\ 0 & * \cdots * \\ \gamma_{1l_1} & * \cdots * \end{bmatrix}, \quad \gamma_{1l_1} \neq 0. \quad \text{(A.7)}$$

In order to show that $\Gamma_r[U_1(s)\bar{D}(s)]$ is non-singular, the property

$$\deg(\rho_j - \frac{\gamma_{1j}}{\gamma_{1l_1}} s^{\delta_{rj}[\bar{D}(s)]-\delta_{rl_1}[\bar{D}(s)]} \rho_{l_1}) \leq \deg \rho_j, \quad j \neq l_1 \quad \text{(A.8)}$$

of the row operation (A.5) can be used. Consider one of the row operations (A.5) for some j and represent it by the unimodular matrix $V_j(s)$. Then, the inequality (A.8) leads to

$$\sum_{i=1}^{m} \delta_{ri}[V_j(s)\bar{D}(s)] \leq \sum_{i=1}^{m} \delta_{ri}[\bar{D}(s)] = \deg\det \bar{D}(s) = n, \quad \text{(A.9)}$$

as $\bar{D}(s)$ is row reduced. Since $\sum_{i=1}^{m} \delta_{ri}[V_j(s)\bar{D}(s)] \geq \deg\det V_j(s)\bar{D}(s)$ is always satisfied (see, *e.g.*, [36]) the relation

$$\sum_{i=1}^{m} \delta_{ri}[V_j(s)\bar{D}(s)] \geq \deg\det \bar{D}(s) = n \quad \text{(A.10)}$$

also holds, because the unimodular matrix $V_j(s)$ satisfies $\det V_j(s) = c \neq 0$. Because of (A.9) and (A.10) one has $\sum_{i=1}^{m} \delta_{ri}[V_j(s)\bar{D}(s)] = n$, which implies $\det \Gamma_r[V_j(s)\bar{D}(s)] \neq 0$ (see, *e.g.*, [36]). Since also the second row operation (A.6) does not influence the row reducedness, all polynomial matrices resulting from the row operations (A.5) and (A.6) are row reduced, *i.e.*, the result $\det \Gamma_r[U_1(s)\bar{D}(s)] \neq 0$ is proven. Furthermore, (A.8) and $\sum_{i=1}^{m} \delta_{ri}[V_j(s)\bar{D}(s)] = \sum_{i=1}^{m} \delta_{ri}[\bar{D}(s)] = n$ show that the row operations (A.5) and (A.6) do not change the row degrees up to a permutation. By applying the same row operations to the first $m-1$ rows of the second column of $U_1(s)\bar{D}(s)$ one obtains

$$\Gamma_r[U_2(s)U_1(s)\bar{D}(s)] = \begin{bmatrix} 0 & 0 & * \cdots * \\ \vdots & \vdots & \vdots \quad \vdots \\ & 0 & * \cdots * \\ 0 & \gamma_{2l_2} & * \cdots * \\ \gamma_{1l_1} & * & * \cdots * \end{bmatrix}, \quad \gamma_{1l_1}, \gamma_{2l_2} \neq 0. \tag{A.11}$$

In (A.11) γ_{2l_2} is the non-vanishing element in the first $m-1$ rows and the second column of $\Gamma_r[U_1(s)\bar{D}(s)]$, which corresponds to the lowest row degree among all non-vanishing elements in these rows and column. Note that such an element $\gamma_{2l_2} \neq 0$ always exists since $\det \Gamma_r[U_1(s)\bar{D}(s)] \neq 0$ (see (A.7)). By applying this procedure to all columns the described algorithm stops at

$$\Gamma_r[U_L(s)\bar{D}(s)] = \begin{bmatrix} 0 & & & \gamma_{ml_m} \\ & & \cdots & * \\ & \gamma_{2l_2} & * & * \\ \gamma_{1l_1} & * & * & * \end{bmatrix}, \quad \gamma_{1l_1}, \gamma_{2l_2}, \ldots, \gamma_{ml_m} \neq 0, \tag{A.12}$$

where

$$U_L(s) = U_m(s)U_{m-1}(s) \cdots U_1(s). \tag{A.13}$$

Moreover, there exists a permutation j_i of the index set $i = (1, 2, \ldots, m)$ such that

$$\delta_{ri}[U_L(s)\bar{D}(s)] = \delta_{rj_i}[\bar{D}(s)], \quad i = 1, 2, \ldots, m. \tag{A.14}$$

The property (A.3) follows from the fact that in every step of the procedure the polynomial matrices are row reduced. In order to prove (A.4) consider the polynomial matrix

$$P(s) = \bar{D}'_\kappa(s) \begin{bmatrix} I_{m-\kappa} & 0 \\ 0 & sI_\kappa \end{bmatrix}, \tag{A.15}$$

with $\bar{D}'_\kappa(s)$ as defined in (A.1) and $\bar{D}'(s) = U_L(s)\bar{D}(s)$. The polynomial matrix $\bar{D}'_\kappa(s)$ can be written as

$$\bar{D}'_\kappa(s) = \bar{D}'(s) \begin{bmatrix} I_{m-\kappa} & 0 \\ 0 & s^{-1}I_\kappa \end{bmatrix} - SP\left\{\bar{D}'(s) \begin{bmatrix} I_{m-\kappa} & 0 \\ 0 & s^{-1}I_\kappa \end{bmatrix}\right\} \tag{A.16}$$

(see (2.24)). By inserting (A.16) in (A.15) the polynomial matrix $P(s)$ takes the form

$$\begin{aligned} P(s) &= \left\{\bar{D}'(s) \begin{bmatrix} I_{m-\kappa} & 0 \\ 0 & s^{-1}I_\kappa \end{bmatrix} - SP\left\{\bar{D}'(s) \begin{bmatrix} I_{m-\kappa} & 0 \\ 0 & s^{-1}I_\kappa \end{bmatrix}\right\}\right\} \\ &\quad \cdot \begin{bmatrix} I_{m-\kappa} & 0 \\ 0 & sI_\kappa \end{bmatrix} \\ &= \bar{D}'(s) - \bar{D}'(0) \begin{bmatrix} 0_{m-\kappa} & 0 \\ 0 & s^{-1}I_\kappa \end{bmatrix} \begin{bmatrix} 0_{m-\kappa} & 0 \\ 0 & sI_\kappa \end{bmatrix} \\ &= \bar{D}'(s) - \bar{D}'(0) \begin{bmatrix} 0_{m-\kappa} & 0 \\ 0 & I_\kappa \end{bmatrix}, \end{aligned} \tag{A.17}$$

which shows that

$$\Gamma_r[P(s)] = \Gamma_r[U_L(s)\bar{D}(s)], \tag{A.18}$$

$$\delta_{ri}[P(s)] = \delta_{ri}[U_L(s)\bar{D}(s)], \quad i = 1, 2, \ldots, m. \tag{A.19}$$

This is a consequence of $\delta_{ri}[\bar{D}'(s)] = \delta_{ri}[U_L(s)\bar{D}(s)] \geq 1$, which follows from $\delta_{ri}[\bar{D}(s)] \geq 1$, $i = 1, 2, \ldots, m$, and (A.14). In view of (A.18) and (A.3) one has $\det \Gamma_r[P(s)] = \det \Gamma_r[U_L(s)\bar{D}(s)] \neq 0$ implying

$$\deg \det P(s) = \sum_{i=1}^{m} \delta_{ri}[P(s)], \tag{A.20}$$

and

$$\deg \det U_L(s)\bar{D}(s) = \sum_{i=1}^{m} \delta_{ri}[U_L(s)\bar{D}(s)] = n, \tag{A.21}$$

since $\deg \det U_L(s)\bar{D}(s) = \deg \det \bar{D}(s) = n$. Hence, (A.19), (A.20) and (A.21) show that

$$\deg \det P(s) = n. \tag{A.22}$$

Then, (A.15) leads to

$$\deg \det P(s) = \deg \det \bar{D}_{\kappa}'(s) + \kappa = n, \tag{A.23}$$

and consequently

$$\deg \det \bar{D}_{\kappa}'(s) = n - \kappa. \tag{A.24}$$

The special form of $\Gamma_r[U_L(s)\bar{D}(s)]$ in (A.12) and the assumption $\delta_{ri}[\bar{D}(s)] \geq 1$, $i = 1, 2, \ldots, m$, imply that the first κ row degrees of $U_L(s)\bar{D}(s)$ are reduced by one when computing $\bar{D}_{\kappa}'(s)$. This yields $\sum_{i=1}^{m} \delta_{ri}[\bar{D}_{\kappa}'(s)] = n - \kappa$, so that with (A.24) the property (A.4) of Theorem A.1 is shown. □

Example A.1. Computing a unimodular matrix such that $\bar{D}_{\kappa}'(s)$ is row reduced
The results of Theorem A.1 can be demonstrated by considering the polynomial matrix

$$\bar{D}(s) = \begin{bmatrix} s^3 & 0 & 0 \\ s & s & 0 \\ s^2 & s^2 & s^2 \end{bmatrix}, \tag{A.25}$$

which satisfies the assumptions of Theorem A.1. The polynomial matrix

$$\bar{D}_1(s) = \begin{bmatrix} s^3 & 0 & 0 \\ s & s & 0 \\ s^2 & s^2 & s \end{bmatrix} \tag{A.26}$$

is not row reduced since

$$\Gamma_r[\bar{D}_1(s)] = \begin{bmatrix} 1 & 0 & 0 \\ 1 & 1 & 0 \\ 1 & 1 & 0 \end{bmatrix} \tag{A.27}$$

is singular. A row-reduced $\bar{D}_1(s)$ can be obtained by the algorithm in the proof of Theorem A.1. In its first step one chooses a non-vanishing element in the first column of

$$\Gamma_r[\bar{D}(s)] = \begin{bmatrix} 1 & 0 & 0 \\ 1 & 1 & 0 \\ 1 & 1 & 1 \end{bmatrix} \tag{A.28}$$

with respect to the lowest row degree $\delta_{r2}[\bar{D}(s)] = 1$, which is $\gamma_{12} = 1$. Then, the elements $\gamma_{11} = \gamma_{13} = 1$ can be eliminated by the row operations $\rho_1 \Rightarrow \tilde{\rho}_1 = \rho_1 - s^2\rho_2$ and $\rho_3 \Rightarrow \tilde{\rho}_3 = \rho_3 - s\rho_2$. By interchanging the second and third row, *i.e.*, by using the elementary row operation $\rho_2 \Leftrightarrow \tilde{\rho}_3$ one obtains the polynomial matrix

$$U_1(s)\bar{D}(s) = \begin{bmatrix} 1 & -s^2 & 0 \\ 0 & -s & 1 \\ 0 & 1 & 0 \end{bmatrix} \bar{D}(s) = \begin{bmatrix} 0 & -s^3 & 0 \\ 0 & 0 & s^2 \\ s & s & 0 \end{bmatrix}, \tag{A.29}$$

with

$$\Gamma_r[U_1(s)\bar{D}(s)] = \begin{bmatrix} 0 & -1 & 0 \\ 0 & 0 & 1 \\ 1 & 1 & 0 \end{bmatrix} \tag{A.30}$$

in the form (A.7). Finally, by interchanging the first and the second row of $U_1(s)\bar{D}(s)$, *i.e.*, $\tilde{\rho}_1 \Leftrightarrow \rho_2$, the polynomial matrix

$$U_2(s)U_1(s)\bar{D}(s) = \begin{bmatrix} 0 & 1 & 0 \\ 1 & 0 & 0 \\ 0 & 0 & 1 \end{bmatrix} U_1(s)\bar{D}(s) = \begin{bmatrix} 0 & 0 & s^2 \\ 0 & -s^3 & 0 \\ s & s & 0 \end{bmatrix} \tag{A.31}$$

is obtained. Thus, the unimodular matrix $U_L(s)$ is

$$U_L(s) = U_2(s)U_1(s) = \begin{bmatrix} 0 & -s & 1 \\ 1 & -s^2 & 0 \\ 0 & 1 & 0 \end{bmatrix}, \tag{A.32}$$

so that

$$\Gamma_r[U_L(s)\bar{D}(s)] = \begin{bmatrix} 0 & 0 & 1 \\ 0 & -1 & 0 \\ 1 & 1 & 0 \end{bmatrix} \tag{A.33}$$

has the form (A.12). Therefore, the polynomial matrix

$$\bar{D}_1'(s) = \begin{bmatrix} 0 & 0 & s \\ 0 & -s^3 & 0 \\ s & s & 0 \end{bmatrix} \tag{A.34}$$

is row reduced because its highest row-degree-coefficient matrix

$$\Gamma_r[\bar{D}_1'(s)] = \begin{bmatrix} 0 & 0 & 1 \\ 0 & -1 & 0 \\ 1 & 1 & 0 \end{bmatrix} \tag{A.35}$$

is non-singular. It is easy to check that also $\bar{D}_2(s)$ and $\bar{D}_3(s)$ are row reduced.

A.2 Proof of Theorem 4.1

First, it is shown that

$$\Pi\left\{\tilde{D}(s)Y(s)C(sI-A)^{-1}[L_1 \quad \Psi_2]\right\} = \left(\tilde{D}(s)Y(s)C-K\right)(sI-A)^{-1}[L_1 \quad \Psi_2] \tag{A.36}$$

holds. Pre-multiplying (4.40) by $D^{-1}(s)$ and postmultiplying the result by $\tilde{D}(s)$ leads to

$$\tilde{D}(s)Y(s)N(s)D^{-1}(s) + \tilde{D}(s)X(s) = \tilde{D}(s)D^{-1}(s), \tag{A.37}$$

or with (2.10), (2.11) and (2.23)

$$\left(\tilde{D}(s)Y(s)C-K\right)(sI-A)^{-1}B = I_p - \tilde{D}(s)X(s). \tag{A.38}$$

The right-hand side of (A.38) is a polynomial matrix. It is first verified that, if the pair (A, B) is controllable, then the polynomial matrix

$$P(s) = \tilde{D}(s)Y(s)C - K \tag{A.39}$$

on the left-hand side of (A.38) must have the form

$$P(s) = \hat{N}(s)(sI-A), \tag{A.40}$$

where $\hat{N}(s)$ is a polynomial matrix. Consider the right-to-left matrix fraction conversion

$$P(s)(sI-A)^{-1} = \bar{L}^{-1}(s)\bar{P}(s) \tag{A.41}$$

yielding the left coprime MFD $\bar{L}^{-1}(s)\bar{P}(s)$, so that the Bezout identity

$$\bar{L}(s)\bar{X}_1(s) + \bar{P}(s)\bar{Y}_1(s) = I_p \tag{A.42}$$

is satisfied (see Theorem 1.2). Controllability of (A, B) implies that the MFD $(sI-A)^{-1}B$ is left coprime (see Remark 1.11). Therefore, solutions $\bar{X}_2(s)$ and $\bar{Y}_2(s)$ of the Bezout identity

$$(sI-A)\bar{X}_2(s) + B\bar{Y}_2(s) = I_n \tag{A.43}$$

exist. From (A.41) follows

$$\bar{P}(s)(sI-A) = \bar{L}(s)P(s). \tag{A.44}$$

Multiplying (A.43) by $\bar{P}(s)$ from the left and by $\bar{Y}_1(s)$ from the right and using (A.44) leads to

$$\bar{L}(s)P(s)\bar{X}_2(s)\bar{Y}_1(s) + \bar{P}(s)B\bar{Y}_2(s)\bar{Y}_1(s) = \bar{P}(s)\bar{Y}_1(s). \tag{A.45}$$

Rearranging (A.42) gives

$$\bar{P}(s)\bar{Y}_1(s) = I_p - \bar{L}(s)\bar{X}_1(s), \tag{A.46}$$

and when inserting this in (A.45) one obtains

$$\bar{L}(s)\bar{X}_3(s) + \bar{P}(s)B\bar{Y}_3(s) = I_p, \tag{A.47}$$

with $\bar{X}_3(s) = \bar{X}_1(s) + P(s)\bar{X}_2(s)\bar{Y}_1(s)$ and $\bar{Y}_3(s) = \bar{Y}_2(s)\bar{Y}_1(s)$. The Bezout identity (A.47) shows that the MFD $\bar{L}^{-1}(s)\bar{P}(s)B$ is left coprime. In view of (A.39) and (A.38) the MFD $\bar{L}^{-1}(s)\bar{P}(s)B = P(s)(sI - A)^{-1}B$ (see (A.41)) is a polynomial matrix implying that the polynomial matrix $\bar{L}(s)$ is unimodular. Then, the MFD (A.41) is also a polynomial matrix $\hat{N}(s)$, *i.e.*,

$$\bar{L}^{-1}(s)\bar{P}(s) = P(s)(sI - A)^{-1} = \hat{N}(s). \tag{A.48}$$

Solving (A.48) for $P(s)$ proves (A.40). Postmultiplying

$$\tilde{D}(s)Y(s)C - K = \hat{N}(s)(sI - A) \tag{A.49}$$

(see (A.39) and (A.40)) by $(sI - A)^{-1}[L_1 \quad \Psi_2]$ yields

$$\tilde{D}(s)Y(s)C(sI-A)^{-1}[L_1 \quad \Psi_2] = K(sI-A)^{-1}[L_1 \quad \Psi_2] + \hat{N}(s)[L_1 \quad \Psi_2]. \tag{A.50}$$

This shows that

$$SP\left\{\tilde{D}(s)Y(s)C(sI - A)^{-1}[L_1 \quad \Psi_2]\right\} = K(sI - A)^{-1}[L_1 \quad \Psi_2], \tag{A.51}$$

because

$$SP\left\{\hat{N}(s)[L_1 \quad \Psi_2]\right\} = 0. \tag{A.52}$$

Inserting (A.51) in

$$\begin{aligned}\tilde{D}(s)Y(s)C(sI - A)^{-1}[L_1 \quad \Psi_2] &= \Pi\left\{\tilde{D}(s)Y(s)C(sI - A)^{-1}[L_1 \quad \Psi_2]\right\} \\ &\quad + SP\left\{\tilde{D}(s)Y(s)C(sI - A)^{-1}[L_1 \quad \Psi_2]\right\}\end{aligned} \tag{A.53}$$

(see (2.24)) yields

$$\begin{aligned}&\tilde{D}(s)Y(s)C(sI - A)^{-1}[L_1 \quad \Psi_2] \\ &= \Pi\left\{\tilde{D}(s)Y(s)C(sI - A)^{-1}[L_1 \quad \Psi_2]\right\} + K(sI - A)^{-1}[L_1 \quad \Psi_2].\end{aligned} \tag{A.54}$$

Solving (A.54) for the polynomial part gives (A.36). By inserting (3.65) in (4.41) one obtains

$$\begin{aligned}\bar{V}(s) &= \Pi\left\{\tilde{D}(s)Y(s)\bar{D}^{-1}(s)\tilde{\bar{D}}(s)\right\} \\ &= \Pi\left\{\tilde{D}(s)Y(s)\begin{bmatrix} I_{m-\kappa} & 0 \\ 0 & 0_\kappa \end{bmatrix} + \tilde{D}(s)Y(s)C(sI - A)^{-1}[L_1 \quad \Psi_2]\right\} \\ &= \tilde{D}(s)Y(s)\begin{bmatrix} I_{m-\kappa} & 0 \\ 0 & 0_\kappa \end{bmatrix} + \Pi\left\{\tilde{D}(s)Y(s)C(sI - A)^{-1}[L_1 \quad \Psi_2]\right\},\end{aligned} \tag{A.55}$$

and with (A.36) this finally gives

$$\bar{V}(s) = \tilde{D}(s)Y(s)\begin{bmatrix} I_{m-\kappa} & 0 \\ 0 & 0_\kappa \end{bmatrix} + \Big(\tilde{D}(s)Y(s)C - K\Big)(sI - A)^{-1}[L_1 \quad \Psi_2]. \tag{A.56}$$

A comparison of (3.61), (3.62) and (3.63) yields the following results

$$\tilde{\bar{D}}^{-1}(s)\bar{D}(s) = \begin{bmatrix} I_{m-\kappa} - C_1\Theta(sI - F)^{-1}TL_1 & \vdots \\ -C_2A\Theta(sI - F)^{-1}TL_1 & \end{bmatrix} \tag{A.57}$$

$$\begin{matrix} C_1\left\{-I - \Theta(sI - F)^{-1}T(A - L_1C_1)\right\}\Psi_2 \\ C_2\left\{sI - A - A\Theta(sI - F)^{-1}T(A - L_1C_1)\right\}\Psi_2 \end{matrix}\Bigg], \tag{A.58}$$

and

$$\tilde{\bar{D}}^{-1}(s)\bar{N}(s) = \begin{bmatrix} C_1\Theta(sI - F)^{-1}TB \\ C_2\left\{I + A\Theta(sI - F)^{-1}T\right\}B \end{bmatrix}. \tag{A.59}$$

Inserting the Relations (A.56) and (A.57) in (4.42) the transfer matrix $N_C^*(s)$ obtains the form

$$\begin{aligned} N_C^*(s) &= \tilde{D}(s)Y(s) - \bar{V}(s)\tilde{\bar{D}}^{-1}(s)\bar{D}(s) \\ &= \tilde{D}(s)Y(s) - \left\{\tilde{D}(s)Y(s)\begin{bmatrix} I_{m-\kappa} & 0 \\ 0 & 0_\kappa \end{bmatrix} + \Big(\tilde{D}(s)Y(s)C - K\Big)(sI - A)^{-1}[L_1 \quad \Psi_2]\right\} \\ &\quad \begin{bmatrix} I_{m-\kappa} - C_1\Theta(sI-F)^{-1}TL_1 & C_1\{-I - \Theta(sI-F)^{-1}T(A-L_1C_1)\}\Psi_2 \\ -C_2A\Theta(sI-F)^{-1}TL_1 & C_2\{sI - A - A\Theta(sI-F)^{-1}T(A-L_1C_1)\}\Psi_2 \end{bmatrix} \\ &= \tilde{D}(s)Y(s) - \tilde{D}(s)Y(s)\begin{bmatrix} I_{m-\kappa} - C_1\Theta(sI - F)^{-1}TL_1 & \vdots \\ 0 & \end{bmatrix. \\ &\qquad \begin{matrix} -C_1\left\{I_n + \Theta(sI - F)^{-1}T(A - L_1C_1)\right\}\Psi_2 \\ 0_\kappa \end{matrix}\Bigg] - T(s). \end{aligned} \tag{A.60}$$

In (A.60) the transfer matrix $T(s)$ has the form

$$\begin{aligned} T(s) = \Big(\tilde{D}(s)Y(s)C - K\Big)(sI - A)^{-1}\big[L_1 - (L_1C_1 + \Psi_2C_2A)\Theta(sI-F)^{-1}TL_1 \,\vdots \\ \left\{sI - \Psi_2C_2A - L_1C_1 - (L_1C_1 + \Psi_2C_2A)\Theta(sI-F)^{-1}T(A - L_1C_1)\right\}\Psi_2\big], \end{aligned} \tag{A.61}$$

where $\Psi_2C_2\Psi_2s = s\Psi_2$ has been used because

$$C_2\Psi_2 = I_\kappa \tag{A.62}$$

(see (3.20)). When using the two relations

$$\Psi_2C_2 = I_n - \Theta T, \tag{A.63}$$

(see (3.7)) and

$$C_2L_1 = 0_{\kappa,m-\kappa}, \tag{A.64}$$

(see (3.19)) which imply

$$L_1 = (\Psi_2 C_2 + \Theta T)L_1 = \Theta T L_1, \tag{A.65}$$

and

$$\Psi_2 C_2 A = A - \Theta T A, \tag{A.66}$$

the transfer matrix $T(s)$ takes the form

$$T(s) = \\ \Big(\tilde{D}(s)Y(s)C - K\Big)(sI - A)^{-1}\big[\Theta T L_1 - \{A - \Theta T(A - L_1 C_1)\}\Theta(sI - F)^{-1}TL_1 \vdots \\ \{sI - A + \Theta T(A - L_1 C_1) - (A - \Theta T[A - L_1 C_1])\,\Theta(sI - F)^{-1}T(A - L_1 C_1)\}\,\Psi_2\big]. \tag{A.67}$$

Observing that $F = T(A - L_1 C_1)\Theta$ (see (4.20)) one can show that the equalities

$$[sI - \Theta T(A - L_1 C_1)]\Theta(sI - F)^{-1}TL_1 = \Theta T L_1, \tag{A.68}$$

and

$$[sI - \Theta T(A - L_1 C_1)]\Theta(sI - F)^{-1}T(A - L_1 C_1) = \Theta T(A - L_1 C_1) \tag{A.69}$$

are satisfied. Substituting in the first column of $T(s)$ the matrix $\Theta T L_1$ as defined by (A.68) and in the second column of $T(s)$ the matrix $\Theta T(A - L_1 C_1)$ as defined by (A.69) gives

$$T(s) = \Big(\tilde{D}(s)Y(s)C - K\Big)\,(sI - A)^{-1}\,\big[(sI - A)\Theta(sI - F)^{-1}TL_1 \vdots \\ \{sI - A + (sI - A)\Theta(sI - F)^{-1}T(A - L_1 C_1)\}\,\Psi_2\big], \tag{A.70}$$

which is equivalent to

$$T(s) = \Big(\tilde{D}(s)Y(s)C - K\Big)\,\big[\Theta(sI - F)^{-1}TL_1 \vdots \\ \{I + \Theta(sI - F)^{-1}T(A - L_1 C_1)\}\,\Psi_2\big]. \tag{A.71}$$

By inserting (A.71) in (A.60) and observing that $C_2\Theta = 0_{\kappa, n-\kappa}$ (see (3.6)) and $C_2\Psi_2 = I_\kappa$ (see (A.62)) one obtains

$$N_C^*(s) = \tilde{D}(s)Y(s) - \tilde{D}(s)Y(s)\begin{bmatrix} I_{m-\kappa} & 0 \\ 0 & I_\kappa \end{bmatrix} + K[\Theta(sI - F)^{-1}TL_1 \vdots \\ \Theta(sI - F)^{-1}T(A - L_1 C_1)\Psi_2 + \Psi_2], \tag{A.72}$$

and this finally gives

$$\begin{aligned} N_C^*(s) &= K\Theta(sI - F)^{-1}[TL_1 \quad T(A - L_1 C_1)\Psi_2] + [0 \quad K\Psi_2] \\ &= \Delta^{-1}(s)N_C(s), \end{aligned} \tag{A.73}$$

because of (4.23).

Inserting the Relations (A.56) and (A.59) in (4.43) the denominator transfer matrix of the compensator obtains the form

$$
\begin{aligned}
D^*_{\tilde{C}}(s) &= \tilde{D}(s)X(s) + \bar{V}(s)\tilde{\bar{D}}^{-1}(s)\bar{N}(s) \qquad\qquad (A.74)\\
&= \tilde{D}(s)X(s) + \left[\tilde{D}(s)Y(s)\begin{bmatrix} I_{m-\kappa} & 0 \\ 0 & 0_\kappa \end{bmatrix} + \left(\tilde{D}(s)Y(s)C - K\right)(sI-A)^{-1}[L_1 \;\; \Psi_2]\right.\\
&\qquad\qquad \left.\cdot\begin{bmatrix} C_1\Theta(sI-F)^{-1}TB \\ C_2B + C_2A\Theta(sI-F)^{-1}TB \end{bmatrix}\right]\\
&= \tilde{D}(s)X(s) + \tilde{D}(s)Y(s)\begin{bmatrix} C_1\Theta(sI-F)^{-1}TB \\ 0_{\kappa,p} \end{bmatrix}\\
&+ \left(\tilde{D}(s)Y(s)C - K\right)(sI-A)^{-1}[(L_1C_1 + \Psi_2C_2A)\Theta(sI-F)^{-1}TB + \Psi_2C_2B].
\end{aligned}
$$

By using (A.65), (A.66) and (A.63) this takes the form

$$
\begin{aligned}
D^*_{\tilde{C}}(s) = \tilde{D}(s)X(s) &+ \tilde{D}(s)Y(s)\begin{bmatrix} C_1\Theta(sI-F)^{-1}TB \\ 0_{\kappa,p} \end{bmatrix} \qquad (A.75)\\
&+ \left(\tilde{D}(s)Y(s)C - K\right)(sI-A)^{-1}[B - \Theta TB\\
&\qquad + (\Theta TL_1C_1 + A - \Theta TA)\Theta(sI-F)^{-1}TB].
\end{aligned}
$$

Observing that $\Theta TB = \Theta\{sI - T(A - L_1C_1)\Theta\}(sI-F)^{-1}TB$ this can be written as

$$
\begin{aligned}
D^*_{\tilde{C}}(s) = \tilde{D}(s)X(s) &+ \tilde{D}(s)Y(s)\begin{bmatrix} C_1\Theta(sI-F)^{-1}TB \\ 0_{\kappa,p} \end{bmatrix} \qquad (A.76)\\
&+ \left(\tilde{D}(s)Y(s)C - K\right)(sI-A)^{-1}[B - \{s\Theta - \Theta T(A - L_1C_1)\Theta\\
&\qquad + \Theta T(A - L_1C_1)\Theta - A\Theta\}(sI-F)^{-1}TB],
\end{aligned}
$$

and after a trivial elimination one obtains

$$
\begin{aligned}
D^*_{\tilde{C}}(s) = \tilde{D}(s)X(s) &+ \tilde{D}(s)Y(s)\begin{bmatrix} C_1\Theta(sI-F)^{-1}TB \\ 0_{\kappa,p} \end{bmatrix}\\
&+ \left(\tilde{D}(s)Y(s)C - K\right)(sI-A)^{-1}[B - (sI-A)\Theta(sI-F)^{-1}TB].
\end{aligned}
\qquad (A.77)
$$

Inserting

$$
\begin{aligned}
\tilde{D}(s)X(s) &= \tilde{D}(s)D^{-1}(s) - \tilde{D}(s)Y(s)N(s)D^{-1}(s) \qquad (A.78)\\
&= I_p + K(sI-A)^{-1}B - \tilde{D}(s)Y(s)C(sI-A)^{-1}B
\end{aligned}
$$

(see (4.40), (2.23) and (4.13)) in (A.77) finally leads to

$$
D^*_{\tilde{C}}(s) = I_p + K\Theta(sI-F)^{-1}TB, \qquad (A.79)
$$

because of $C_2\Theta = 0$. Pre-multiplying (4.27) by $\Delta^{-1}(s)$ and substituting (4.24) yields

$$
\Delta^{-1}(s)D_C(s) = I_p + K\Theta(sI-F)^{-1}TB, \qquad (A.80)
$$

and this shows that (A.79) is indeed equivalent to $D^*_{\tilde{C}}(s) = \Delta^{-1}(s)D_C(s)$.

Thus, (4.42) and (4.43) are verified. The Relations (4.45) and (4.46) follow directly by introducing (4.44) in (4.42) and (4.43), which completes the proof. □

References

1. Ackermann, J.: Sampled-data Control Systems: Analysis and Synthesis, Robust System Design. Springer, Berlin Heidelberg New York (1985)
2. Anderson, B.D.O., Kučera, V.: Matrix fraction construction of linear compensators. IEEE Transactions on Automatic Control **30**, 1112–1114 (1985)
3. Anderson, B.D.O., Moore, J.B.: Optimal Filtering. Prentice Hall, Englewood Cliffs, NJ (1979)
4. Anderson, B.D.O., Moore, J.B.: Optimal Control – Linear Quadratic Methods. Prentice Hall, Englewood Cliffs, NJ (1989)
5. Antsaklis, P.J., Michel, A.N.: Linear Systems. McGraw-Hill, New York (1997)
6. Aström, K.J., Wittenmark, B.: Computer Controlled Systems: Theory and Design. Prentice Hall, Englewood Cliffs, NJ (1997)
7. Bryson, A., Johansen, D.: Linear filtering for time-varying systems using measurements containing colored noise. IEEE Transactions on Automatic Control **10**, 4–10 (1965)
8. Burl, J.B.: Linear Optimal Control. Addison-Wesley, Menlo Park (1999)
9. Chen, C.T.: Linear System Theory and Design. Holt, Rinehart and Winston, New York (1984)
10. Davison, E.J.: The robust control of a servomechanism problem for linear time-invariant multivariable systems. IEEE Transactions on Automatic Control **21**, 25–34 (1976)
11. Davison, E.J., Goldenberg, A.: Robust control of a general servomechanism problem: The servo-compensator. Automatica **11**, 461–471 (1975)
12. Descusse, J., Dion, J.: On the structure at infinity of linear square decoupled systems. IEEE Transactions on Automatic Control **27**, 971–974 (1982)
13. Deutscher, J.: On parametric state feedback design in the frequency domain. In: Proceedings of the IFAC World Congress. Barcelona, Spain (2002)
14. Deutscher, J.: A linear differential operator approach to flatness-based tracking for linear and non-linear systems. International Journal of Control **76**, 266–276 (2003)
15. Deutscher, J.: Flatness-based Control of Multivariable Systems Using Linear Differential Operator Representations (in German). Shaker Verlag, Aachen (2004)
16. Fahmy, M., O'Reilly, J.: On eigenstructure assignment in linear multivariable systems. IEEE Transactions on Automatic Control **27**, 690–693 (1982)

17. Falb, P.L., Wolovich, W.A.: Decoupling in the design and synthesis of multivariable control systems. IEEE Transactions on Automatic Control **12**, 651–659 (1967)
18. Fliess, M., Levine, J., Martin, P., Rouchon, P.: Flatness and defect of nonlinear systems: Introductory theory and examples. International Journal of Control **61**, 1327–1361 (1995)
19. Franklin, G.F., Powell, J.D., Emami-Naeini, A.: Feedback Control of Dynamic Systems. Prentice Hall, Upper Saddle River, NJ (2002)
20. Friedland, B.: On the properties of reduced-order Kalman filters. IEEE Transactions on Automatic Control **34**, 321–324 (1989)
21. Gelb, A.: Applied Optimal Estimation. The M.I.T. Press, Cambridge, Mass. (1996)
22. Goodwin, G.C., Graebe, S.F., Salgado, M.E.: Control System Design. Prentice Hall, Upper Saddle River, NJ (2001)
23. Haddad, W., Bernstein, B.: The optimal projection equations for reduced-order state estimation: The singular measurement noise case. IEEE Transactions on Automatic Control **32**, 1135–1139 (1987)
24. Hippe, P.: The design of observer-based controllers in view of a reduced sensitivity to errors in its realization parameters (in German). Ph.D. Thesis, Universität Erlangen-Nürnberg (1976)
25. Hippe, P.: Parameterization of full-order compensator in the frequency domain. International Journal of Control **48**, 1583–1603 (1988)
26. Hippe, P.: Design of reduced-order optimal estimators directly in the frequency domain. International Journal of Control **50**, 2599–2614 (1989)
27. Hippe, P.: Parameterization of reduced-order compensator in the frequency domain. International Journal of Control **49**, 1143–1162 (1989)
28. Hippe, P.: Design of observer based compensators: the polynomial approach. Kybernetika **27**, 125–150 (1991)
29. Hippe, P.: Design of MIMO compensators for systems with unmeasurable disturbances: the polynomial approach. Automatica **28**, 1003–1009 (1992)
30. Hippe, P.: Windup prevention for unstable systems. Automatica **39**, 1967–1973 (2003)
31. Hippe, P.: Windup in Control – Its Effects and Their Prevention. Springer, Berlin Heidelberg New York London (2006)
32. Hippe, P., Deutscher, J.: Problems in the computational solution of the stationary discrete-time LQG problem. In: Proceedings of the IFAC Conference on System Structure and Control, 147–151. Nantes, France (1998)
33. Hippe, P., Deutscher, J.: Time domain design of the reduced-order H_∞ filter. In: Proceedings of the 1st IFAC/IEEE Symposium on System Structure and Control, 010. Prague, Czech Republic (2001)
34. Hippe, P., Wurmthaler, C.: Optimal reduced-order estimators in the frequency domain: the discrete-time case. International Journal of Control **52**, 1051–1064 (1990)
35. Johnson, C.D.: Accommodation of external disturbances in linear regulator and servomechanism problems. IEEE Transactions on Automatic Control **16**, 635–644 (1971)
36. Kailath, T.: Linear Systems. Prentice Hall, Englewood Cliffs, NJ (1980)
37. Kailath, T.: Linear Estimation. Prentice Hall, Upper Saddle River, NJ (2000)
38. Khalil, H.K.: Nonlinear Systems. Prentice Hall, Englewood Cliffs, NJ (1996)

39. Koussiouris, T.: A frequency domain approach to the block decoupling problem II. Pole assignment while block decoupling a minimal system by state feedback and a constant non-singular input transformation and the observability of the block decoupled system. International Journal of Control **32**, 443–464 (1980)
40. Kreindler, E., Jameson, A.: Conditions for nonnegativeness of partitioned matrices. IEEE Transactions on Automatic Control **17**, 147–148 (1972)
41. Kučera, V.: Analysis and Design of Discrete Linear Control Systems. Prentice Hall, New York London Toronto Sydney (1991)
42. Kwakernaak, H., Sivan, R.: Linear Optimal Control Systems. Wiley-Interscience, New York London Sidney Toronto (1972)
43. Liu, G.P., Patton, R.J.: Eigenstructure Assignment for Control System Design. John Wiley & Sons, Chichester (1998)
44. Lohmann, B.: Partial decoupling of non-minimum phase systems by constant state feedback. In: Proceedings of the IFAC Workshop on System Structure and Control, 269–272. Prague, Czech Republic (1989)
45. Lohmann, B.: Complete and Partial Input–output Decoupling of Dynamical Systems in the State Space (in German), vol. 244. VDI Verlag, Düsseldorf (1991)
46. Luenberger, D.G.: Observers for multivariable systems. IEEE Transactions on Automatic Control **11**, 190–197 (1966)
47. Moness, M., Amin, M.: Minimal order precompensators for decoupling linear multivariable systems. International Journal of Control **47**, 1925–1936 (1988)
48. Morgan, B.: The synthesis of linear multivariable systems by state feedback. IEEE Transactions on Automatic Control **9**, 405–411 (1964)
49. Morse, A., Wonham, W.: Status of noninteracting control. IEEE Transactions on Automatic Control **16**, 568–581 (1971)
50. O'Halloran, W.F., Uttam, B.J.: Derivation of observers for continuous linear stochastic systems from the discrete formulation. In: Proceedings of the 6th Asilomar Conference, pp. 323–328. Pacific Grove, California (1972)
51. O'Reilly, J.: Observers for Linear Systems. Academic Press, London (1983)
52. O'Reilly, J., Fahmy, M.M.: The minimum number of degrees of freedom in state feedback control. International Journal of Control **41**, 749–768 (1985)
53. PolyX, Ltd.: The polynomial toolbox for MATLAB®. Version 2.5 released in 2000, Prague, Czech Republic (2000)
54. Roppenecker, G.: Pole assignment by state feedback (in German). Regelungstechnik **29**, 228–233 (1981)
55. Roppenecker, G.: Modal approach to parametric state observer design. Electronic Letters **22**, 118–120 (1986)
56. Roppenecker, G.: On parametric state feedback design. International Journal of Control **43**, 793–804 (1986)
57. Roppenecker, G.: Time Domain Design of Linear Control Systems (in German). Oldenbourg Verlag, München (1990)
58. Rosenbrock, H.H.: State Space and Multivariable Theory. Nelson, London (1970)
59. Schmitt, G.: Frequency domain evaluation of circle criterion, Popov criterion and off-axis cricle criterion in the MIMO case. International Journal of Control **72**, 1299–1309 (1999)
60. Shaked, U.: Solution in the z-domain to the discrete-time stationary Kalman filtering problem for systems with perfect measurements. IEEE Transactions on Automatic Control **31**, 1156–1159 (1986)

61. Skogestad, S., Postlethwaite, I.: Multivariable Feedback Control. John Wiley & Sons, Chichester (1996)
62. Uttam, B.J., O'Halloran, W.F.: On observers and reduced-order optimal filters for linear stochastic systems. In: Proceedings of the Joint American Control Conference, pp. 842–846. Stanford, California (1972)
63. Vidyasagar, M.: Nonlinear Systems Analysis. Prentice Hall, Englewood Cliffs, NJ (1993)
64. White, B.: Eigenstructure assignment: a survey. Proceedings of the Institute of Mechanical Engineers **209**, 1–11 (1995)
65. Willems, J.C., Trentelman, H.L.: On quadratic differential forms. SIAM Journal on Control and Optimization **36**, 1703–1749 (1998)
66. Williams, T., Antsaklis, P.: Decoupling. In: W. Levine (ed.) The Control Handbook, pp. 795–804. CRC Press, Boca Raton, Florida (1996)
67. Wolovich, W.A.: Linear Multivariable Systems. Springer, Heidelberg New York London (1974)
68. Youla, D.C.: On the factorization of rational matrices. IRE Transactions on Information Theory **7**, 172–189 (1961)

Index

FSC
www.fsc.org
MIX
Papier | Fördert
gute Waldnutzung
FSC® C083411

Kolibri 360 GmbH
Ferdinand-Jühlke-Straße 7
99095 Erfurt, Deutschland
produktsicherheit@kolibri360.de